ADVANCES IN

Immunology

VOLUME 50

ADVANCES IN
Immunology

EDITED BY

FRANK J. DIXON

Scripps Clinic and Research Foundation
La Jolla, California

ASSOCIATE EDITORS

K. Frank Austen
Leroy E. Hood
Jonathan W. Uhr
Tadamitsu Kishimoto
Fritz Melchers

VOLUME 50

ACADEMIC PRESS, INC.
Harcourt Brace Jovanovich, Publishers
San Diego New York Boston
London Sydney Tokyo Toronto

This book is printed on acid-free paper. ∞

Copyright © 1991 BY ACADEMIC PRESS, INC.
All Rights Reserved.
No part of this publication may be reproduced or transmitted in any form or by any means, electronic or mechanical, including photocopy, recording, or any information storage and retrieval system, without permission in writing from the publisher.

Academic Press, Inc.
San Diego, California 92101

United Kingdom Edition published by
ACADEMIC PRESS LIMITED
24-28 Oval Road, London NW1 7DX

Library of Congress Catalog Card Number: 61-17057

ISBN 0-12-022450-X (alk. paper)

PRINTED IN THE UNITED STATES OF AMERICA
91 92 93 94   9 8 7 6 5 4 3 2 1

CONTENTS

Selective Elements for the Vβ Region of the T Cell Receptor: Mls and the Bacterial Toxic Mitogens

CHARLES A. JANEWAY, JR.

I. Introduction	1
II. The Working Model for Vβ Selective Elements	4
III. Vβ Selective Elements	6
IV. The Molecular Basis of Vβ Selective Element Action	16
V. The Cellular Basis of Vβ Selective Element Action	29
VI. The Impact of Vβ Selective Elements on T Cell Development and Function	32
VII. Vβ Selective Elements are Encoded by Endogenous or Extrinisic Retroviruses	37
VIII. Unanswered Questions About Vβ Selective Elements	42
IX. Conclusions	47
References	48

Programmed Cell Death in the Immune System

J. JOHN COHEN

I. Introduction and Definitions	55
II. Mechanisms of Programmed Cell Death	56
III. Programmed Cell Death in Immunologic Systems	63
IV. Triggering and Regulation of Programmed Cell Death	71
V. Conclusions	76
References	76

Avian T Cell Ontogeny

MAX D. COOPER, CHEN-LO H. CHEN, R. PAT BUCY, AND CRAIG B. THOMPSON

I. Introduction	87
II. Avian T Cell Differentiation Antigens	88
III. Thymocyte Precursors: Origin and Thymic Attraction	89
IV. Diversification in the Thymus	90
V. T Cell Migration to the Periphery	96
VI. Functional Capabilities of TCR1, TCR2, and TCR3 Cells	101
VII. Experimental Manipulation of T Cell Development	102
VIII. T Cell Tumors	105

IX. TCR Genes	108
X. Concluding Remarks	112
References	114

Structural and Functional Chimerism Results from Chromosomal Translocation in Lymphoid Tumors

T. H. RABBITTS AND T. BOEHM

I. Introduction	119
II. Mechanism of Translocation and Inversion	121
III. Timing of Chromosome Translocation and Inversion	122
IV. Consequences of the Formation of Chromosomal Abnormalities on Adjacent Oncogenes	124
V. Effects of Chromosomal Abnormalities on Lymphoid Cells	132
VI. The Development of T Cell Leukemia: A Paradigm for Clonal Evolution and Tumor Development	141
References	142

Interleukin-2, Autotolerance, and Autoimmunity

GUIDO KROEMER, JOSÉ LUIS ANDREU, JOSÉ ANGEL GONZALO, JOSÉ C. GUTIERREZ-RAMOS, AND CARLOS MARTÍNEZ-A.

I. Introduction	147
II. Physiology of Interleukin-2 and Its Receptor	149
III. IL-2 and Autotolerance	165
IV. IL-2 in Autoimmunity—Phenomenology	171
V. IL-2 in Autoimmunity—*In Vivo* Interventions	187
VI. Role of IL-2-Induced Cytokines in Autoimmunity	210
VII. Concluding Remarks	214
References	217

Histamine Releasing Factors and Cytokine-Dependent Activation of Basophils and Mast Cells

ALLEN P. KAPLAN, SESHA REDDIGARI, MARIA BAEZA, AND PIOTR KUNA

I. Introduction	237
II. Purification of Histamine Releasing Factors	238
III. The Cell Source(s) of HRF	243
IV. Mechanism of Action of Human HRFs	244
V. Cytokines and Histamine Release	247
VI. Modulation of Histamine Release by Cytokines	248
VII. Clinical Considerations	250
References	254

Immunologic Interactions of T Lymphocytes with Vascular Endothelium

JORDAN S. POBER AND RAMZI S. COTRAN

I. Introduction	261
II. Specific Antigen Presentation by Vascular Endothelial Cells	261
III. Costimulator Activities of Vascular Endothelial Cells	267
IV. Antigen-Independent Recruitment of T Lymphocytes into Tissues by Endothelial Cells	274
V. Summary	288
References	288

Adoptive Transfer of Human Lymphoid Cells to Severely Immunodeficient Mice: Models for Normal Human Immune Function, Autoimmunity, Lymphomagenesis, and AIDS

DONALD E. MOSIER

I. Introduction	303
II. Transfer of Normal or Autoimmune Human Cells to Severely Immunodeficient Mice	304
III. EBV-Associated Lymphoproliferative Disease in hu-PBL-SCID Mice	310
IV. Other Tumor Models in SCID Mice	313
V. HIV Infection of hu-PBL-SCID Mice	314
VI. Biosafety Issues: Viral Pseudotypes in HIV-Infected hu-PBL-SCID Mice	317
VII. Summary	319
References	320

INDEX	327
CONTENTS OF RECENT VOLUMES	339

Selective Elements for the Vβ Region of the T Cell Receptor: Mls and the Bacterial Toxic Mitogens

CHARLES A. JANEWAY, Jr.

Section of Immunobiology, Yale University School of Medicine and Howard Hughes Medical Institute, New Haven, Connecticut 06510

I. Introduction

In 1973, Festenstein described a novel immunological phenomenon, a strong primary murine T lymphocyte proliferative response elicited by stimulator cells from a strain identical at the major histocompatibility complex (MHC). The genetic locus that induced this response was called minor lymphocyte stimulating, or *Mls*, to distinguish it from the MHC, the only known inducer of such responses. This phenomenon appears to violate many of the normal rules of T cell responses. The frequency of responding T cells is very high (~20%), the response is not MHC restricted, and it is unidirectional. These unusual characteristics led originally to the suggestion the *Mls* must encode a mitogen for T cells; others thought that *Mls* stood for "makes little sense."

The finding that T cells responding to the *Mls*-encoded product, Mls, are identifiable on the basis of the Vβ gene segment encoding their receptors converted this curiosity into a subject of intense interest (Kappler *et al.*, 1988; MacDonald *et al.*, 1988a). This finding built on earlier studies showing that the response to Mls involved the same cell surface structures as the response to antigen, namely MHC class II, CD4, and the T cell receptor (TCR) (for review, see Janeway *et al.*, 1989b). The determination of responsiveness by Vβ gene segment usage readily explained the very high precursor frequency of responding T cells, because 5–15% of normal T cells can express receptors encoded by a single Vβ gene segment. Furthermore, mice expressing stimulatory alleles at *Mls* were shown to maintain self-tolerance to Mls by eliminating T cells whose receptors are encoded by responsive Vβ gene segments, explaining the unidirectional nature of the response (Kappler *et al.*, 1988; MacDonald *et al.*, 1988a). However, these findings have raised new questions. How can Mls act only on the Vβ-encoded portion of the TCR, when responses to antigen:self MHC complexes or to nonself MHC are determined by all of the variable elements of the TCR? Why does *Mls* exist in the mouse genome? What is the nature of the *Mls* gene product and how does it produce its effects? In this review, I will not answer these questions. Rather, I hope to provide a clear description of the literature in this rapidly moving field of investigation, to suggest a working model to account for these data, and to raise the questions that need to be answered in order to incorporate *Mls* into the immunological mainstream.

In doing so, I wish to broaden the consideration of Mls to include two other sets of molecules with similar functional effects. The first are bacterial proteins that mimic all of the properties of Mls, the staphylococcal enterotoxins and their relatives (Janeway *et al.*, 1988a, 1989b; Marrack and Kappler, 1990; Purdie *et al.*, 1991). The second are products of genes that produce some but not all of the effects of *Mls* in the mouse. I will group these together with *Mls* as Vβ selective elements (Vbse), because their most striking shared characteristic is their selective effect on T cells whose receptors are encoded by particular Vβ gene segments. I will suggest that Vbse be used to generate a new nomenclature for this scattered family of functionally related genes, because *Mls* is clearly too narrow a definition. In the end, however, it is clear that a molecular definition of Vbse is required before one can say that *Mls* stands for "makes lotsa sense"; if the molecular definition is not obtained, then Vbse may end up being derided as a "very bad second effort." Fortunately, a molecular definition is now in the offing (Janeway, 1991).

A. Vβ SELECTIVE ELEMENTS

The Vbse are defined by their ability to affect the behavior of T cell populations, defined primarily by the expression of products of particular Vβ gene segments in the T cell receptor. This characteristic was first defined in T cells expressing Vβ17a that are stimulated by B cells expressing the MHC class II molecule I-E. It was subsequently shown that this response was directed not solely at I-E, but also required a second genetic element (Marrack and Kappler, 1988a). This second gene is apparently nonpolymorphic, so that it only elicits responses when T cells derive from I-E-negative strains. Subsequently, T cells bearing Vβ6 and Vβ8.1 were shown to dominate the response to Mls^a (Kappler *et al.*, 1988; MacDonald *et al.*, 1988a), whereas T cells bearing Vβ3 dominated the response to Mls^c (see below for a discussion of *Mls* genetics and nomenclature)(Abe *et al.*, 1988; Fry and Matis, 1988; Pullen *et al.*, 1988, 1989b). Following these findings, T cells bearing TCRs encoded by a variety of Vβ gene segments have been shown to be over- or underrepresented in the mature peripheral T cell repertoire, or in certain immune responses, suggesting the action of some genetic element distinct from MHC class II that plays the major role in Vβ selection.

Initially, the prime candidate for these Vbse was a self peptide, because it was believed that all T cell responses involve TCR binding to peptide:self MHC complexes (Marrack and Kappler, 1988a,b). However, this hypothesis raised a difficult problem, which was that all known responses to peptides depend on all of the variable elements of the TCR, most particularly the hypervariable sequences at the junctions of the V gene segments with the D and J gene segments (Davis and Bjorkman, 1988). An alternative formulation of the effect of Vbse was subsequently proposed, based on the response to the staphylococcal enterotoxins (SEs) that closely resembles the response to Mls (Janeway *et al.*, 1988a, 1989a,b). These molecules were shown to bind directly to MHC class II molecules without modification or processing (Fischer *et al.*, 1989; Fraser, 1989; Mollick *et al.*, 1989), to stimulate T cells bearing certain Vβ-encoded receptors (Fleischer *et al.*, 1989; Janeway *et al.*, 1989b; Kappler *et al.*, 1989a; White *et al.*, 1989; Yagi *et al.*, 1990), and to lose their biological potency if they were degraded with chemical or enzymatic treatment (Fraser, 1989). This analysis led to the proposal of what I will refer to as "the working model" (see below), in which the Vbse bind to the outer surface of the MHC class II molecule and the outer surface of the TCR in the region encoded by the Vβ gene segment (Fleischer *et al.*, 1989; Janeway *et al.*, 1989; Marrack and Kappler, 1990).

The existence of extrinsic substances that mimic the action of Vbse has led to further interest in this class of compounds, and rapid progress is being made now in analyzing these responses. Many bacterial toxic mitogens act like Vbse, and at least some of the pathological effects of these toxins are due to T cell activation (Marrack *et al.*, 1990). This provides further impetus to understanding the molecular, genetic, and evolutionary bases of these phenomena.

B. Questions to Be Addressed about Vβ Selective Elements

There are many unanswered questions about Vbse, as well as some disagreement about basic observations among laboratories. In this review, I will try to answer certain basic questions, point out where answers are simply unknown, and guess at possible answers. The main questions I wish to address are as follows:

1. What is the molecular basis of Vbse stimulation of T cells? Exactly how do the T cell receptor and its associated coreceptors, MHC class

II, and the Vbse come together? Is the working model valid, or are other interpretations of the data more consistent with the currently available facts?

2. What is the molecular nature of Vbse? This has been the single greatest problem in this field. We have known nothing of the nature of Vbse gene products, not even whether they are structural proteins or enzymes that modify structural proteins. Why is this so? Why are there no antibodies specific for any Vbse? Only recently have data been advanced suggesting a surprising answer to this question.

3. What is the cellular distribution of Vbse *in vivo?* Because Vbse have not been identified serologically or chemically, we do not know the distribution of these structures. This imposes severe constraints on understanding their behavior.

4. What is the biological function of the endogenous Vbse? These genes appear to be numerous, dispersed in the genome, and significantly polymorphic. It seems likely that they were maintained to perform some beneficial biological function. To date, the only known effects are on the immune system. Assuming this to be the case, do the Vbse exist to eliminate cells potentially reactive to bacterial toxic mitogens or susceptible to viral infections, protecting their hosts from infection, as has been suggested? Or do Vbse have an endogenous function that is independent of the existence of the bacterial toxic mitogens or retroviruses?

5. What is the evolutionary origin of Vbse? Did the mouse develop Vbse to counteract the effects of bacterial toxic mitogens or of retroviruses? Are there Vbse in vertebrate species other than the mouse?

This list of questions is extensive, reflecting two facts about Vbse, the fascination they have held for immunologists and the frustration of not knowing precisely what they are for many years. I believe this intellectual logjam will give way soon, now that the structure of these genes is being elucidated. Once a pure product is produced, and studies similar to those initiated with the staphylococcal enterotoxins have taken place, much of the mystery about Vbse should evaporate.

II. The Working Model for Vβ Selective Elements

The working model for Vbse attempts to account for the major data available about T cell responses to Vbse, binding data on SEs, and the assumed structure of the TCR (Fleischer *et al.*, 1989; Janeway, 1990; Janeway *et al.*, 1989a,b; Marrack and Kappler, 1990). It also acknowledges the important role of CD4 in T cell activation by both Mls and SEs. It is shown in schematic form in Fig. 1.

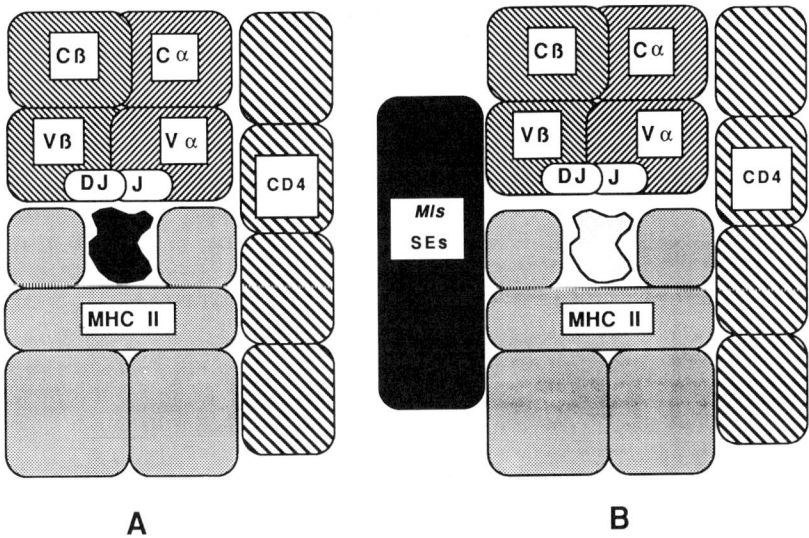

FIG. 1. The "working model" for Mls and bacterial toxic mitogen action. Distances between elements as well as intensity of shading reflect their relative importance in controlling the interaction. (A) Antigen recognition by the T cell receptor involves all the variable elements of the T cell receptor binding to a complex ligand consisting of an antigenic peptide bound to a self MHC class II molecule. The CD4 coreceptor is also involved in binding this same ligand and participates in signaling. (B) The response to Mls and to bacterial toxic mitogens is believed to involve binding of Mls or the bacterial toxic mitogen to both MHC class II molecules and the lateral face of the Vβ domain of the T cell receptor. In the mouse, CD4 also plays a critical role in this response, although this is less clear in responses of human T cells (see text). Direct contacts between the hypervariable elements of the T cell receptor and peptide:MHC are not controlling the specificity of this response.

The major elements of the working model should be contrasted to the elements in conventional models for TCR : peptide : MHC interaction, also shown in Fig. 1 (Davis and Bjorkman, 1988). In the recognition of conventional peptides, the TCR is believed by many to make contact with the surface of a peptide:MHC complex much as an antibody binds to protein antigen. The six hypervariable loops defined in antibody molecules have analogs in the TCR Vα and Vβ regions. The Vβ gene segment encodes the analog of the immunoglobulin heavy-chain CDR1 and CDR2 loops. Exchanging these loops, singly or together, changes the specificity of peptide:MHC recognition by the T cell receptor (M. M. Davis, personal communication).

By contrast, the working model postulates that the critical interactions for responses to Mls or SEs take place away from these hypervariable loops at the site of binding of the Vbse to the side of the Vβ-encoded region of the TCR (Choi et al., 1990; Pullen et al., 1990b; Cazenave, 1990). Similarly, the peptide-binding groove of the MHC molecule is not occupied by the Vbse (Dellabona et al., 1989). In consequence, the fine specificity of the interaction of the TCR with the peptide:MHC complex is unimportant in the action of strong Vbse, such as Mls-1 or SEs. However, it should be added that certain combinations of TCR and MHC:peptide may not be permissive for the action of Vbse as shown here. This constraint makes analysis of such interactions complex. Finally, one can envisage a role for both Vbse and conventional TCR:MHC:peptide interactions in T cell receptor ligation. In this case, the Vbse would serve to enhance recognition of peptide:MHC complexes by the T cell receptor or vice versa, and these effects might broaden the specificity of recognition as well. Thus, the Vbse could participate in all T cell receptor:ligand interactions. In this case, the difference between peptide:MHC recognition and responses to Vbse disappears, the only remaining distinction being which element plays the dominant role in T cell activation. In a syngeneic system, this will always be the antigenic peptide. This aspect of Vbse function has been termed "coligand" function (Janeway et al., 1989a).

The working model will be used to interpret and analyze the data about Vbse in this review. Thus, it is important that the major aspects of this model be clear. The most important points have to do with the interactions of the molecular elements of the model, particularly of the Vbse with both MHC class II molecules and the Vβ-encoded portion of the T cell receptor. However, the other contacts of the T cell receptor with the peptide:MHC complex, and of the coreceptor with the MHC molecule and the T cell receptor, are also important. If the working model is correct, it demonstrates that the molecular complexity of T cell antigen recognition extends well beyond the interaction of the TCR with its peptide:MHC ligand.

III. Vβ Selective Elements

Vbse and bacterial toxic mitogens have been identified primarily by the potent stimulation of T cells in *in vitro* cultures. This system of identification works well for bacterial toxic mitogens, which are extrinsic products that can be added to such cultures. However, this system of identification only allows the detection of polymorphic Vbse in mice. Recently, several Vbse have been described on the basis of their impact on the representation of particular Vβ gene segment encoded receptors on the T cells of mice, even where T cell stimulation has not been observed. In this section, the available information about murine and exogenous Vbse will be detailed. I will also suggest a generalized nomenclature for Vbse in mice, list the various nomenclatures that have been used, and give the strain distribution of these Vbse.

A. Mls

The primary T cell proliferative response to Mls has already been described. Using this response and the proliferative response of cloned T cell lines to Mls, the strongly stimulatory polymorphic Vbse have been characterized by a number of laboratories. These studies have shown that Vbse in the mouse are genetically complex, and this complexity is compounded by the impact of MHC polymorphism on responses (see below). Here, I will review the initial description of Mls and its nomenclature. I will then describe how subsequent studies have modified our view of Mls and required a change in nomenclature to the current standard, which will probably require further modifications to account for new data on retroviral coding of Vbse.

1. The Original Definition of Mls

Mls was originally defined as the ability to stimulate a T cell proliferative response in mixed lymphocyte culture in the absence of MHC polymorphism. Using this criterion, Festenstein identified four phenotypes, which he termed Mls^a, Mls^b, Mls^c, and Mls^d (Festenstein, 1972, 1973, 1974; Festenstein and Berumen, 1984; Festenstein et al., 1971, 1977; Festenstein and Kimura, 1988). T cells from mice of the Mls^b phenotype were responders to all other genotypes, whereas T cells from mice of the Mls^d phenotype did not respond to stimulators from mice of any other genotype. The studies were made more complex by the need to match at the MHC, which Festenstein achieved by making F_1 hybrid mice between responder and stimulator strains, allowing MHC-disparate strains to be compared for the Mls phenotype. The genes and the major strains typed for each phenotype are shown in Table I.

TABLE I
THE ORIGINAL CLASSIFICATION OF
Mls GENOTYPES

Genotype	Representative mouse strains
Mls^a	AKR/J, DBA/2, DBA/1
Mls^b	BALB/c, C57BL/6, C57BL/10, CBA/CaJ
Mls^c	C3H/HeJ, A/J
Mls^d	CBA/J

Festenstein also mapped Mls^a of DBA/2 mice to chromosome 1, using a series of recombinant inbred strains (Festenstein et al., 1977). He then produced a congenic line of mice, B10.D2-Mls^a, which confines the Mls^a trait to a small region of chromosome 1 (Festenstein and Berumen, 1984). Subsequent mapping studies using a variety of techniques have shown that Mls^a is close to Spna, which encodes the spectrin α chain (Kingsmore et al., 1989; Moseley and Seldin, 1989; Moseley et al., 1989; Seldin et al., 1988a,b; Watson et al., 1990). A map of chromosome 1 is shown in Fig. 2, with the approximate chromosomal location of the polymorphic gene controlling the Mls^a trait shown. Note that the endogenous retroviral insertion Mtv-7 is not separable from Mls^a and it is now thought to encode this structure (Frankel et al., 1991). It was initially assumed that all Mls loci were alleles of Mls^a, but this has proved not to be the case. Rather, multiple

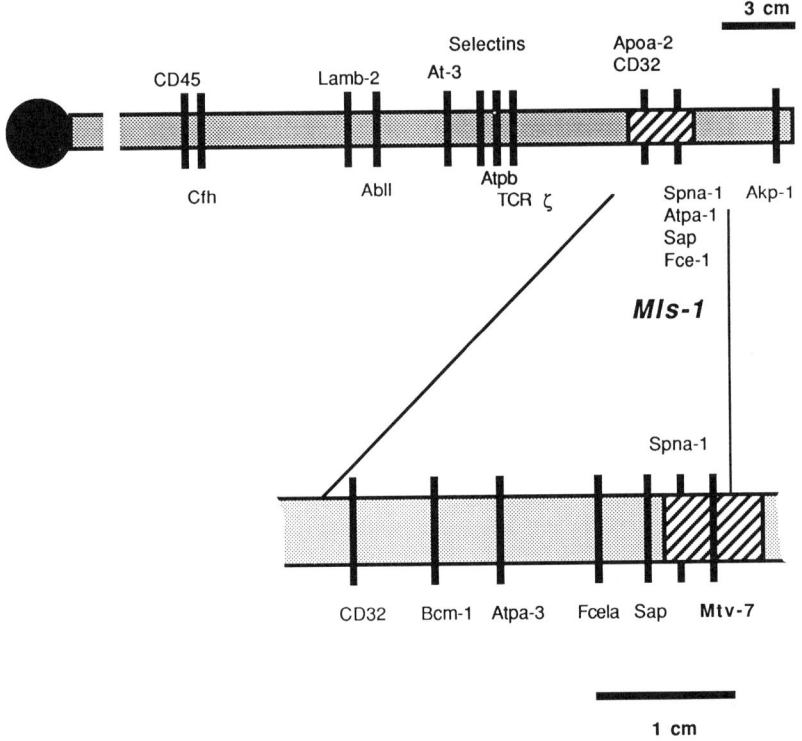

FIG. 2. Map of murine chromosome 1 with the position of Mls-1 and of various genes, including Mtv-7, now believed to encode the product of the Mls-1 locus.

loci control the traits known as *Mls*, and they map to several distinct chromosomal locations (see below).

2. The Existence of Multiple Mls Loci

Two observations have led to a reevaluation of the genetics of *Mls*. The first was the finding that T cells from strains that were typed *Mls^a* or *Mls^c* would respond to stimulator cells from strains typed *Mls^d*, but T cells from *Mls^{axc}* F$_1$ hybrids would not respond to *Mls^d* stimulator cells. Furthermore, *Mls^d* T cells would not respond to either *Mls^a* or *Mls^c* stimulator cells (Table II). This suggested that *Mls^d* was a trait compounded of two separate genes, *Mls^a* and *Mls^c* (Abe et al., 1987a,b,c, 1989; Abe and Hodes, 1988, 1989; Abromson-Leeman et al., 1988; Click and Adelmann, 1988b; Click et al., 1987; Ryan et al., 1987b).

The two-gene control of the Mls^d phenotype, and the independence of the loci controlling Mls^a and Mls^c, were shown by making cloned T cell lines specific for either Mls^a or Mls^c (Abe et al., 1987a,b,c, 1989; Abe and Hodes, 1988, 1989). These cloned T cell lines revealed several important findings (Table III). First, the cloned T cell lines fully discriminated Mls^a and Mls^c, but both sets of cloned T cell lines responded to Mls^d. Second, *Mls^{axc}* F$_1$ stimulator cells stimulated both sets of cloned lines. When F$_2$ mice were generated, *Mls^a* and *Mls^c* segregated independently of one another. Third, when *Mls^d* mice were crossed with an *Mls^b* strain, and the F$_1$ mice were backcrossed to *Mls^b* mice, the offspring typed Mls^a, Mls^b, Mls^c, or Mls^{a+c}. Thus, there are clearly at least two loci that control the classical Mls traits of Festenstein, and they are located on two different chromosomes, because they segregate independently of one another.

These findings led to the proposal of a new nomenclature (Janeway et al., 1988b). This nomenclature designates two loci, called *Mls-1*

TABLE II
EVIDENCE FOR COMPLEXITY IN *Mls* GENETICS[a]

Responder strain	Mls	Stimulator strain Mls				
		b	a	c	d	axc
B10.BR	b	—	+++	+	++++	++++
AKR/J	a	—	—	+	+	+
C3H/HeJ	c	—	++	—	++	++
CBA/J	d	—	—	—	—	—
(AKR × C3H)F$_1$	axc	—	—	—	—	—

[a] The table shows the intensity of the primary mixed lymphocyte response among responder T cells of *H-2^k* mice of various *Mls* genotypes and stimulator cells from the same *H-2^k* strains.

TABLE III
THE RESPONSES OF CLONED T CELL LINES DEFINE TWO
INDEPENDENT Mls LOCI[a]

Stimulator cells	Mls	Specificity of cloned T cell line for	
		Mls^a	Mls^c
B10.BR	b	−	−
AKR/J	a	++	−
C3H/HeJ	c	−	++
CBA/J	d	++	++
(CBA/J × B10.BR)F$_1$	bxd	++	++
(CBA/J × B10.BR)F$_2$	bxd	++	++ (25%)
		++	− (25%)
		−	++ (25%)
		−	− (25%)
(AKR/J × C3H/HeJ)F$_1$	axc	++	++
(AKR/J × C3H/HeJ)F$_2$	axc	++	++ (25%)
		++	− (25%)
		−	++ (25%)
		−	− (25%)

[a] Data obtained by typing with multiple cloned T cell lines of each specificity show that Mls^a and Mls^c segregate independently and the products are recognized by different T cells.

and Mls-2. At each locus, there is a stimulatory allele called a, and a nonstimulatory or responder allele called b (Table IV). Mice with the a allele at a given locus delete T cells able to respond to the a allele at that locus, whereas mice with the b allele retain these T cells, allowing them to generate the Mls response. Mls-1 maps to chromosome 1 in DBA/2 mice, and Mls-1^a is identical to Mls^a, and so on as shown in Table IV. This nomenclature has since become standard in the field, although not all investigators agree with its use. Furthermore, recent studies examining a widening number of strains of mice have defined new Mls-like stimulatory loci that will need molecular definition and, ultimately, naming (Abromson-Leeman et al., 1988; Janeway and Katz, 1985; Vacchio et al., 1990). The recent discovery that Vbse are encoded by retroviral insertions (see below) may allow a more rational approach to this problem.

B. NON-Mls Vβ SELECTIVE ELEMENTS IN THE MURINE GENOME

Although Mls allelic polymorphisms generate potent unidirectional T cell responses by acting selectively on T cells whose receptors are

TABLE IV
THE CURRENT NOMENCLATURE FOR Mls LOCI AND STRAIN DISTRIBUTION[a]

Old nomenclature	New nomenclature		Strain distribution
Mls	Mls-1	Mls-2	
a	a	b	AKR/J, MA/MyJ, RF/J, LT/Ch
b	b	b	C57BL/6, C57BL/10, CBA/CaJ, MRL/mp, SJL/J
c	b	a	C3H/HeJ, BALB/c, A/J
d	a	a	CBA/J, C58/J, CE/J, DBA/2J, NZB, DBA/1, D1.LP, PL/J, SM/J

[a] Mice congenic with C57BL/10, BALB/c, and A/J all type identically at Mls.

encoded by responding Vβ gene segments, not all Vbse induce a primary mixed lymphocyte reaction among MHC-identical T cells or even stimulate T cell hybrids that bear an appropriate Vβ-encoded TCR. Thus, although Mls fits the definition of a Vbse, not all Vbse fit the definition of Mls. In this section, the other murine Vbse that have been defined will be described.

1. Murine Vbse That Do Not Stimulate a Primary T Cell Response among MHC-Identical Cells

The first Vbse studied was thought initially to be the MHC class II molecule I-E, the murine homologue to HLA-DR (Kappler et al., 1987a,b). Many mouse strains lack expression of I-E due to defects in the gene encoding the Eα chain (Mathis et al., 1983). Kappler et al. (1987a) discovered that most T cells bearing receptors encoded by a particular Vβ gene segment, called Vβ17a, responded to stimulator cells only if they express the I-E molecule. Furthermore, strains that express I-E molecules delete T cells expressing TCRs encoded by Vβ17a, whereas strains that lack I-E molecules generate cells expressing Vβ17a-encoded TCRs at varying frequency. Although T cells expressing Vβ17a-encoded receptors respond to spleen cells from mice that express I-E molecules, they do not respond to fibroblasts transfected with these same I-E molecules, nor to I-E-expressing macrophages derived from bone marrow culture (Marrack and Kappler, 1988a). From this, it is clear that some element that apparently lacks relevant polymorphism is responsible for responsiveness of Vβ17a-bearing T cells to I-E. Presently, it is believed that a Vbse is required for this selective effect of I-E molecules on Vβ17a expression, and that this gene is selectively expressed in B cells. The gene encoding this Vbse has not been mapped, as no polymorphisms in this gene have been detected. The absence of polymorphism also prevents this Vbse from fulfilling the classical definition of Mls, the stimulation of MHC-identical T cells.

Once it was established that Vbse can mediate clonal deletion of those developing T cells whose receptors are encoded by a particular Vβ gene segment, several surveys of Vβ gene expression using both antibodies to the TCR (Bill et al., 1989; Happ et al., 1989; Kanagawa et al., 1989; Pullen et al., 1989a,b, 1990a; Tomonari and Lovering, 1988; Vacchio et al., 1990; Woodland et al., 1990), panels of T cell hybrids (Bill et al., 1989), RNase protection assays (Okada and Weissman, 1989), and the polymerase chain reaction (Choi, 1989) were conducted. The clearest findings from these surveys are that expression of I-E leads to the deletion of T cells expressing TCRs encoded by Vβ5, Vβ11, and Vβ12 in most but not all strains of mice. Using these I-E-positive strains of mice, several groups have conducted mapping studies of the Vbse assumed to be responsible for this effect. These studies are summarized in Table V. Additional data have come from rare T cell hybrids that respond to these I-E:Vbse combinations; such responses appear to correlate perfectly with the data obtained by studying clonal deletion (Palmer et al., 1989; Woodland et al., 1990).

It can be seen from Table V that several different Vbse located on several distinct chromosomes mold the T cell receptor repertoire by deleting developing T cells. Some of these Vbse are only observably active when combined with I-E, whereas I-E-negative strains express the corresponding Vβ-encoded TCR normally. It may be asked why no Vbse associated with the MHC class II molecule I-A, equivalent to human HLA-DQ, have been detected. There are two possible reasons for the failure to observe I-A-associated Vbse. The first is that there are

TABLE V
PROPOSAL FOR A GENERAL NOMENCLATURE FOR Mls AND OTHER Vβ SELECTIVE ELEMENTS[a]

Vbse	Proposed nomenclature	Vβ gene segments affected	Chromosomal location
Mls-1	Vbse-1	6,8.1,9	1
Mls-2	Vbse-2	3	4
Mls-3	Vbse-3	3	12
?	Vbse-4	17a	—
?	Vbse-5	5	—
?	Vbse-6	11	—
?	Vbse-7	7	—
?	Vbse-8	14	—

[a] All of these Vbse require class II MHC to induce deletion of various Vβ-encoded TCR.

no I-A-negative strains of mice, so nonpolymorphic Vbse acting together with I-A go undetected. The second is that Vbse only act together with I-E. The latter is unlikely, because Mls-1, at least, can be stimulatory on I-E-negative stimulator cells, although I-E-expressing strains usually stimulate responses to Mls-1 and especially Mls-2 more efficiently than do I-E-negative strains (see below for further discussion of MHC effects). Thus, at least one Vbse is functional in conjunction with I-A molecules.

The situation is even more complex in that some Vbse appear to augment the representation of T cells bearing TCRs encoded by particular $V\beta$ gene segments. Thus, one cannot use the simple criterion of loss of T cells in identifying Vbse. For instance, it has been shown that an I-E-associated Vbse actually enhances representation of T cells expressing $V\beta6$-encoded TCRs in some cases (MacDonald et al., 1988c). This is almost certainly due to positive selection of T cells bearing such receptors (see below). Again, the Vbse involved has not been identified.

2. A Suggested General Nomenclature for Vbse

There is no simple solution to the problem of nomenclature of the Vbse at present. The two Mls nomenclatures are both in use, and neither nomenclature encompasses the Mls-like Vbse that delete T cells bearing TCRs encoded by particular $V\beta$ gene segments, or those that positively select T cells on the same basis. Furthermore, where such genes are nonpolymorphic, they cannot be mapped or assigned an allele designation. Thus, what is clearly needed is a nomenclature that allows all Vbse to be encompased. This nomenclature should ideally reflect relationships between individual genes, their function, and the T cells on which they act.

The proposed nomenclature is shown in Table V. It uses Vbse to denote the general trait, assuming that all Vbse are related. Second, it uses numbers to denote individual loci. It retains the designation Vbse-1 for Mls-1, and Vbse-2 for Mls-2. As there is some evidence for a third locus that comprises Mls^c, and as this has been referred to as Mls-3, this is called Vbse-3. Third, it retains the allele designation a when the gene has an effect (stimulation, T cell deletion), and b for those strains in which no effect is observed. Thus, Mls^a or Mls-1^a becomes Vbse-1^a.

An alternative nomenclature would list the Vbse on the basis of the $V\beta$ gene segments on which they act, such that Mls-1 would be Vbse-6, 8.1,9 etc. However, this nomenclature is cumbersome and also assumes that all $V\beta$ effects are identified when the gene is discovered.

Thus, although this nomenclature would be informative as to the specificity of each gene, it is rejected because it is too cumbersome.

It is hoped that this nomenclature is sufficiently broad to encompass all possible changes in the understanding of the genetics of these molecules. It's advantage is that it designates genes on the basis of their most striking property, selective effects on T cells defined on the basis of Vβ gene segment usage in the TCR. Second, by simply adding a number for each newly defined locus, any new locus can be defined. Third, by using only the designations a and b, we can denote the two main types of Vbse so far observed, those that can act and those whose effect is not observed. As this is only a proposed nomenclature, the current standard nomenclature shown in Table IV will be used in this review. However, the term Vbse will be used to designate all such genetic elements as a class.

3. Null Alleles of Vbse

What is the nature of the null or b alleles in this system? Are these genes not expressed, or do they have no impact on T cells? This question may be answerable by the new molecular definition of Vbse; null alleles lack the retroviral insertions thought to encode Vbse. Thus, null alleles really should be null.

Nevertheless, there are hints in the data that there are silent or nonpolymphic Vbse products that do have effects on T cells. First, Vbse without the ability to stimulate mixed lymphocyte reactions (MLRs) between MHC-identical strains (Bill et al., 1988) or delete developing T cells have been shown to have effects on T cell development (Benoist and Mathis, 1989; Bill and Palmer, 1989). Such effects could be due to absence of Vbse polymorphism, as in the case of the Vbse that affects Vβ17a expression. Second, Click has presented data that he interprets as T cell responses to Mls-1^b. One can criticize these latter experiments for their failure to remove T cells from the stimulator cell population and B cells from the responder cell population, allowing a possible "back-reaction." In a back-reaction, the T cells in the stimulator population cannot proliferate, because they are irradiated or treated with mitomycin C, but they can secrete growth factors that drive the proliferation of B cells in the responder population. Thus, these experiments need to be examined closely for evidence of responding T cells expressing TCRs with restricted Vβ usage. Should such cells be found in the responding population, then it will be clear that "silent" Vbse are not totally undetectable (Click and Adelmann, 1988a, 1989; Click et al., 1984, 1985).

C. The Bacterial Toxic Mitogens

It has been known for many years that a series of bacterial toxins are potent T cell mitogens (Peavy *et al.*, 1970). The demonstration that the staphylococcal enterotoxins stimulate T cells by binding to MHC class II molecules and act on the CD4:TCR complex of some but not all cloned T cell lines suggested a strong similarity to *Mls* (Carlsson *et al.*, 1988; Fischer *et al.*, 1989; Fleischer and Schrezenmeier, 1988; Fraser, 1989; Janeway *et al.*, 1988a; Mollick *et al.*, 1989; White *et al.*, 1989). This similarity was extended by demonstrating that T cells responding to a given bacterial toxic mitogen are identifiable on the basis of Vβ gene segment usage in their TCRs (Janeway *et al.*, 1989b; Kappler *et al.*, 1989a; Marrack and Kappler, 1990; White *et al.*, 1989; Yagi *et al.*, 1990). Thus, the T cell stimulatory bacterial toxic mitogens behave as useful analogs of Vbse. That bacterial toxic mitogens can also function in clonal deletion has been shown *in vivo* and in thymic organ culture (White *et al.*, 1989; Yagi *et al.*, 1991).

The most extensively studied of the bacterial mitogens have been the staphylococcal enterotoxins, the distantly related toxic-shock syndrome toxin (TSST), the streptococcal exfoliating toxins, and the *Mycoplasma arthritidis* T cell mitogen (Table VI). The structure of many of these substances has been defined by protein and nucleic acid sequencing, and excellent recent reviews about their toxic (Bergdoll,

TABLE VI
The Bacterial Toxic Mitogens and the Murine Vβ Gene Segments Associated with Responses to Them[a]

Bacterial toxic mitogen (abbreviation)	Responding murine Vβ gene segments
Staphylococcal enterotoxin (SE)	
SEA	1, 3, 10, 11, 17
SEB	7, 8.1, 8.2, 8.3, 17
SEC1	(3),[b] 8.2, 8.3, 11, 17
SEC2	(3), 8.2, 10, 17
SEC3	(3), 7, 8.1, 8.2
SED	(3), 7, 8.1, 8.2, 8.3, 11, 17
SEE	11, 15, 17
Toxic shock syndrome toxin (TSST)	3, 15, 17
Exfoliating toxin (ExFT)	3, 10, 11, 15, 17
Mycoplasma arthritidis mitogen (MAM)	6, 8.1, 8.2, 8.3

[a] Data about murine Vβ associations are from references given in text.
[b] Vβ 3 responses may reflect contamination in enterotoxin preparations.

1970, 1979), molecular (Wannamaker an Schlievert, 1988), and immunologic (Marrack and Kappler, 1990; Purdie et al., 1991) properties are available.

The most important feature of the bacterial toxic mitogens is their availability as purified proteins of known molecular structure. The working model of Vbse action is based in large part on the assumption that bacterial toxic mitogens are molecular mimics of Vbse. However, it must be realized that however striking the similarities between bacterial toxic mitogens and endogenous Vbse, these may not be homologous structures; indeed, the recent molecular identification of Vbse suggests that these are analogs, not homologues. Thus, each property discovered for one class of Vbse must be confirmed for the other.

IV. The Molecular Basis of Vβ Selective Element Action

The working model of Vbse action is based on a series of studies in which T cell responses to endogenous Vbse or bacterial toxic mitogens have been measured and the role of individual molecules determined. To date, three structures in addition to the Vbse have been unequivocally identified as contributing to this response: the MHC class II molecule, the CD4 coreceptor, and the $\alpha:\beta$ heterodimeric T cell receptor for antigen:MHC. However, evidence for involvement of class I MHC molecules, the CD8 coreceptor, a putative receptor structure for Mls, and the $\gamma:\delta$ heterodimeric T cell receptor has been obtained in a limited number of studies and will also be mentioned here.

The involvement of a given molecule in the response to Vbse can be shown in several ways. First, genetic polymorphisms that affect the response to Vbse have identified some of the molecules involved. Second, the ability of antibodies to a molecule to inhibit the response has been taken as evidence for that molecule's involvement. Third, enrichment of expression of a given structure in the responding cells suggests but does not directly document that structure's involvement in the response. Fourth, in vivo or in vitro ontogenetic deletion of cells expressing a particular structure implies its involvement in the response. Fifth, the binding of a bacterial toxic mitogen to a molecule suggests its involvement in the response. Sixth, when the loss of a molecule is associated with a loss of response, that molecule is implicated in the response. In such a system, conferring the ability to respond by means of gene transfection is definitive evidence for that structure's involvement in the response. This last technique has the particular advantage that the transferred gene can be manipulated, allowing one to pinpoint the critical parts of the structure. This approach has been notably successful in studying the role of the T cell receptor and the MHC class II molecule in these responses. These approaches have given a consistent picture of the molecular basis of the T cell response to Vbse, although much work remains to be done.

A. THE INVOLVEMENT OF MHC MOLECULES

The MHC class II molecules are the transport system that delivers peptide fragments of antigens from cellular vesicles to the cell surface. The complex of peptide and MHC class II molecule is recognized by the TCR of CD4 T cells. The MHC class I molecules similarly deliver peptides from the cell cytoplasm to the surface, where the complex is recognized by CD8 T cells. The MHC molecules are also the major stimulators of mixed lymphocyte reactions; ~3% of T cells will respond to a given nonself MHC molecule and its associated peptides (Wilson et al., 1968). It was originally suggested that *Mls* might be an unlinked gene that encodes an MHC class II-like molecule. If this were the case, then the response to Mls would differ from all other T cell responses in being independent of molecules encoded in the MHC on chromosome 17 of the mouse. However, early studies showed that some responses to Mls were strongly influenced by the MHC class II genotype of the stimulator cell (Gress et al., 1981; Janeway et al., 1980; Jones and Janeway, 1982; Peck et al., 1977), suggesting that Mls was like other antigens in all respects except the unusually high number of responding T cells (~20%; see below).

The exact role of MHC molecules had to be determined by detailed studies of specific responses. Although there was initially a controversy over whether or not responses to Mls were MHC restricted (Gress et al., 1981; Janeway et al., 1980; Jones and Janeway, 1982; Molnar-Kimber and Sprent, 1980; Molnar-Kimber et al., 1980; Peck et al., 1977; Webb et al., 1981), it soon became apparent that MHC class II molecules were critically involved in responses to both endogenous Vbse and bacterial toxic mitogens, because polymorphisms in MHC class II influenced the T cell response to Vbse (Cole et al., 1981; Fleischer and Schrezenmeier, 1988; Janeway et al., 1988a; Marrack and Kappler, 1990). However, this genetic effect was quite distinct from classical MHC-restricted antigen recognition, in which only a single MHC allelic form is able to present a given peptide. Rather, in the response to Vbse, most MHC class II molecules appear to be competent for response, with one or a few allelic forms failing to provide stimulation. The details of individual responses, and the means by which they were studied, are given in this section.

1. The Involvement of I-E Molecules

The most apparent influence of MHC polymorphism on responses to Vbse is seen for the MHC class II I-E molecule, the murine homologue of HLA-DR. The studies that were initially thought to demonstrate MHC restriction in T cell responses to Mls actually showed that I-E-positive strains of mice could stimulate a response to Mls-2^a whereas I-E-negative strains could not (Peck et al., 1977). Many strains of mouse lack expression of the I-E molecule due to a deletion in the promoter region of the Eα chain (Mathis et al., 1983). Cells from such strains either do not stimulate a response to Mls-2^a or do so far less effectively than do congenic mice that differ only by expression of I-E. Second, anti-I-E antibodies are highly effective at inhibiting the responses of both normal T cells and Mls-specific T cell clones, and they do so by acting on the stimulator cell (Katz and Janeway, 1985). Third, some cloned T cell lines fail to respond to bacterial toxic mitogens unless the stimulator cell expresses I-E, as seen by comparing I-E-positive with I-E-negative cells (Yagi et al., 1991). This effect depends upon the Vβ gene segment expressed by the T cell clone. We have recently observed that T cells stimulated with staphylococcal enterotoxin B (SEB) are enriched for expression of Vβ8.1 and Vβ8.2 only when the antigen-presenting cell expresses I-E, and most Vβ8.2-expressing cloned T cell lines will respond to SEB presented by the I-E-positive strain B10.D2, but not by the MHC-congenic, I-E-negative strain B10.GD. By contrast, T cells expressing Vβ3 are stimulated by either staphylococcal enterotoxin A (SEA) or SEB to a similar degree, whether I-E or only I-A is expressed on the stimulating cell. This result may be affected by contaminants in SEB preparations. Cells that do not express MHC class II molecules do not stimulate responses to SEs. Direct binding of labeled SEB to mouse B lymphoma cells can be partially inhibited with anti-I-E monoclonal antibodies, but not by anti-class I MHC monoclonal antibodies (Yagi et al., 1990). Finally, as noted in Table V, many of the Vbse delete developing T cells only in I-E-positive strains of mice. Thus, all these data point to a dominant role of I-E molecules in responses to Vbse in mice.

In humans, no endogenous Vbse have been defined. However, T cell responses to the bacterial toxic mitogens have been examined for influences of HLA polymorphism and directly for interaction of SEs with MHC class II molecules, to which they bind with high affinity. These studies show that HLA polymorphism does not detectably affect responsiveness to SEs, although detailed studies may reveal effects on the Vβ repertoire expression in the future. As in the mouse, L cell transfectants expressing HLA-DR, -DP, or -DQ will allow SEA, SEB, and TSST to stimulate T cells, whereas class II-negative human cells or nontransfected L cells will not (Purdie et al., 1991). Direct binding of SEA, SEB, and TSST to HLA-DR has been reported by a number of groups (Fischer et al., 1989; Fraser, 1989; Mollick et al., 1989). Up to 60% of the SEA bound by human B cells is bound to HLA-DR, as determined by precipitation of labeled SEA from lysates of B cells with monoclonal anti-HLA-DR antibody (Fraser, 1989). Thus, the human homologue of murine I-E appears to be the dominant, but not the exclusive, receptor for SEs in humans. Taken together, these data demonstrate that I-E molecules and their human homologues, HLA-DR molecules, are the major cell surface receptor for Vbse and bacterial toxic mitogens.

Why should I-E molecules be particularly important in Vbse phenomena? There are several possible answers to this question. There appear to be structural features of I-E molecules that favor interactions with Vbse; David and co-workers have observed such effects in MHC class II transgenic mice, in which Eα chains allow Vbse-mediated clonal deletion of T cells in mice lacking Eβ chains (Anderson and David, 1989). In this case, the Aα:Aβ complex is not able to mediate deletion, whereas the Eα:Aβ complex does so even though it is poorly expressed on the cell surface. These results suggest that the Eα chain is a principal site of Vbse interaction. However, it must be remembered that the binding site for Vbse may involve binding to combinatorial determinants involving both chains, or conformational changes in Aβ induced by pairing with Eα. Chemical cross-linking followed by immunoprecipitation coupled with site-directed mutagenesis may be required to identify the binding site of Vbse to I-E and HLA-DR molecules.

Similar studies have not been performed with HLA-D hybrid molecules. However, competitive binding studies carried out by Fraser give interesting results (Purdie et al., 1991). SEA can compete with SEB and TSST for binding to fixed B cells. However, SEB and TSST do not compete with one another for binding, and neither can displace binding of SEA, presumably because SEA has a higher affinity. These results suggest that SEA binds to a site overlapping the site bound by SEB on one side and TSST on the other, but that TSST and SEB bind to nonoverlapping sites. However, other explanations for this finding are possible (see Section VIIIA.). Further studies of binding of all the toxins to purified molecules and hybrid complexes should map the binding sites on HLA-DR and I-E molecules with accuracy, thus enhancing our knowledge of the structure of this molecular complex.

2. The Involvement of I-A Molecules

The I-A molecule is the homologue of HLA-DQ in humans, and it is found in all strains of mice. It plays a dominant role in the presentation of peptides derived from most but not all test antigens and in the mixed lymphocyte response between strains of mice that differ at the MHC. However, its contribution to responses to Vbse and bacterial toxic mitogens appears to be less important than the contribution of I-E molecules (see above). Nevertheless, there is decisive evidence for the ability of I-A molecules to present both Vbse and bacterial toxic mitogens to T cells. First, some strains of mice lack I-E molecules but still reveal Vbse-mediated T cell stimulation and clonal deletion of developing T cells bearing TCR encoded by responsive $V\beta$ gene segments (Jones and Janeway, 1982; Kappler et al., 1988; Lynch et al., 1985; MacDonald et al., 1988a). Second, anti-I-A antibodies inhibit normal and cloned T cell responses to many Vbse (Debreuil et al., 1982; Janeway and Katz, 1985; Janeway et al., 1980). Interestingly, both the response of a given cloned T cell line and of unselected normal T cells to Mls-1^a can be partially blocked by either anti-I-A or anti-I-E, although anti-I-E is generally the more potent inhibitor (Janeway et al., 1983; Janeway and Katz, 1985). Furthermore, mixtures of anti-I-A and anti-I-E antibodies are required to fully inhibit stimulation by Vbse.

Some cloned T cell lines that respond to peptides presented by I-A can also cross-react with Mls-1^a stimulator cells. The response to Mls-1^a is often more effectively inhibited by anti-I-E than by anti-I-A antibodies (Janeway et al., 1983; Janeway and Katz, 1985). Thus, neither the restriction specificity nor the isotype of the MHC class II molecule recognized by the T cell receptor in its physiological response to peptide:MHC class II appears to affect the TCR's interaction with MHC class II molecules mediated by Vbse. These findings strongly imply that the T cell receptor's interaction with MHC class II molecules during responses to Vbse is independent of specific antigenic peptide or specific MHC recognition. Although it seems likely that there is intimate contact of the TCR with the MHC molecule during such responses, there is no conclusive evidence that this is so.

When murine cells are used in responses to bacterial toxic mitogens, one finds that certain I-A molecules are less effective than I-E molecules in stimulation, but that many I-A molecules can stimulate with various bacterial toxic mitogens. Although I-A^d can be shown to bind SEB to an extent similar to I-E, this complex is a weak stimulator of T cells, especially those expressing Vβ8-encoded TCR (Yagi et al., 1991). The response to MAS is absent in I-E-negative strains of mice (Bekoff et al., 1987; Cole et al., 1981, 1982a,b).

The most striking disparity between I-A and I-E is in T cell deletion. T cells expressing TCR encoded by several different Vβ gene segments are deleted in I-E-positive but not I-E-negative strains of mice (Bill et al., 1989; Kanagawa et al., 1989; Kappler et al., 1987a,b; Tomonari and Lovering, 1988; Vacchio et al., 1990; Woodland et al., 1990). No comparable effect has yet been observed for I-A. However, this may simply reflect the absence of I-A-negative mice at the present time. Gene knockout experiments may reveal nonpolymorphic Vbse that act together with I-A to delete developing T cells. Alternatively, one may argue that deletion of developing T cells by Vbse that work in concert with I-E are allowable, just as the absence of I-E is allowable, but that Vbse that delete developing T cells in association with I-A are not permissible just as I-A-negative mice have not been observed. The Vbse that associate with I-A may have too low an affinity for the Vβ site on the TCR to cause frank deletion, as this effect would be too deleterious to the host.

As noted above, SEB and TSST binding to HLA-DQ, the human homologue of I-A, has been observed using transfected mouse L cells, and such cells are competent to stimulate human T cell proliferative responses (Scholl et al., 1989a,b, 1990). Thus, HLA-DQ is clearly competent to present bacterial toxic mitogens. To date, no effect of polymorphism in HLA-DQ has been observed in such responses. HLA-DQ binds less toxin than HLA-DR and appears to be less critical in responses. SEA binds less avidly to HLA-DQ as well.

These findings demonstrate that murine I-A molecules and their human HLA-DQ counterparts are receptors for Vbse and for bacterial toxic mitogens, albeit less efficient ones than I-E and HLA-DR. The implications of these differences are difficult to decipher, as no clear cut distinction can yet be made between the roles of I-A and I-E in murine T cell responses to either conventional antigens or Vbse. In keeping with the absence of effects of MHC polymorphism on responses to Vbse and bacterial toxic mitogens, one very illuminating study shows that mutations in Aα chains that affect presentation of peptide to a T cell hybrid do not affect presentation of SEs to the same T cell hybrid. Furthermore, SEs do not compete with peptide for presentation (Dellabona et al., 1989). These data indicate that SEs do not interact efficiently with the peptide-binding groove, but must have a binding site elsewhere in the I-A molecule.

3. Incompetent MHC Class II Molecules

As noted above, some MHC class II molecules fail to present Vbse to responsive T cells. The most clearly defined example is the I-A^q molecule, which does not present Mls-1^a to T cells (Jones and Janeway, 1982; Lynch et al., 1985). Furthermore, some I-A molecules present SEs very poorly to certain T cells, particularly as noted above in the case of SEB presentation to T cells whose TCR is encoded by Vβ8.1 or Vβ8.2 (Yagi et al., 1991). The fact that the same I-A molecule does present SEB to T cells expressing Vβ3 suggests that SEB binds effectively to these molecules, and this can be shown by direct binding studies also (Yagi et al., 1990, 1991). Thus, the failure of this response is likely to involve the interaction of the TCR with the MHC class II molecule, or with the complex of SEB and the MHC class II molecule. MHC class II molecules could be incompetent to bind Vbse, or they could be incompetent to present the bound Vbse to a particular T cell receptor, either because the T cell receptor cannot interact simultaneously with both the Vbse and the MHC class II molecule for steric reasons, because the MHC class II molecule contributes too little binding energy to the interaction, or because bound peptides prevent docking of the T cell receptor with the MHC molecule in a manner that allows the Vbse to bind the Vβ-encoded portion of the receptor. Studies of incompetent MHC alleles will be very valuable in determining the precise interaction of the receptor with the MHC molecule that leads to stimulation by Vbse. An alternative interpretation of these studies is given in Section IX.

Although there are I-A molecules that are incompetent to present certain Vbse to certain T cell receptors, it is fair to state that there are no known examples of I-E or HLA-DR molecules that fail to present Vbse or bacterial toxic mitogens. Furthermore, it has been shown that Eα paired with the Aβ chain of an incompetent I-A molecule generates a competent allele, implicating polymorphic regions of the Aα chain in interaction with Vbse (Anderson and David, 1989). Because it is known that Aα chains are polymorphic whereas Eα chains are not, these data strongly suggest that the failure of some I-A molecules to present Vbse may be due to differences in the Aα chains, and the universal competence of Eα-containing MHC class II molecules points to a binding site on Eα. This may be an important clue in searching for Vbse-binding sites on MHC class II molecules. The detailed analysis of various types of incompetent MHC class II molecules may reveal much about the interaction of the T cell receptor, MHC class II, and Vbse leading to T cell activation.

B. The Involvement of T Cell Receptors and Coreceptors

Conventional protein antigens are recognized as a complex of the specific antigenic peptide bound in the major peptide-binding groove on the outer aspect of the MHC class II molecule (for review, see Davis and Bjorkman, 1988). This groove and its associated peptides are apparently not involved in the recognition of MHC class II molecules presenting Vbse (see above). The Vβ-encoded portion of the TCR is the critical element in inducing T cell responses to Vbse. In addition, the CD4 coreceptor plays a critical role in most murine T cell responses to peptides presented by MHC class II, increasing sensitivity to stimulation by at least 100-fold (Janeway et al., 1989c). In this section, the contributions of the elements of the T cell receptor to responses to Vbse will be explored.

1. The Involvement of the T Cell Receptor

The earliest indication that the T cell receptor might be involved in the recognition of Vbse was the finding that responses to Mls were clonal in nature (Jones and Janeway, 1982; Molnar-Kimber et al., 1980; Webb et al., 1981). However, as the precursor frequency was so high, it was difficult to believe that the same receptor that recognized antigen could also recognize Mls. A similar percentage of T cells specific for a wide range of protein antigens also responded to Mls (Janeway et al., 1983; Janeway and Katz, 1985; Katz and Janeway, 1985; Webb and Sprent, 1986). However, the finding that antibodies to the receptors on these T cell clones would block both antigen and Mls recognition (Katz and Janeway, 1985) made it more likely that the TCR was directly involved in Mls recognition.

In contrast to these findings, studies using T cell hybrids between dual reactive cloned T cells and the AKR lymphoma BW5147 showed independent segregation of reactivity to Mls and reactivity to antigen or allogeneic MHC molecules. These studies led to the proposal of an alternative hypothesis, that the T cell has a separate receptor for Mls (Webb et al., 1988; Webb and Sprent, 1986). It was initially proposed that there was no role for the TCR in the response to Mls, but later this hypothesis was modified to state that the functioning of the putative receptor for Mls is dependent on the T cell receptor for antigen but distinct from that receptor. This possibility is still not ruled out by available data. For instance, there could be a T cell molecule that associates with the TCR and forms part of the binding site for Mls. If this protein were present in all T cells, but only able to function when the Vbse also binds to Vβ, then it could go undetected in most studies.

An alternative explanation for the results of Webb and co-workers is that BW5147 has endogenous αTCR and βTCR chains that can combine with the donor clone receptors to generate Mls reactivity. In this case, one could readily explain the selective loss of antigen reactivity. However, it is much more difficult to accommodate a loss of reactivity to Mls with retention of reactivity to antigen in such hybrids. It is these latter results that give the Webb findings their unique interest.

2. The Involvement of the Vβ Region of the T Cell Receptor

Whether or not there are other entities on the T cell surface required to recognize Mls, it is abundantly clear that responses to all Vbse and to bacterial toxic mitogens involve the T cell receptor. There is no other way to explain all of the data. The defining characteristic of the response to a given Vbse is the involvement of T cells marked by expression of TCR encoded by a restricted set of Vβ gene segments (Abe et al., 1988; Bill et al., 1988, 1989; Janeway et al., 1989b; Kappler et al., 1987a,b, 1988, 1989a; MacDonald et al., 1988b; Marrack and Kappler, 1988b, 1990; Palmer et al., 1989; Purdie et al., 1991; Woodland et al., 1990). Antibodies to the T cell receptor inhibit responses to Vbse or bacterial toxic mitogens under conditions wherein responses to other classes of T cell activators, such as concanavalin A or anti-CD3, are not inhibited (Janeway et al., 1988a, 1989b; Katz and Janeway, 1985). Loss of the T cell receptor is associated with loss of reactivity, and reactivity can be transferred to recipient cells by transfection with the genes encoding the TCR, provided that the β chain of the responsive clone is expressed on the surface of the recipient cell line (Kaye, 1989). Of particular interest is the finding that in most cases, the β chain alone will generate reactivity to Vbse regardless of the α chain with which it is paired, although details of reactivity may differ. These findings point so strongly to the region of the β chain encoded by the Vβ gene segment that it is to this portion of the molecule that attention has turned in attempting to map sites critical for activation by Vbse.

Although it is true that Vbse show marked specificity for that region of the TCR encoded by the Vβ gene segment, it is also true that not all T cells bearing such receptors are necessarily affected (Kappler et al., 1988). Furthermore, elements other than this part of the receptor may play a role in some phenomena. For instance, a series of cloned T cell lines expressing Vβ8.2 have recently been observed to differ in their MHC class II molecule requirements for stimulation by SEB (Yagi et al., 1991). A more common observation is that some, but not all, T cells expressing a particular Vβ-encoded receptor are deleted during intrathymic ontogeny (Kappler et al., 1988; Tomonari and Lovering, 1988).

Some of the residual cells can respond to antireceptor antibody. This suggests that more subtle aspects of the interaction between the TCR, MHC class II, and the Vbse may involve β chain VDJ junctions or the α chain of the receptor. Under conditions of limited stimulation, these differences may be critical.

3. Mapping Relevant Sites on the T Cell Receptor Vβ Region

Having identified the Vβ-encoded segment of the TCR as the critical site of action of Vbse, Kappler, Marrack and co-workers sought pairs of closely related Vβ that differed in their reactivity to Vbse. A wild mouse Vβ8.2 was found that conferred reactivity to Mls-1^a to all cloned T cells on which it was expressed, whereas most cloned T cells lines expressing Vβ8.2-encoded receptors derived from laboratory mice lacked this reactivity (Pullen et al., 1990b). Pullen et al. systematically altered the amino acids of the laboratory Vβ8.2 gene to ask which changes conferred Mls-1^a reactivity on cells expressing the mutant β chain. This study pinpointed amino acids 22, 70, and 71 as critical for this response. Using a proposed model structure for the TCR, these residues are found on the lateral face of the Vβ domain, well away from the three hypervariable loops in the model structure. Importantly, the one residue that differed between the wild mouse and laboratory mice was found on a CDR loop did *not* affect the response to Mls-1^a. The mutations that altered responses to Mls-1^a had no influence on the response of cells bearing these receptors to the peptide:self MHC complex for which the original clone showed specificity, indicating the integrity of the receptor and of the signaling apparatus, and the lack of involvement of hypervariable residues in reactivity to Mls-1^a (Pullen et al., 1990b). Cazenave and co-workers have recently identified another wild mouse in which a Vβ did not show the expected deletion by I-E, even when crossed with laboratory strains that could give this deletion (Cazenave, 1990). This Vβ differed from that of laboratory mice by two amino acids at positions in close proximity to the site defined by Pullen et al. (1990b).

A parallel approach was taken by Choi and co-workers using closely related human Vβ13 genes that differ in reactivity to SEC2 and SEC3. The amino acids between positions 67 and 77 containing amino acid differences at 8/11 positions between these Vβ13 genes were exchanged. This exchange generated a TCR that could respond to SEC2 and SEC3 from one that could not. In this case, it was not possible to detect antigen reactivity (Choi et al., 1990). However, this experiment did prove that differences in the three CDRs believed to contact the MHC molecule were *not* required to confer reactivity to SEs, although they may play a role in the response. Thus, both murine Vβ responses to Mls-1^a and human T cell responses to SEs involve the region of the TCR believed to be on its lateral face and involving the loop between amino acids 67 and 77. These results are summarized in Fig. 3.

An alternative approach was taken by M. M. Davis and co-workers (personal communication). They transplanted whole CDR loops from a responsive to a nonresponsive βTCR chain. The resultant cells were tested for reactivity to SEB. This analysis implicated the hypervariable loops in the response. Unfortunately, given the unknown effects of such manipulations on the conformation of the region that appears to interact with SEB, the likely importance of TCR:MHC contacts in responses to SEB, and the variability of the data, it is premature to suggest that the hypervariable loops directly contact Vbse. Nevertheless, these data emphasize the complexity of this response, and demonstrate that much remains to be discovered in this system.

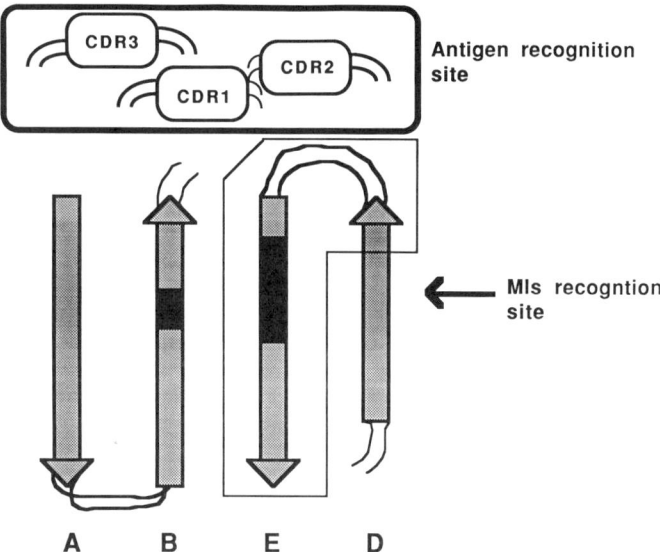

FIG. 3. Schematic view of the Vβ domain of the T cell receptor. At the top are the three hypervariable loops believed to make contact with the peptide:MHC ligand. The recognition site for Mls-1 is shown in black on β strands B (amino acid residue 22) and E (amino acid residues 70 and 71) according to Pullen et al. (1990b). The box denotes the region of Vβ that is responsible for reactivity to SEC 2 in human Vβ13 (Choi et al., 1990). Changes at residues 22, 70, and 71 affect reactivity to Mls-1 but do not affect responses to antigen:MHC.

Taken together, these data are most consistent with the consensus model, stating that the critical contact sites between Vbse or bacterial toxic mitogens and the Vβ region of the TCR lie on the lateral face of the TCR Vβ domain and are encoded in the adjacent loops involving amino acids 67–77 and amino acid 22. The involvement of the TCR loops equivalent to CDRs 1, 2, and 3 in these responses requires further analysis. Are the changes observed in this region consistent, or do they reflect particular MHC and Vbse combinations. I favor this latter interpretation, because it is readily explained by the consensus model. However, conformational changes in the Vβ domain may also explain these results.

4. The Involvement of Coreceptors in the Response

The T cell response to conventional protein antigens involves not only the TCR but also the coreceptors CD4 or CD8, depending on the class of MHC molecule being recognized (Janeway et al., 1989c). This molecular association is established during positive selection of T cells in the thymus (von Boehmer et al., 1989). When murine T cells are stimulated by Mls-1^a, one finds predominantly CD4 T cells in the responding population (Janeway et al., 1980). Furthermore, anti-CD4 antibody added to such cultures can completely inhibit the response (Janeway and Katz, 1985). Purified CD4 T cells respond strongly to these stimuli (Janeway et al., 1980), whereas purified CD8 T cells do not (Webb and Sprent, 1989a). However, this does not mean that CD8 T cells lack the ability to recognize Vbse. First, the CD8 T cell blasts derived from such cultures consist of cells expressing the same Vβ genes encoding their receptors as do those of the responding CD4 T cells, demonstrating that the TCR on the CD8 T cell is required for the response (Webb and Sprent, 1989a). Second, a CD8 T cell clone specific for influenza virus presented by class I MHC also responded to Mls-1^a stimulator cells, a response not inhibited by anti-CD8 (Braciale and Braciale, 1981). Third, Vbse delete both CD4 and CD8 T cells bearing TCRs encoded in the selected Vβ gene segment (Kappler et al., 1987b, 1988; MacDonald et al., 1988a). Nevertheless, it has been argued that CD8 T cells respond differently from CD4 T cells to Vbse, because it has been shown experimentally that anti-CD4 *in vivo* inhibits the negative selection of CD8 T cells expressing these receptors (Fowlkes et al., 1988; MacDonald et al., 1988b). These data taken together show that CD4 is clearly involved in the T cell response to Vbse, but they also suggest that CD4 is not essential for this response.

In the response to bacterial toxic mitogens, there is a significant difference in the mouse and the human situation. Responses of both normal and cloned murine T cells to SEs are strongly inhibited by anti-CD4 antibody (Janeway et al., 1988a, 1989b). However, both cloned and normal murine CD8 T cells have also been shown to respond to SEB in vitro (Herrmann et al., 1990). Furthermore, antibodies to the coreceptors have no effect on the human T cell response to SEs (Purdie et al., 1991). Whether this reflects the much greater potency of SEs on human as opposed to murine T cells, bypassing the requirement for coreceptor function, or whether this is a fundamental difference in the mode of T cell activation by SEs in the two species remains to be determined.

C. THE INTERACTION OF T CELL RECEPTORS, MHC MOLECULES, AND Vβ SELECTIVE ELEMENTS: EVIDENCE FOR THE WORKING MODEL

The above data strongly support the working model for Vbse function. In particular, the finding that alterations of a TCR that conferred reactivity to Mls-1^a did not affect antigen recognition is strong evidence for the independent siting of these two events (Pullen et al., 1990b). Likewise, the finding that alterations in the MHC molecule that did not affect SE presentation nevertheless altered peptide recognition is strong support for the model (Dellabona et al., 1989). However, there are some anomalies that cannot be ignored, and a convincing explanation for all of these is not yet available.

First, it should be emphasized that responses to Vbse clearly depend on more than just the MHC molecule, the Vβ gene segment encoded portion of the TCR, and the Vbse. As mentioned above, there are combinations of MHC and TCR that do not yield the response expected if only these elements of the consensus model of Vbse are required for response. In particular, the combination of MHC molecules known to bind Vbse and TCRs known to respond to that Vbse does not always yield a response (Yagi et al., 1991). Second, not all cloned lines of cells expressing a particular Vβ are stimulated by the same Vbse bound by the same MHC molecule. This was evident in the earliest study of this phenomenon, both at the level of hybrids that did not respond to I-E or to Mls-1^a, and also at the level of cells not being deleted or anergized during development (Kappler et al., 1987a, 1988; Yagi et al., 1991). Presumably, other receptor elements prevented effective interaction of these TCRs with the MHC class II:Vbse complex. These findings can be accommodated into the consensus model by stating that the TCR:MHC interaction must be permissive for interaction, and most but not all such combinations are permissive. Second, as mentioned above, Webb and co-workers have data suggesting an extra element on the T cell involved in the response to Mls-1[a], although the nature and existence of this second element is mysterious (Webb et al., 1988). Likewise, Yagi et al. have evidence that L cells and B cells may differ in the ability of a single MHC class II molecule to present a given SE to a given set of TCR (Yagi and Janeway, 1991). Third, Webb and colleagues have recently shown that T cells responding to Mls do not generate a transient increase in intracellular Ca^{2+}, whereas the same cells do generate this response when stimulated by peptide:MHC class II (O'Rouke et al., 1990). Fourth, the amounts of SEA required to stimulate T cells in some individual humans is extremely low, well below one molecule per responding cell (J. D. Fraser, personal communication). Finally, despite early encouraging results (J. Yagi and C. A. Janeway, Jr., unpublished data), conclusive data for direct binding of SEs to the TCR are not yet available.

These data could be reconciled if the model is modified as follows. First, some Vbse may be present on the responding T cell rather than the stimulating B cell. Their function could be determined by their interaction with the Vβ-encoded region of the TCR. Second, the SEs may be effective mimics of Vbse polymorphisms not because they are Vbse homologues, but rather because they bind to and alter the function of existing, nonpolymorphic Vbse (Janeway et al., 1991). These possible modifications of the working model are shown in Fig. 4.

V. The Cellular Basis of Vβ Selective Element Action

The molecular basis of Vbse action has been discussed in terms of the working model. However, before one can understand these structures fully, the cellular nature of the response must also be described. Here, the crucial issues are which cells can respond to Vbse and which cells can elicit that response. Some of this material has been covered in the previous section as part of the argument for involvement of various molecules.

A. The Responding Cells

The response to Vbse is mediated by T cells; to date, only T cells expressing the α/βTCR have been detected in responses to both endogenous and bacterial Vbse. However, certain cloned lines of human T cells expressing a particular γ/δTCR would kill class II-bearing target cells pretreated with SEA (Rust et al., 1990). These cells do not proliferate in response to this stimulus. The precise meaning of this finding is unclear at present. This same subset of γ/δ T cells responds strongly to a structure on the human Burkitt's lymphoma line Daudi

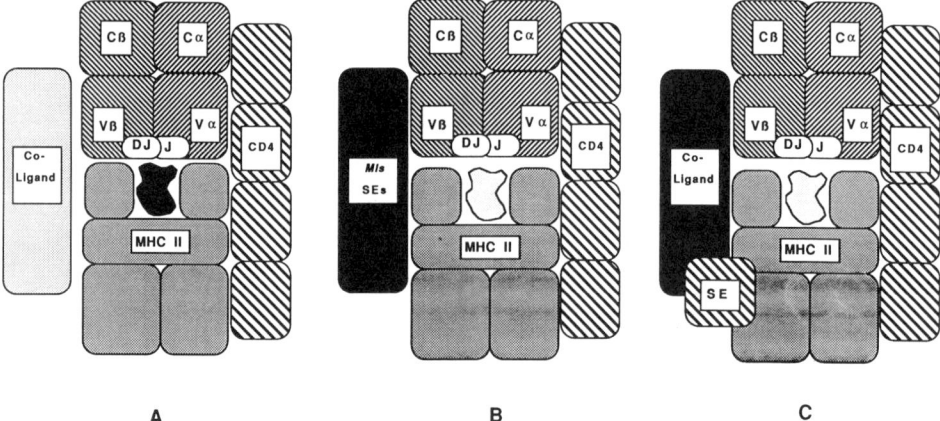

FIG. 4. A modification of the working model to account for anomalies seen in various studies. (A) Several studies suggest that Mls-like proteins participate in antigen recognition, serving to stabilize or orient the T cell receptor to its MHC class II ligand. We have termed such structures coligands. (B) The current working model, as in Fig. 1B. (C) A modified model of stimulation by the staphylococcal enterotoxins (SE) in which the SE binds to MHC class II and the coligand, allowing the coligand to interact with the $V\beta$ domain of the T cell receptor with sufficient strength to stimulate the T cell. Neither direct contacts of the hypervariable portions of the T cell receptor with the peptide:MHC class II complex nor the specific peptide are critical to this response.

and to mycobacteria of certain strains (Fisch et al., 1990). For the present discussion, we will confine ourselves to T cells with α/βTCRs, although the γ/δ T cells may be telling us an important lesson.

1. The Response of CD4 T Cells

As stated earlier, all Vbse described to date depend on class II MHC molecules and stimulate predominantly CD4 T cells. This applies to endogenous Vbse and bacterial toxic mitogens. At least in the mouse, anti-CD4 profoundly inhibits this response. Thus, the CD4 T cell is the predominant cell that proliferates in response to Vbse.

2. The Response of CD8 T Cells

As stated above, purified CD8 T cells in mice will not proliferate in response to Vbse, but the CD8 T cell blasts that appear after stimulation of whole spleen T cells are apparently Vbse selected because they bear the appropriate TCR. The failure to obtain CD8 T cell proliferation in response to Vbse, when class I MHC differences can stimulate CD8 T cell responses, may reflect differences in the antigen-presenting cells (APCs) that express these two determinants or differences in signal transduction pathways (O'Rourke et al., 1990).

B. THE STIMULATING CELLS

Steinman and colleagues have shown that conventional allogeneic mixed lymphocyte responses are driven primarily or exclusively by dendritic cells (Steinman et al., 1986). These cells can stimulate proliferative responses of either CD4 or CD8 T cells. Thus, it was expected that dendritic cells would also be the predominant stimulating cell in responses to Vbse. However, this appears not to be the case. Rather, the B cell is the predominant cell driving this response.

1. The Role of B Cells in the Response

Early data on cell separation showed that T cell depletion of spleen stimulator cells enriched for stimulator cells in responses to Mls-1^a, whereas T cells purified by depletion of B cells removes stimulatory activity for responses to Mls-1^a (von Boehmer and Sprent, 1974). Furthermore, highly purified B cell preparations will stimulate such responses (Webb et al., 1985, 1989). Further enhancement of these responses is seen using cells stimulated with anti-IgD *in vivo* (Ryan et al., 1983, 1987a, 1988). Finally, spleen cells from mice carrying the *xid* defect that affects the maturation of a subset of B cells has been reported to prevent stimulation by Mls-1^a (Ahmed et al., 1977). Thus, it appears that B cells are the predominant stimulator of proliferative responses to Vbse.

2. The Role of Dendritic Cells in the Response

The role of the dendritic cell in the response to Mls-1^a is more problematic. Early studies reported that dendritic cells were excellent stimulators of such responses (Sunshine et al., 1985), but more recent studies have shown conclusively that dendritic cells lack the ability to initiate such responses (Webb et al., 1989). Not only were purified dendritic cells shown to be lacking the ability to stimulate responses to Vbse, but B cell-depleted mice were also shown to lack this ability (Webb et al., 1989). In both cases, APC function of the dendritic cells was normal for stimulation of responses across MHC differences. This poses some problems in accounting for thymic selection by Vbse, as discussed below, because few B cells are found in the thymus, yet deletion driven by Vbse is prominent.

3. The Transfer of Vbse between Cells

As mentioned above, T cells do not stimulate primary proliferative responses to Vbse. Because murine T cells do not express class II MHC, it may be that T cells express the Vbse but cannot present it to other T cells in a stimulatory form. Furthermore, it has been reported that small numbers of T cells, especially CD8 T cells, can transfer tolerance to Vbse, suggesting that the T cell may express Vbse but transfer them to other cells *in vivo* (Webb et al., 1990; Webb and Sprent, 1990). If this were the case, one might expect to find other aspects of this transfer *in vivo* or *in vitro*, such as induction of stimulatory capacity in mixed chimeras or in tissue culture. Indeed, evidence for transfer of Vbse was first reported in tissue culture experiments (De Kruyff *et al.*, 1986) and later supported by *in vivo* experiments (Kappler *et al.*, 1988). These experiments took advantage of nonpermissive Ia alleles on B cells to demonstrate transfer of Vbse activity. This experiment has been repeated with varying success in many laboratories; it succeeds less often than it fails. This may reflect the fact that Vbse function most effectively if a single cell produces both the Vbse and the permissive class II MHC molecule. Alternatively, it may be that transfer is determined in part by the genotype of the actual cells used. Thus, successful transfer has been observed from cells of MHC genotype $H\text{-}2^q$ and $H\text{-}2^d$ but not $H\text{-}2^k$. (It should be noted that in the experiments of de Kruyff *et al.*, $H\text{-}2^k$ mice were used, but in this case, it is also possible to explain the results by other mechanisms.) Data showing transfer of Vbse could explain the disparate results regarding dendritic cells in stimulation of responses to Vbse. Thus, Sunshine and co-workers used DBA/2 ($H\text{-}2^d$) dendritic cells, and Webb and co-workers used CBA/J ($H\text{-}2^k$) dendritic cells. R. M. Steinman and K. Inaba (personal communication) have used both types and find only $H\text{-}2^d$ dendritic cells stimulate such responses. Taken together, these data suggest that dendritic cells do not make Vbse but can acquire Vbse from B cells in some situations, and that this transfer may be determined by MHC genotype. Alternatively, it may be that although functionally indistinguishable, the *Mls-1* genes of CBA/J and DBA/2 mice are distinct in this property.

VI. The Impact of Vβ Selective Elements on T Cell Development and Function

The current interest in Vbse stems largely from their ability to shape the repertoire of T cells during intrathymic ontogeny. In particular,

Vbse clearly cause major changes in the mature repertoire of receptors by deleting or inactivating developing T cells having TCRs encoded by certain Vβ gene segments, essentially independently of the rest of the TCR (Marrack and Kappler, 1988b). More subtle effects are seen in positive selection, or "overselection," as it is sometimes called. Finally, there is evidence for a role of Vbse in the functioning of T cells in the periphery. Cells differing for Vbse can also differ in antigen-presenting capability, can induce inactivation or deletion of T cells, and can activate profound suppression of a wide range of immune responses. Whether the endogenous nonpolymorphic Vbse do the same thing is not clear. In this section, I will present the evidence regarding the role of Vbse in these processes in mice; to date, no evidence exists for human endogenous Vbse, so no statements can be made. However, Vbse play a role in most or all T cell interactions with ligand; this is unlikely to be a function restricted to mice.

A. THE ROLE OF Vβ SELECTIVE ELEMENTS IN INTRATHYMIC DEVELOPMENT

Five critical events occur during intrathymic development of T cells. First, the cells express receptors and coreceptors, generating large CD4/CD8 double-positive CD3-low thymocytes. Second, those cells whose TCRs are able to recognize antigens presented by self MHC molecules are selected for further maturation, a process called positive selection. It appears to be accompanied by increased TCR expression and the third critical event, determination of whether the cell will express CD4 or CD8. TCRs restricted to class I MHC molecules select CD8 expression, whereas TCRs restricted to class II MHC molecules select CD4 expression. Furthermore, it appears that CD4 T cells are also signaled to become precursors of lymphokine-secreting cells, and CD8 cells are signaled to become precursors of cytolytic T cells, although the means of this commitment event are unclear. Fourth, cells whose TCRs would lead to activation by self MHC or by self peptides bound to self MHC molecules expressed on professional APCs are eliminated by a process of apoptosis, called negative selection. Finally, cells surviving these processes are exported to the periphery (for review, see von Boehmer et al., 1989). Strong evidence exists for a role for Vbse in negative selection, while somewhat weaker evidence supports a role for Vbse in positive selection. These processes appear to be more sensitive indicators of Vbse action than is peripheral T cell activation.

1. Positive Selection of Vβ Selective Elements

Several groups have reported the presence of mature T cells expressing TCRs encoded by a particular Vβ gene segment at higher than anticipated levels in mice of certain MHC genotypes. For instance, I-E-positive mice appear to express higher than expected levels of Vβ6-expressing cells (MacDonald et al., 1988c), while H-2^q mice express higher than expected levels of Vβ17a-bearing cells (Kappler et al., 1987b; Marrack et al., 1991; Marrack and Kappler, 1988b). This overselection by MHC molecules may reflect positive selection events. Whether this is due to a role of Vbse, or whether it reflects utilization of the particular Vβ to recognize a particular MHC molecule, is not yet clear in these systems. However, the data are sufficiently striking, and the involvement of I-E in some of these systems is sufficiently reminiscent of the activation and deletion characteristic of Vbse, that it seems likely that Vbse will play a role in this process. To date, a polymorphism in Vbse that leads to this overselection has not been observed; such a finding would provide conclusive evidence for a role for Vbse in positive selection in the thymus.

2. Negative Selection by Vβ Selective Elements

There is no doubt whatsoever that Vbse participate in negative selection of autoreactive T cells by a process of clonal deletion. Numerous studies document this effect. This is one of the two diagnostic criteria for identification of Vbse. In addition to polymorphic Vbse, there are a number of apparently nonpolymorphic Vbse that act in conjunction with certain MHC molecules to cause deletion of T cells whose receptors are encoded by particular Vβ gene segments. While it is accepted that bone marrow-derived cells play a leading role in clonal deletion in the thymus, there are few B cells in this site. Most of the bone marrow-derived cells in the thymic medulla are macrophages and dendritic cells that are purported to lack Vbse expression. Thus, it is a striking observation that neonatal tolerance to Mls-1^a is most efficiently induced by injection of CD8 T cells (Webb and Sprent, 1990). Taken together with the data on transfer of Vbse cited above, this suggests that Vbse can be expressed by many cell types, but act only after transfer to cells expressing appropriate MHC molecules.

A number of Vbse:MHC combinations appear to cause deletion of T cells expressing various Vβ-encoded receptors, but these same Vbse:MHC combinations cannot cause stimulation of most cells bearing TCRs encoded by the same Vβ (Palmer et al., 1989). This suggests that deletion has a much lower threshold than activation. To test this, the SEs were used in *in vitro* models of clonal deletion and clonal activation. It was found that clonal deletion in thymic organ culture was 30- to 100-fold more sensitive to SEB than was clonal activation of mature T cells from the spleen (Yagi and Janeway, 1990). Although the two sets of experiments were carried out under different conditions, this is the best current estimate for the difference in the activation threshold of these two processes. This difference is probably required to avoid autoreactivity of activated T cells to APCs expressing increased levels of class II MHC molecules and/or Vbse in the periphery.

B. THE IMPACT OF Vβ SELECTIVE ELEMENTS IN THE PERIPHERY

Vbse were originally described by their role in mixed lymphocyte reactions between peripheral lymphocytes. More recent data on the *in vivo* effects of Vbse suggest that they participate in a variety of processes that again influence the utilization of TCRs based on the Vβ gene segment expressed.

1. Clonal Inactivation by Vβ Selective Elements

T cells from $Mls\text{-}1^b$ mice injected with $Mls\text{-}1^a$ cells fail to respond to the original donor cells (Rammensee *et al.*, 1989). This is in part due to deletion of the cells bearing relevant Vβ-encoded TCRs; CD4 T cells are especially affected (Webb *et al.*, 1990). In addition, remaining cells bearing the relevant Vβ-encoded TCRs appear to be immunologically inert, unable to proliferate in response to stimulation (Kawabe and Ochi, 1990; Rammensee *et al.*, 1989; Webb *et al.*, 1990). The exact mechanism of this effect is not yet known. A similar effect is seen when SEB is injected *in vivo*. After an initial clonal expansion, CD4 T cells bearing the selected Vβ are lost, but CD8 T cells are somewhat expanded (Kawabe and Ochi, 1991). As CD8 T cells respond poorly to Vbse, one has an impression of T cell anergy (Rammensee *et al.*, 1989). In addition, a role for CD8 suppressor cells has not been ruled out. Why a signal that elicits strong T cell activation in tissue culture should induce deletion and/or inactivation *in vivo* is also not known. This is a very interesting problem in T cell biology.

An alternative approach to this problem is to examine mature peripheral T cells expressing Vβ-encoded receptors that should lead to a response to self Vbse. Such cells are self-tolerant, and many if not all are anergic, as determined by their failure to respond to the self Vbse or to anti-Vβ antibodies that activate cells in mice lacking the Vbse (Blackman *et al.*, 1990a; Fowlkes *et al.*, 1988). Thus, not all self-tolerance to Vbse is due to clonal deletion; at least a fraction is due to clonal anergy.

A similar effect has been observed using mice transgenic for I-E molecules placed in pancreatic β cells by linking the genes to the rat insulin-1 promoter (Burkly et al., 1989). These mice are tolerant to I-E, but have not deleted cells expressing TCRs associated with I-E recognition, such as Vβ5 and Vβ17a. These T cells are tolerant to I-E, and they cannot be stimulated by antibodies directed at the TCRs, although the same antibodies do activate T cells from normal control mice. Thus, I-E expressed on pancreatic β cells appears to induce clonal anergy in a manner strongly suggesting involvement of Vbse (Blackman et al., 1990a; Fowlkes et al., 1988). The cellular source of the Vbse in this system has not been ascertained.

2. Enhancement of Responses to Antigen by Vβ Selective Elements

Initially, it was thought that T cells either did or did not respond to Vbse; intermediate effects of Vbse were not observed. However, in examining the effect of Mls-1^a on antigen presentation, it was found that APCs expressing Mls-1^a were able to present antigen to a series of cloned T cell lines about 10-fold more efficiently than MHC-matched cells expressing Mls-1^b (Janeway et al., 1983). This finding was mapped to *Mls-1* using recombinant inbred lines of mice (Hammerling et al., 1988). This effect is observed with some cloned lines of T cells, and these can derive from donors that are either *Mls-1^a* or *Mls-1^b*, implying that the effect acts on T cells whose receptors are not directly reactive to Mls-1^a (Janeway et al., 1983). However, not all investigators have been able to repeat these findings using other cloned lines of T cells (Needleman et al., 1988). The explanation for these differences is not clear; it seems most likely that the *Mls-1^a* product will augment antigen presentation by class II MHC molecules to some but not all TCRs, depending on the Vβ expressed. This needs to be explored in detail to determine whether these effects are indeed Vβ specific.

These findings are important for three reasons. First, they provide a possible function for Vbse in the activation of peripheral T cells, lowering the threshold concentration of antigen by 10-fold. This increased sensitivity to antigen could be decisive in the course of an infection. Second, they suggest that Vbse work together with, not as an alternative to, standard peptide:self MHC recognition, consistent with the working model. Third, these findings suggest that Vbse participate in all T cell responses. That this may be so is further supported by an anomaly we recently described in the response of a cloned T cell line to antigen presented by self MHC (Portoles et al., 1989). The responses of this cloned T cell line could be inhibited by Fab fragments of a variety of antibodies directed at TCR V region determinants. All of these Fab fragments inhibited T cell activation proportionally to their affinity for the TCR. However, two antibodies to a Vβ8.2 framework determinant known to map to the region of Vβ close to Cβ, and hence distal to the antigen recognition site, inhibited far more effectively than predicted from their affinity. This effect was seen using APCs that express Mls-1[b]. As these antibodies are unlikely to affect ligand binding, it may be that their effect is on Vbse recognition of Vβ. If this can be confirmed, it would prove that even in "null" strains of mice Vbse can participate critically in antigen:self MHC recognition.

3. Suppression Induced by Vβ Selective Elements in Vivo

One of the earliest known effects of *Mls* was the induction of suppression upon injection of *Mls*-disparate cells into recipient mice (Festenstein *et al.*, 1989; Halle-Pannenko *et al.*, 1986; Matossian-Rogers and Festenstein, 1976, 1977). Many studies have been carried out on this strange phenomenon without a definitive explanation being achieved. The experimental finding is that injection of spleen or lymph node cells from mice differing at *Mls-1* will lead to the suppression of a variety of different immune responses, such as mixed lymphocyte responses, antibody formation, graft rejection, etc. (Halle-Pannenko *et al.*, 1983, 1985, 1986, 1987; Jacobsson *et al.*, 1975; Lilliehöök *et al.*, 1975). These suppressive phenomena are mimicked by injection of SEs (Holly *et al.*, 1988; Pinto *et al.*, 1978; Poindexter and Schlievert, 1986). It is possible that all of these effects are due to the profound T cell activation that occurs when these substances are injected, leading to the secretion of a wide range of cytokines in a nonspecific fashion. The suppression is entirely lacking in specificity for Mls. This effect must be kept in mind when such injections are carried out, as they may confound interpretation. Thus, purported loss of response to Mls-1[a] must be confirmed by showing that other responses are unaffected, such as responses to non-self MHC molecules.

VII. Vβ Selective Elements Are Encoded by Endogenous or Extrinsic Retroviruses

Genetic mapping of Vbse initially placed *Mls-1* in a position clearly distinct from the endogenous retroviral insertion known as *Mtv-7*. However, recent reanalysis of this question revealed numerous errors in the genetic map, such that *Mtv-7* is now inseparable from *Mls-1* (Eicher and Lee, 1990; Frankel *et al.*, 1991). In addition, a Vbse on chromosome 12 that allows deletion of Vβ5.1, 5.2, and 11 is not separable from Mtv-9 (Woodland et al., 1990). These observations have led to a rapid analysis of the relationship between Vbse and retroviruses, especially those of the mammary tumor virus (Mtv) family. This analysis has led to many data that support the hypothesis that Vbse are encoded by endogenous retroviruses.

A. Endogenous Retroviral Insertions Are Inseparable from Vbse

A number of different Vbse have been mapped in extensive genetic crosses and recombinant inbred strains of mice, and no disparities between the presence of a given Mtv insertion and a given Vbse have been observed in these studies. Table VII summarizes these data as currently understood. Given the strong case implicating these viral insertions as Vbse, it has been suggested that each locus be renamed vSAG (for viral superantigen), followed by the accepted number for the retroviral insertion (e.g., vSAG-7 would be the same as Mtv-7 or Mls-1^a). Whether this will become the accepted nomenclature is not

TABLE VII
Vβ Selective Elements and Their Linkage to Retroviruses

Vbse	Vβ	MHC	Chromosome	Retrovirus	Reference
Mls-1^a	6, 8.1, 9^a	II, E > A	1	Mtv-7	Frankel et al. (1991)
Mls-2^a	3	E >> A	4	Mtv-13	Frankel et al. (1991)
Mls-3^a	3	E >> A	16	Mtv-6	Frankel et al. (1991)
Mls-2-like	3	?	7	Mtv-1	Frankel et al. (1991)
Etc-1 (Dvb11-2)	5.1, 5.2, 11	E	12	Mtv-9	Woodland et al. (1990, 1991) Dyson et al. (1991)
Dvb11-1	11	E	6	Mtv-8	Dyson et al. (1991)
Dvb11-3	11	E	14	Mtv-11	Dyson et al. (1991)
Unnamed	14	E	Extrinsic	MMTV	Marrack et al. (1991)
MAIDS B cell	5	II	Extrinsic	MuLV	Hugin et al. (1991)
Unnamed	17a	E	Not mapped^b	?	Kappler et al. (1987b) Marrack and Kappler (1988a)
Unnamed	12	E	Not mapped	?	Vacchio et al. (1990)
Unnamed	16	E	Not mapped	?	Vacchio et al. (1990)
Unnamed	19a	E	Not mapped	?	R. Hodes, personal communication

^{a} Mls-1^a is often stated to delete Vβ7 as well, but this is not always confirmed (R. Abe, personal communication).
^{b} Nonpolymorphic.

clear yet. It has the advantage of utilizing the present numbering system for the many retroviral insertions. It has the disadvantage that it renames or renumbers all the *Mls* and other loci, and it attributes to them the property of superantigens, even where this has been documented only by deletion rather than by stimulation of cloned T cell lines. Finally, it will not accommodate Vbse that are not encoded by Mtv or other viruses, if such should be discovered.

B. Expression of Mtv in B Cells Correlates with Vbse Expression

Once it was realized that Mtv might encode Vbse, it was possible to examine the expression of mRNA encoded by the appropriate Mtv in established or normal cells that do or do not express a Vbse. Using the polymerase chain reaction (PCR), Mtv-9 expression was shown to correlate with Etc-1 expression in a series of B lymphomas (Woodland *et al.*, 1991), and likewise, Mtv-7 expression correlated with Mls-1^a in a series of B cell hybrids (B. T. Huber, personal communication).

A more emphatic demonstration that Vbse may be encoded by retroviral sequences has come from infection and transfection experiments. Using cloned Mtv genes, it has been found that the 3′ half of Mtv-7 can confer Mls-1^a reactivity on a *B* cell lymphoma, whereas the *env* gene of this virus, contained in the 3′ half, cannot (B. T. Huber, personal communication). This leaves the mysterious open reading frame (orf) in the Mtv long terminal repeat (LTR) as the main candidate for encoding Vbse. This gene does encode a protein that is expressed in some cells, such as phorbol myristate acetate (PMA)-induced EL-4 cells, which express very high levels of mRNA encoded by this gene (Kwon and Weissman, 1984; Racevskis, 1986; Racevskis and Prakash, 1984). Similar studies have been carried out with mouse mammary tumor virus genes (P. Marrack and J. Kappler, personal communication). Thus, it seems likely that the orf gene of Mtv can encode Vbse. Whether this is the only source of Vbse, or whether there are Vbse in other retroviruses or in cellular genes, is not yet known.

C. Mouse Mammary Tumor Virus Vertically Transmits a Vbse

The mouse mammary tumor virus (MMTV) that causes breast cancer in C3H mice has also been shown to encode a Vbse that deletes T cells expressing Vβ14 (Marrack, 1991). This Vbse also maps to the orf gene within the MMTV genome. MMTV is the exogenous form of the endogenous Mtv retroviruses. Apparently, MMTV genomes have inserted into the germ line of mice over the past million years, moving to several different locations and varying in structure (Coffin, 1990). This gives rise to the surprising complexity of Vbse inheritance.

D. Mtv Transcripts Are Highly Expressed in Activated B Cells

As noted earlier, B cells are the main stimulator type in responses to Vbse. Although MMTV and Mtv sequences were thought to be expressed predominantly in mammary gland, it is now known that Mtv is potently activated by B cell stimuli. Corley and co-workers have noted Mtv transcripts in B cells activated with lipopolysaccharide (LPS) (King *et al.*, 1990). In this regard, it is interesting that Vbse should be inducible genes, because it is a common observation that responses to Mls-1^a, although very potent on most days, vary far more than do responses to MHC differences (C. Janeway, personal observation). This is in keeping with the inducible nature of Vbse genes.

E. Mice Transgenic for a Given Mtv Express a Vbse

Acha-Orbea and colleagues have also noted a relationship between Vbse and endogenous retroviruses. Using cloned Mtv genes, they have prepared mice transgenic for an Mtv associated with Vβ14 deletion. These mice show a strict correlation between the presence of the transgenic Mtv and deletion of Vβ14 (H. Acha-Orbea, personal communication). Thus, both *in vivo* and *in vitro*, there is excellent evidence that Mtv encode Vbse.

F. Other Retroviruses May Also Encode Vbse

Although all the data presently available point to Mtv as providing the genes encoding Vbse, one piece of evidence may implicate other retroviruses in Vbse expression. Morse and colleagues have been studying a novel retroviral disease called mouse AIDS, in which a murine leukemia virus (MuLV) causes a syndrome of B cell activation and CD4 T cell depletion (Buller *et al.*, 1987; Cerny *et al.*, 1990a,b; Chattopadhyay *et al.*, 1989; Hartley *et al.*, 1989; Hugin and Morse, 1990; Klinman and Morse, 1989; Mosier *et al.*, 1985; Yetter *et al.*, 1988). The syndrome requires both B cells and CD4 T cells for its effects to occur. Recently, Morse and colleagues have isolated B lymphomas from end-stage disease mice. Those that express retroviral *gag*-encoded fusion proteins on the cell surface stimulate potent syngeneic T cell proliferative responses, whereas those that are surface-gag negative do not. Interestingly, the responding T cells are CD4 T cells that are mainly Vβ5, 11, and 12 positive, but are depleted of T cells expressing other Vβ genes (Hugin *et al.*, 1991). The response is inhibited by anti-CD4, anti-I-A, and an anti-gag monoclonal antibody. This virus is unrelated to Mtv and lacks the orf gene of the Mtv viruses. Nevertheless, it appears to encode a Vbse. Alternatively, it could be that the transformed B cell is expressing an Mtv sequence that is normally silent in the mouse strain from which it was derived, but that becomes active upon transformation, although this leaves the role of the gag protein open. This issue needs to be resolved. Nevertheless, it is wise to keep in mind the possibility that viruses other than the MMTV may encode Vbse.

G. Retroviral Sequences Bear No Homology to Bacterial Toxic Mitogens

Assuming that the retroviral sequences identified to date are indeed encoding Vbse, it can be determined whether endogenous Vbse and bacterial toxic mitogens show any molecular homology. At present, no such homology has been identified. This is interesting, because the two structures cause such highly analogous effects. As other studies have led us to question whether bacterial toxic mitogens are homologues of Vbse or alternatively bind to Vbse and alter their behavior, the lack of homology would tend to favor the latter model.

H. Why Have Vbse Sequences Been Retained in the Mouse?

The finding that Vbse are encoded by oncogenic viruses and also are retained in significant numbers in the mouse genome raises two questions. Why do viruses encode Vbse? And why does the host retain these sequences?

There are at least three possible answers to the first question. It may be that T cell activation by the Vbse encoded by the retrovirus leads to the same type of generalized immunosuppression noted when *Mls*-disparate cells are injected into mice, or when SEs are administered (see above). This should favor spread of the virus. Second, it may be that Vbse elicit polyclonal B and T cell responses that favor viral growth by providing activated cells in which to replicate. Finally, it may be that the virus uses its Vbse to enter specific T cells expressing the corresponding Vβ sequence, similar to the suggestion of Weissman some years ago (McGrath and Weissman, 1979). All of these mechanisms would favor successful parasitization of the host by the virus, which should be a sufficient criterion for retention of this potential.

It has been proposed that Vbse exist primarily to delete T cells that would respond deleteriously to bacterial toxic mitogens (Kappler *et al.*, 1989; Marrack and Kappler, 1990). Very similar arguments can be made about deletion to remove targets of opportunity for retroviruses. By incorporating the retrovirus in defective form, while retaining its Vbse, the host can remove the critical T cell targets of the retrovirus.

Two other mechanisms could achieve the same end: deletion of I-E (Mathias et al., 1983), which removes the necessary MHC molecule for many Vbse, or deletion of a cluster of Vβ genes (Pullen et al., 1990a,b), both of which occur in wild mouse populations. This argument is somewhat mitigated by the finding that wild mice can also mutate Vβ so that they are no longer deleted (Cazenave, 1990). However, the weight of the evidence suggests that some interplay between receptor and exogenous ligand is being modulated by a corresponding interplay between receptor and endogenous ligand. The fact that in this case the two are highly homologous suggests that the relationship may be causal.

VIII. Unanswered Questions About Vβ Selective Elements

A. Is the Working Model Valid?

This review has been written largely from the point of view of the working model. Many of the data fit this paradigm. Does this mean that the working model is correct? The answer is clearly that we do not know. In this section, I want to summarize the data that do not fit the working model and propose some alternatives.

There are three major sets of findings that do not fit the working model. The first is that SEs have not been shown to bind to Vβ of the TCRs that they stimulate. This is surprising, because it is a basic tenet of the working model that they should do so. Similarly, SEs adsorbed to any surface fail to stimulate, unless that surface bears MHC class II molecules. Thus, it has been considered that SEs may modify MHC class II molecules such that they bind more effectively to the TCRs. However, this model does not explain the Vβ selectivity of Vbse, because all other changes in MHC molecules lead to selection for all variable portions of the TCR. Alternatively, binding to the MHC may lead to an alteration in the conformation of SEs, allowing them to bind to Vβ directly, whereas the unmodified SE cannot. This seems unlikely as well, because the SEs appear to be very tightly packed, rigid molecules. A third possibility is that the SEs activate B cells in such a way that they stimulate potent T cell responses. This explanation is unlikely given that fixed APCs (Fleischer and Schrezenmeier, 1988; Yagi et al., 1990), MHC class II transfected fibroblasts (Dellabona et al., 1989; Yagi et al., 1991), or even MHC class II molecules in planar lipid membranes (Bekoff et al., 1987; Cole et al., 1989; Macphail and Stutman, 1990) can present SEs to T cells.

The second major problem with the working model is the failure to detect the products of the Vbse. Numerous attempts have been made to raise antibodies to Vbse without a single reported success. It could be that these molecules vary quantitatively rather than qualitatively, and thus that there are no null strains; however, even heterologous immunizations fail to raise antibodies. Furthermore, if Mtv do encode Vbse, then null alleles really do lack the gene, and should be able to mount an antibody response. It could be that the Vbse induce profound suppression, particularly abrogating antibody responses to themselves. This still deserves exploration. At the present time, it seems anomalous that antibody has not been produced. Perhaps when we know more, we will be able to explain this failure.

The third problem with the working model is the uncanny potency of SEA in some people. It has been estimated that as little as one molecule of SEA per 100 cells can elicit a response in some individual humans. This suggests some catalytic or enzymatic activity imparted by SEs. This property must be explained before one can be satisfied with the working model or its replacement.

We have recently proposed an alternative to the working model to account for these anomalies, two of which apply to SEs rather than to self Vbse. This modification is shown in schematic form in Fig. 4. It proposes that SEs actually bind to self Vbse. This is based on another anomalous finding in my laboratory. SEB can be presented to Vβ3-bearing T cells from B10.A(5R) mice by spleen cells but not by fibroblasts transfected with suitable MHC class II genes. By contrast, both spleen cells and fibroblasts can present SEA to Vβ3-bearing T cells, and SEB to Vβ8-bearing T cells. These data suggest that some element is missing in the fibroblasts, because all the components of the working model are present and functional. Thus, we have proposed that SEs may act not as a homologue of Mls but rather by binding to self Vbse. This new model is much like the working model in all other respects. However, it has several interesting implications. First, it suggests that SEs can only function in the presence of Vbse, and thus Vbse cannot have evolved as a host defense against bacterial toxins, nor can the toxins have evolved from captured host Vbse genes. Indeed, this model accounts for the lack of homology between SE and retroviral genes that may encode Vbse. Second, it would explain the absence of SE binding to Vβ domains of the TCR. Third, it might account for the catalytic effects of SEA by proposing that SEA not only binds to but permanently modifies the Vbse. Fourth, it explains our anomalous findings with fibroblasts by saying that not all cells that can express MHC class II can also express all Vbse. Indeed, it appears that dendritic cells do not express Mls-1[a] (see above) and I-E-transfected fibroblasts do not express the Vbse that acts on Vβ17a. Fifth, it implies that most if not all T cell responses involve Vbse, and that humans have Vbse similar to those in the mouse. Finally, it suggests that structures other than MHC molecules that bind to SEs might be Vbse. The ability of Vbse to act when presented by I-E in planar membranes appears to rule out this model. However, T cells may supply the Vbse in these cases.

Why have such Vbse not been observed on gels? It may be that they are expressed on relatively few MHC class II molecules, or that the Vbse are sufficiently heterogeneous that they cannot be detected on gels. Finally, a serious examination of gels for species other than MHC class II may not yet have been undertaken; we are certainly doing so now.

B. What is the Molecular Nature of Endogenous Vβ Selective Elements?

This is the key to the puzzle. Recent data suggest that Vbse are encoded by endogenous retroviruses of the Mtv family. At least 30 different Mtv genomes have been identified in the mouse genome. Several of these have already been linked to Vbse, and Mtv transfections confer Vbse function on these cells. Thus, it seems certain that a product of these genomes encodes Vbse. However, the nature of this product, its binding properties, and its polymorphism remain to be determined. A critical issue is the presence of Vbse in species other than the mouse. The availability of probes from Mtv may allow homologous genes to be identified in other species, including humans. If Vbse are confined to mice, their general biological significance must be questioned. Alternatively, the retention of these structures in the mouse must also be examined. Is it an accident of evolution, or has strong selection led to the retention of these genes?

C. What Is the Distribution of Vβ Selective Elements?

Vbse are currently defined only by their ability to stimulate primary mixed lymphocyte reactions or responses of cloned T cell lines. This has led to the observation that the B cell is the major cell type expressing Vbse. However, as noted above, one can observe the impact of Vbse on developing T cells when CD8 T cells are used to induce clonal deletion, and on mature T cells when I-E is placed exclusively on pancreatic β cells. This suggests that Vbse may be more widespread than is observed using the indirect assay of immune stimulation. Clearly, a direct assay for Vbse that is independent of their ability to stimulate a T cell proliferative response is what is needed for an accurate answer to this question. In this regard, Mtv are known to be induced by LPS activation of B cells, which also potentiates their ability to activate T cells.

A second aspect of this question is whether Vbse can move from one cell to another. Several observations suggest that this may be so; however, many studies have failed to observe this behavior. It is difficult to explain all the data if Vbse cannot move from cell to cell, but a definitive answer will again require molecular definition.

D. WHAT IS THE BIOLOGICAL FUNCTION OF Vβ SELECTIVE ELEMENTS?

There has been a great deal of interest in the biological function of Vbse, largely stimulated by the finding that they shape the repertoire of T cell receptors. At present, there are three main proposals to account for their biological function, all of which may be correct as they are not mutually exclusive.

It has been suggested that Vbse may be essential for the positive selection of developing T cells for self MHC recognition (Marrack and Kappler, 1988b). This process is at present one of the most mysterious and fascinating areas in immunology. It is widely agreed that interaction between the T cell receptor and a self MHC molecule is required for T cell maturation. The difficulty is to account for the recognition process, because the same MHC molecule in the periphery will stimulate these T cells only if it is modified by a highly specific nonself peptide. Numerical arguments make unlikely the alternative proposal that random self peptides generated by the thymic epithelium can select for a full repertoire of nonself recognition (Kourilsky and Claverie, 1989); there are about 10^{13} peptides of 10 amino acids, so such random peptide generation would never achieve a level of more than one peptide:MHC complex per cell. Unless a T cell can be selected via ligation of a single receptor, it seems more likely that peptides are irrelevant to positive selection, or are permissive but not directly recognized. An alternative hypothesis is to state that Vbse provide the necessary binding energy to drive positive selection, and indeed, as outlined above, some impact of Vbse on this process has been inferred from several studies. However, this proposal also creates a problem. At least for Mls and SEs, virtually any MHC class II molecule will present to any receptor having the appropriate Vβ, whereas positive selection is allele and locus specific and acts on all variable elements of the TCR. Thus, one must modify this proposal by saying that the effects are too weak to select unless there is also a suitable TCR:self MHC interaction.

A second proposal has focused on the most dramatic effect of self Vbse, namely, clonal deletion. This proposal states that Vbse have evolved to eliminate T cells based on $V\beta$ gene segment usage. A possible biological function of this repertoire deletion model is protection of mice from bacterial toxic mitogens that cause disease, at least in part, by stimulating T cell responses that are immunosuppressive and pathogenic (Marrack et al., 1990). Given that the toxins we know about are far more potent in humans than in mice, and that humans have a much lower reproductive rate, it is surprising that similar $V\beta$ deletion effects have not yet been observed in humans if this is the only biological function of Vbse. A modification of this hypothesis states that Vbse have other functions, and that the polymorphism that leads to clonal deletion in the mouse has been selected by the bacterial toxins, with mice that can delete responsive T cells having a selective advantage in the wild. This argument is somewhat mitigated by the recent finding that wild mice also mutate their $V\beta$ gene segments such that they no longer are deleted by Vbse, and their Vbse such that they no longer delete $V\beta$-encoded receptors (Cazenave, 1990). A more recent proposal is that MMTV encodes a Vbse and uses this to infect mice. Deletion of T cells bearing complementary $V\beta$ may protect the host from such viruses (Janeway, 1991; Marrack, 1991; Woodland et al., 1991).

A third proposal, which we call the coligand hypothesis, can encompass all of the above proposals. The coligand hypothesis states that Vbse participate in all TCR:MHC interactions, serving to stabilize and orient this molecular event (Janeway, 1990; Janeway et al., 1989b). The role of the coligand would be to make the TCR:MHC interaction maximally sensitive to foreign peptide, thus lowering the threshold antigen concentration required to observe an immune response. This would clearly favor survival by making the immune response maximally efficient. However, this proposal would also require that $V\beta$ gene segments and Vbse coevolve over time such that coligand function was optimized. A change in one or the other element could lead either to inefficient antigen recognition or to frank stimulation requiring clonal deletion to maintain self-tolerance. The former change would be seen as an underrepresentation of a particular set of $V\beta$-defined receptors in T cell responses, whereas the latter would be seen as clonal deletion. At any one time, most $V\beta$:Vbse pairs would be at equilibrium. Coligands could participate in positive selection, in negative selection, in antigen-induced T cell activation, and in the mediation of effector function. The coligand hypothesis makes several predictions. First, coligands will be found in other species, specifically humans. Second, coligands will be important at all phases of T cell development and function. Third, coligands may be the target of bacterial toxin action, as outlined above in the modification of the working model of Vbse action.

One notable problem with the coligand hypothesis is that it calls for 10 to 20 Vbse, one for every one or two Vβ gene segments. If Vbse bind stably to MHC class II molecules, then they will lower the ligand density available to a given TCR. However, if Vbse bind only to existing MHC:TCR complexes and stabilize them, this problem is eliminated.

IX. Conclusions

The mouse has been shown to express products of several genes mapping to different chromosomes whose effect is to influence the expressed repertoire of T cell receptors solely in terms of the Vβ gene segment utilized in the receptor. Allelic polymorphisms of some of these genes can directly stimulate T cell proliferative responses, and they are known as *Mls* (minor lymphocyte stimulating) genes; others have been detected solely through their ability to alter the developing T cell repertoire. I propose that these structures be called Vβ selective elements, or Vbse. Although much is known about Vbse, the crucial questions in this field remain to be answered. It is my personal belief that Vbse are a consistent component of the TCR:MHC interaction that I refer to as a coligand. The coligand would serve to make recognition of foreign peptides more efficient, thus increasing the sensitivity of the immune system to antigen. Bacterial toxins would act either by mimicking the coligand but binding with a higher affinity, or by binding to the coligand and altering its affinity for Vβ such that the interaction became stimulatory.

It is likely that the recent identification of the products of Vbse genes will allow us to understand the structure, function, and evolutionary origin of these interesting and important molecules, as well as their relationship to the bacterial toxins whose behavior so closely mimics that of the *Mls* product. As long as *Mls* "made little sense" and was thought to be a curiosity of the mouse, little effort was made to identify the structures. It seems certain that the structures and the genes encoding them will be identified in the near future. This should put an end to speculation and answer the key questions about Vbse. It is clearly *the* goal for the future in this field.

Acknowledgments

The author wishes to acknowledge his many collaborators in studies of T cell responses to Mls and the SEs, especially Bruff Peck, Hans Wigzell, Barry Jones, Ethan Lerner, Michael Katz, Pat Conrad, John Tite, Junji Yagi, Steve Buxser, Jan Chalupny, Punu Rath, and Don Murphy. Many people have contributed ideas, discussion, and insight into various problems, especially Ed Palmer, Pippa Marrack, John Kappler, John Fraser, Kim Bottomly, Brigette Huber, Sandy Morse, and Hans Fischer, and numerous individuals have been very helpful in sharing their thoughts, findings, and papers with me to help in preparing this review. Finally, the author wants to thank Susan Morin for invaluable secretarial assistance in preparing the manuscript and bibliography. The author's work on Mls and SEs has been supported by funds from NIH Grant Al-14579.

References

Abe, R., and Hodes, R. J. (1988). *Immunogenetics* **28**, 221–232.
Abe, R., and Hodes, R. J. (1989). *Immunol. Rev.* **109**, 5–28.
Abe, R., Ryan, J., Finkelman, F., and Hodes, R. (1987a). *J. Immunol.* **138**, 373–379.
Abe, R., Ryan, J., and Hodes, R. (1987b). *J. Exp. Med.* **165**, 1113–1129.
Abe, R., Ryan, J. J., and Hodes, R. (1987c). *J. Exp. Med.* **166**, 1150–1155.
Abe, R., Vacchio, M., Fry, A., Fox, B., Matis, L. A., and Hodes, R. J. (1988). *Nature (London)* **335**, 827–830.
Abe, R., Foo-Phillips, M., and Hodes, R. J. (1989). *J. Exp. Med.* **170**, 1059–1073.
Abromson-Leeman, S. R., Laning, J. C., and Dorf, M. E. (1988). *J. Immunol.* **140**, 1726–1731.
Ahmed, A., Scher, I., Smith, A. H., and Sell, K. W. (1977). *J. Immunogenet.* **4**, 201.
Anderson, G. D., and David, C. S. (1989). *J. Exp. Med.* **170**, 1003–1008.
Bekoff, M. C., Cole, B. C., and Grey, H. M. (1987). *J. Immunol.* **139**, 3189–3194.
Benoist, C., and Mathis, D. (1989). *Cell (Cambridge, Mass.)* **58**, 1027–1033.
Bergdoll, M. S. (1970). In "Microbial Toxins" (T. C. Montie, S. Kadis, and S. J. Ajl, eds.), Vol. III, pp. 265–326. Academic Press, New York.
Bergdoll, M. S. (1979). In "Food-Borne Infections and Intoxications" (H. Riemann and F. L. Bryan, eds.), Vol. 2, pp. 443–494. Academic Press, New York.
Bill, J., and Palmer, E. (1989). *Nature (London)* **341**, 649–651.
Bill, J., Appel, V., and Palmer, E. (1988). *Proc. Natl. Acad. Sci. U.S.A.* **85**, 9184.
Bill, J., Kanagawa, O., Woodland, D. L., and Palmer, E. (1989). *J. Exp. Med.* **169**, 1405–1419.
Blackman, M. R., Gerhardt-Burget, H., Woodland, D. L., Palmer, E., Kappler, J., and Marrack, P. (1990a). *Nature (London)* **345**, 540–542.
Braciale, V. L., and Braciale, T. J. (1981). *J. Immunol.* **127**, 859–862.
Buller, R. M., Yetter, R. A., Fredrickson, T. N., and Morse, H. C., III (1987). *J. Virol.* **61**, 383–387.
Burkly, L., Lo, D., Kanagawa, O., Brinster, R. L., and Flavell, R. L. (1989). *Nature (London)* **342**, 562–564.
Carlsson, R., Fischer, H., and Sjögren, H.-O. (1988). *J. Immunol.* **140**, 2484–2488.
Cazenave, P.-A. (1990). *Cell (Cambridge, Mass.)* **63**, 717–728.
Cerny, A., Hügen, A. W., Hardy, R. R., Hayakawa, K., Zinkernagel, R. M., Makino, M., and Morse, H. C., III (1990a). *J. Exp. Med.* **171**, 315–320.
Cerny, A., Hügin, A. W., Holmes, K. L., and Morse, H. C. III (1990b). *Eur. J. Immunol.* **20**, 1577–1581.

Chattopadhyay, S. K., Morse, H. C., III, Makino, M., Ruscetti, S. K., and Hartley, J. W. (1989). *Proc. Natl. Acad. Sci. U.S.A.* **86**, 3862–3866.
Choi, Y. (1989). *Proc. Natl. Acad. Sci. U.S.A.* **86**, 8941.
Choi, Y., Herman, A., DiGiusto, D., Wade, T., Marrack, P., and Kappler, J. (1990). *Nature (London)* **346**, 471–473.
Click, R. E., and Adelmann, A. (1988a). *J. Immunogenet.* **15**, 39–47.
Click, R. E., and Adelmann, A. (1988b). *Immunogenetics* **28**, 412–416.
Click, R. E., and Adelmann, A. (1989). *Immunogenetics* **29**, 155–160.
Click, R. E., Schneider, D., Sitzmann, L. A., and Azar, M. M. (1984). *Immunogenetics* **20**, 301–310.
Click, R. E., Adelmann, A. M., and Azar, M. M. (1985). *J. Immunol.* **134**, 2948–2952.
Click, R. E., Cahill, G., Schneider, D., Adelmann, A., Tarquino, J. J., and Peck, A. B. (1987). *J. Immunol.* **139**, 321–325.
Coffin, J. M. (1990). In "Virology" (B. N. Fields and D. M. Knipe, eds.), pp. 1437–1500. Raven Press, New York.
Cole, B. C., Daynes, R. A., and Ward, J. R. (1981). *J. Immunol.* **127**, 1931.
Cole, B. C., Sullivan, G. J., Daynes, R. A., Sayed, I. A., and War, J. R. (1982a). *J. Immunol.* **128**, 2013.
Cole, B. C., Daynes, R. A., and Ward, J. R. (1982b). *J. Immunol.* **129**, 1352–1359.
Cole, B. C., Kartchner, D. R., and Welles, D. J. (1989). *J. Immunol.* **142**, 4131.
Davis, M. M., and Bjorkman, P. J. (1988). *Nature (London)* **334**, 395–402.
Debreuil, P. C., Caillol, D. H., and Lemonnier, F. A. (1982). *J. Immunogenet.* **9**, 11.
De Kruyff, R. H., Ju, S.-T., Laning, J., Cantor, H., and Dorf, M. E. (1986). *J. Immunol.* **137**, 1109–1114.
Dellabona, P., Peccoud, J., Benoist, C., and Mathis, D. (1989). *Cold Spring Harbor Symp. Quant. Biol.* **54**, 375–381.
Dyson, P. J., Knight, A. M., Fairchild, S., Simpson, E., and Tomonari, K. (1991). *Nature (London)* **349**, 531–532.
Eicher, E. M., and Lee, B. K. (1990). *Genetics* **125**, 431–446.
Festenstein, H. (1972). In "Symposium Standardization of HL-A Reagents," Vol. 18, pp. 298–299. Munksgaard, Copenhagen.
Festenstein, H. (1973). *Transplant. Rev.* **15**, 62–88.
Festenstein, H. (1974). *Transplantation* **18**, 555–557.
Festenstein, H., and Berumen, L. (1984). *Transplantation* **37**, 322–324.
Festenstein, H., and Kimura, S. (1988). *J. Immunogenet.* **15**, 183–196.
Festenstein, H., Sachs, J. A., and Oliver, R. T. D. (1971). In "Immunogenetics of the H-2 System" (A. Lengerova and M. Voitiskova, eds.), pp. 170–177. Karger, Basel.
Festenstein, H., Bishop, C., and Taylor, B. (1977). *Immunogenetics* **5**, 357–361.
Festenstein, H., Kimura, S., and Biasi, B. (1989). *Immunol. Rev.* **107**, 29–59.
Fisch, P., Malkovsky, M., Kovats, S., Sturm, E., Braakman, E., Klein, B. S., Voss, S. D., Morrissey, L. W., DeMars, R., Welch, W. J., Bolhuis, R. L. H., and Sondell, P. M. (1990). *Science* **250**, 1269–1273.
Fischer, H., Dohlsten, M., Lindwall, M., Sjögren, H.-O., and Carlsson, R. (1989). *J. Immunol.* **142**, 3151–3157.
Fleischer, B., and Schrezenmeier, H. (1988). *J. Exp. Med.* **167**, 1697–1707.
Fleischer, B., Schrezenmeier, H., and Conradt, P. (1989). *Cell. Immunol.* **120**, 92.
Fowlkes, B. J., Schwartz, P. H., and Pardoll, D. M. (1988). *Nature (London)* **334**, 620–623.
Frankel, W. N., Rudy, C., Coffin, J. M., and Huber, B. T. (1991). *Nature (London)* (in press).

Fraser, J. D. (1989). *Nature (London)* **339**, 221–223.
Fry, A. M., and Matis, L. A. (1988). *Nature (London)* **335**, 830–832.
Gress, R. E., Wesley, M. N., and Hodes, R. J. (1981). *J. Immunol.* **127**, 1763–1766.
Halle-Pannenko, O., Pritchard, L. L., and Rappaport, H. (1983). *Transplantation* **36**, 60–68.
Halle-Pannenko, O., Pritchard, L. L., Bruley-Rosset, M., Berumen, L., and Motta, R. (1985). *Immunol. Rev.* **88**, 59–85.
Halle-Pannenko, O., Pritchard, L. L., Festenstein, H., and Berumen, L. (1986). *J. Immunogenet.* **13**, 437–450.
Halle-Pannenko, O., Pritchard, L. L., and Bruley-Rosset, M. (1987). *Eur. J. Immunol.* **17**, 1751–1755.
Hammerling, U. M., Toulon, M., Chun, M., Chun, M., Palfree, S., and Hoffman, M. K. (1988). *J. Immunol.* **140**, 2543.
Happ, M. P., Woodland, D. W., and Palmer, E. (1989). *Proc. Natl. Acad. Sci. U.S.A.* **86**, 6293–6296.
Hartley, J. W., Fredrickson, T. N., Yetter, R. A., Makino, M., and Morse, H. C., III (1989). *J. Virol.* **63**, 1223–1231.
Herrmann, T., Maryanski, J. L., Romero, P., Fleischer, B., and MacDonald, H. R. (1990). *J. Immunol.* **144**, 1–6.
Holly, M., Lin, Y.-S., and Rodgers, T. J. (1988). *Immunology* **64**, 643.
Hugin, A. W., Vacchio, M. S., and Morse, H. C., III (1991). *Science* (in press).
Jacobsson, J., Lilliehöök, B., and Blomgren, H. (1975). *Scand. J. Immunol.* **4**, 181–191.
Janeway, C. A., Jr. (1990). *Cell (Cambridge, Mass.)* **63**, 659–661.
Janeway, C. A., Jr. (1991). *Nature (London)* **349**, 459–461.
Janeway, C. A., Jr., and Katz, M. (1985). *J. Immunol.* **134**, 2057–2063.
Janeway, C. A., Jr., Lerner, E. A., Jason, J. M., and Jones, B. (1980). *Immunogenetics* **10**, 481–497.
Janeway, C. A., Jr., Conrad, P. J., Tite, J. P., Jones, B., and Murphy, D. B. (1983). *Nature (London)* **306**, 80–82.
Janeway, C. A., Jr., Chalupny, J., Conrad, P. J., and Buxser, S. (1988a). *J. Immunogenet.* **15**, 161–168.
Janeway, C. A., Jr., Fischer-Lindahl, K., and Hämmerling, U. (1988b). *Immunol. Today* **9**, 125–126.
Janeway, C. A., Jr., Dianzani, U., Portoles, P., Rath, S., Reich, E.-P., Rojo, J., Yagi, J., and Murphy, D. B. (1989a). *Cold Spring Harbor Symp. Quant. Biol.* **54**, 657–666.
Janeway, C. A., Jr., Yagi, J., Conrad, P., Katz, M., Vroegop, S., and Buxser, S. (1989b). *Immunol. Rev.* **107**, 61–88.
Janeway, C. A., Jr., Rojo, J., Saizawa, K., Dianzani, U., Portoles, P., Tite, J., Haque, S., and Jones, B. (1989c). *Immunol. Rev.* **109**, 77–92.
Janeway, C. A., Jr., Yagi, J., and Rath, S. (1991). *Behring Inst. Mitt.* **88**, 177–182.
Jones, B., and Janeway, C. A., Jr. (1982). *Immunogenetics* **16**, 243–255.
Kanagawa, O., Palmer, E., and Bill, J. (1989). *Cell. Immunol.* **119**, 412–426.
Kappler, J., Wade, T., White, J., Kushnir, E., Blackman, M., Bill, J., Roehm, R., and Marrack, P. (1987a). *Cell (Cambridge, Mass.)* **49**, 263–271.
Kappler, J. W., Roehm, N., and Marrack, P. (1987b). *Cell (Cambridge, Mass.)* **49**, 273–280.
Kappler, J. W., Staerz, U., White, J., and Marrack, P. (1988). *Nature (London)* **332**, 35–40.
Kappler, J. W., Kotzin, B., Herron, L., Gelfand, E. W., Bigler, R. D., Boylston, A., Carrel, S., Posnett, D. N., Choi, Y., and Marrack, P. (1989a). *Science* **244**, 811–813.

Kappler, J. W., Pullen, A., Callahan, J., Choi, Y., Herman, A., White, J., Potts, W., Wakeland, E., and Marrack, P. (1989b). *Cold Spring Harbor Symp. Quant. Biol.* **54**, 401–408.
Katz, M. E., and Janeway, C. A., Jr. (1985). *J. Immunol.* **134**, 2064–2070.
Kawabe, Y., and Ochi, A. (1990). *J. Exp. Med.* **172**, 1065–1070.
Kawabe, Y., and Ochi, A. (1991). *Nature (London)* **349**, 245–248.
Kaye, J. (1989). *Nature (London)* **336**, 580.
King, L. B., Lund, F. E., White, D. A., Sharma, S., and Corley, R. B. (1990). *J. Immunol.* **144**, 3218–3227.
Kingsmore, S. F., Watson, M. L., Howard, T. A., and Seldin, M. F. (1989). *EMBO J.* **8**, 4073–4080.
Klinman, D. M., and Morse, H. C., III (1989). *J. Immunol.* **142**, 1144–1149.
Kourilsky, P., and Claverie, J. M. (1989). *Cell (Cambridge, Mass.)* **56**, 327–329.
Kwon, B. S., and Weissman, S. M. (1984). *J. Virol.* **52**, 1000–1004.
Lilliehöök, B., Jacobsson, H., and Blomgren, H. (1975). *Scand. J. Immunol.* **4**, 209–216.
Lynch, D. H., Grees, R. E., Needleman, B. W., Rosenberg, S. A., and Hodes, R. J. (1985). *J. Immunol.* **134**, 2071–2078.
MacDonald, H. R., Schneider, R., Lees, R. K., Howe, R. C., Acha-Orbea, H., Festenstein, H., Zinkernagel, R. M., and Hengartner, H. (1988a). *Nature (London)* **332**, 39–45.
MacDonald, H. R., Hengartner, H., and Pedrazzini, T. (1988b). *Nature (London)* **335**, 174–176.
MacDonald, H. R., Lees, R. K., Schneider, R., Zinkernagel, R. M., and Hengartner, H. (1988c). *Nature (London)* **336**, 471–472.
Macphail, S., and Stutman, O. (1990). Submitted for publication.
Marrack, P. (1991). *Nature (London)* **349**, 524–526.
Marrack, P., and Kappler, J. (1988a). *Nature (London)* **332**, 840–843.
Marrack, P., and Kappler, J. (1988b). *Immunol. Today* **9**, 308–314.
Marrack, P., and Kappler, J. (1990). *Science* **248**, 705–711.
Marrack, P., Blackman, M., Kushnir, E., and Kappler, J. (1990). *J. Exp. Med.* **171**, 455–464.
Marrack, P., Blackman, M., Choi, K., and Kappler, J. (1991). Submitted for publication.
Mathis, D. J., Benoist, C., Williams, V. E., II, Kanter, M., and McDevitt, H. O. (1983). *Proc. Natl. Acad. Sci. U.S.A.* **80**, 237.
Matossian-Rogers, A., and Festenstein, H. (1976). *J. Exp. Med.* **143**, 456–461.
Matossian-Rogers, A., and Festenstein, H. (1977). *Transplantation* **23**, 316–321.
McGrath, M. S., and Weissman, I. L. (1979). *Cell (Cambridge, Mass.)* **17**, 65–75.
Mollick, J. A., Cook, R. G., and Rich, R. R. (1989). *Science* **244**, 817–820.
Molnar-Kimber, K. L., and Sprent, J. (1980). *J. Exp. Med.* **151**, 407–417.
Molnar-Kimber, K. L., Webb, S. R., Sprent, J., and Wilson, D. B. (1980). *J. Immunol.* **125**, 2643–2645.
Moseley, W. S., and Seldin, M. F. (1989). *Genomics* **5**, 899–905.
Moseley, W. S., Watson, M. L., Kingsmore, S. F., and Seldin, M. F. (1989). *Immunogenetics* **30**, 378–382.
Mosier, D. E., Yetter, R. A., and Morse, H. C., III (1985). *J. Exp. Med.* **161**, 766–784.
Needleman, B. W., Lynch, D. H., and Hodes, R. J. (1988). *J. Immunol.* **141**, 3760–3767.
Okada, C. Y., and Weissman, I. L. (1989). *J. Exp. Med.* **169**, 1703–1719.
O'Rourke, A. M., Mescher, M. F., and Webb, S. R. (1990). *Science* **249**, 171–174.
Palmer, E., Woodland, D. L., Happ, M. P., Bill, J., and Kanagawa, O. (1989). *Cold Spring Harbor Symp. Quant. Biol.* **54**, 135–146.

Peavy, D. L., Adler, W. H., and Smith, R. T. (1970). *J. Immunol.* **105**, 1453-1458.
Peck, A. B., Janeway, C. A., and Wigzell, H. (1977). *Nature (London)* **266**, 840-842.
Pinto, M., Torten, M., and Birnbaum, S. C. (1978). *Transplantation* **25**, 320.
Poindexter, N. J., and Schlievert, P. M. (1986). *J. Infect. Dis.* **153**, 772.
Portoles, P., Rojo, J. M., and Janeway, C. A., Jr. (1989). *J. Mol. Cell. Immunol.* **4**, 129-137.
Pullen, A. M., Marrack, P., and Kappler, J. W. (1988). *Nature (London)* **335**, 796.
Pullen, A. M., Kappler, J. W., and Marrack, P. (1989a). *Immunol. Rev.* **107**, 125-139.
Pullen, A. M., Marrack, P., and Kappler, J. W. (1989b). *J. Immunol.* **142**, 3033-3037.
Pullen, A. M., Potts, W., Wakeland, E. K., Kappler, J. and Marrack, P. (1990a). *J. Exp. Med.* **171**, 49-62.
Pullen, A. M., Wade, T., Marrack, P., and Kappler, J. W. (1990b). *Cell (Cambridge, Mass.)* **61**, 1365-1374.
Purdie, K. J., Hudson, K., and Fraser, J. D. (1991). *CRC Crit. Rev. Immunol.* (in press).
Racevskis, J. (1986). *J. Virol.* **58**, 441-449.
Racevskis, J., and Prakash, O. (1984). *J. Virol.* **51**, 604-610.
Rammensee, H., Kroschewski, R., and Frangoulis, B. (1989). *Nature (London)* **339**, 541.
Rust, C. J. J., Verreck, F., Vietor, H., and Konig, F. (1990). *Nature (London)* **346**, 572.
Ryan, J. J., Mond, J. J., Finkelman, F. D., and Scher, I. (1983). *J. Immunol.* **130**, 2534-2541.
Ryan, J. J., Miner, D. W., Mond, J. J., Finkelman, F. D., and Woody, J. N. (1987a). *J. Immunol.* **138**, 2392-2401.
Ryan, J. J., Mond, J. J., and Finkleman, F. D. (1987b). *J. Immunol.* **138**, 4085-4092.
Ryan, J. J., Thompson, C. B., Mond, J. J., and Finkelman, F. D. (1988). *J. Immunogenet.* **15**, 121-133.
Scholl, P. R., Diez, A., Mourad, W., Parsonnet, J., Geha, R. S., and Chatila, T. (1989a). *Proc. Natl. Acad. Sci. U.S.A.* **86**, 4210.
Scholl, P. R., Diez, A., and Geha, R. S. (1989b). *J. Immunol.* **143**, 2583.
Scholl, P. R., Diez, A., Karr, R., Sekaly, R. P., Trowsdale, J., and Geha, R. (1990). *J. Immunol.* **144**, 226.
Seldin, M. F., Morse, H. C., LeBoeuf, R. C., and Steinberg, A. D. (1988a). *Genomics* **2**, 48-56.
Seldin, M. F., Abe, R., Steinberg, A. D., Hodes, R. J., and Morse, H. C. (1988b). *J. Immunogenet.* **15**, 59-66.
Steinman, R. M., Van-Voohis, W. C., and Spalding, D. M. (1986). In "Handbook of Experimental Immunology" (D. M. Weir, L. A. Herzenberg, C. Blackwell, and L. A. Herzenberg, eds.), pp. 49.1-49.9. Blackwell, Oxford.
Sunshine, G. H., Mitchell, T. J., Czitrom, A. A., Edwards, S., Glasebrook, A., Kelso, A., and MacDonald, H. R. (1985). *Cell. Immunol.* **91**, 60.
Tomonari, K., and Lovering, E. (1988). *Immunogenetics* **28**, 445-451.
Vacchio, M. S., Ryan, J. J., and Hodes, R. J. (1990). *J. Exp. Med.* **172**, 807-813.
von Boehmer, H., and Sprent, J. (1974). *Nature (London)* **249**, 363-365.
von Boehmer, H., Teh, H. S., and Kisielow, P. (1989). *Immunol. Today* **10**, 57-61.
Wannamaker, L. W., and Schlievert, P. M. (1988). In "Handbook of Bacterial Toxins" (C. M. Hardegree and A. T. Tu, eds.), Vol. 4 p. 267. Dekker, New York.
Watson, M. L., Kingsmore, S. F., Johnston, G. I., Siegelman, M. H., Le Beau, M. M., Lemons, R. S., Bora, N. S., Howard, T. A., Weissman, I. L., McEver, R. P., and Seldin, M. F. (1990). *J. Exp. Med.* **173**, 263-271.
Webb, S. R., and Sprent, J. (1986). *Int. Rev. Immunol.* **1**, 151-182.
Webb, S. R., and Sprent, J. (1989a). *J. Exp. Med.* **171**, 953-958.
Webb, S. R., and Sprent, J. (1989b). *Immunol. Rev.* **107**, 141-158.

Webb, S. R., and Sprent, J. S. (1990). *Science* **248**, 1643–1646.
Webb, S. R., Molnar-Kimber, K., Bruce, J., Sprent, J., and Wilson, D. B. (1981). *J. Exp. Med.* **154**, 1970–1974.
Webb, S. R., Li, J. H., Wilson, D. B., and Sprent, J. (1985). *Eur. J. Immunol.* **15**, 92–96.
Webb, S. R., Okamoto, A., and Sprent, J. (1988). *J. Immunol.* **141**, 1828–1834.
Webb, S. R., Okamoto, A., Ron, Y., and Sprent, J. (1989). *J. Exp. Med.* **169**, 1–12.
Webb, S. R., Morris, C., and Sprent, J. (1990). *Cell (Cambridge, Mass.)* **63**, 1249–1256.
White, J., Herman, A., Pullen, A. M., Kubo, R., Kappler, J., and Marrack, P. (1989). *Cell (Cambridge, Mass.)* **56**, 27–35.
Wilson, D. B., Blyth, J. L., and Nowell, P. C. (1968). *J. Exp. Med.* **128**, 1157–1181.
Woodland, D. L., Happ, M. P., Bill, J., and Palmer, E. (1990). *Science* **247**, 964–967.
Woodland, D. L., Happ, M. P., Gollob, K., and Palmer, E. (1991). *Nature (London)* **349**, 529–530.
Yagi, J., and Janeway, C. A., Jr. (1990). *Int. Immunol.* **2**, 83–89.
Yagi, J., and Janeway, C. A., Jr. (1991). Submitted for publication.
Yagi, J., Baron, J., Buxser, S., and Janeway, C. A., Jr. (1990). *J. Immunol.* **144**, 892–901.
Yagi, J., Rath, S., and Janeway, C. A., Jr. (1991). Submitted for publication.
Yetter, R. A., Buller, R. M. L., Lee, J. S., Elkins, K. L., Mosier, D. E., Fredrickson, T. N., and Morse, H. C., III (1988). *J. Exp. Med.* **168**, 623–635.

This article was accepted for publication on 4 March 1991.

Programmed Cell Death in the Immune System

J. JOHN COHEN

Department of Microbiology and Immunology, University of Colorado Medical School, Denver, Colorado 80262

I. Introduction and Definitions

There are two conceptually different ways in which a eukaryotic cell can die. One, most familiar to ordinary people, is *accidental* death, such as might happen to a cell when its blood supply is cut off or on exposure to certain poisons or to physical injury. The other is *programmed* cell death (PCD), death that is required or desirable for the proper development or function of an organ, cell system, or individual as a whole (1–5). A striking example, seen in embryonic development, is generically called morphogenetic death (6–8). This refers to cell death that helps form the ultimate shape of limbs and organs, as when two structures fuse or when a solid structure divides; the formation of digits in the primitive limb bud is due to death of web-space cells, precisely programmed in time and space (7,9). The central and peripheral nervous systems provide other examples, in which neurons degenerate if they fail to make the proper trophic connections at the right time (10); a well-studied experimental model is the death of those dorsal root ganglion cells that should have innervated an amputated fetal limb (11,12). Another category of programmed cell death is that which follows removal of a growth factor or hormone. Seasonal or cyclic involution of sex organs in response to declining levels of gonadotrophins or steroids, and involution of secondary structures following gonadectomy (13), are familiar examples. The third category may be normal turnover, for example, in the skin. Here basal cells divide and give rise to other basal cells and cells that begin to differentiate into granular keratinocytes. Once this decision is made the granular cells have about a week to live; this is true terminal differentiation, with the terminus being death. Similar fates await most of the leukocytes.

In this review the premise to be examined is that, in many different examples of PCD, regardless of the cell type involved and regardless of the stimulus, the cells undergo similar biochemical and morphological events, and that PCD often follows a final common pathway. Evidence that there is not only programmed cell death, but a cell death program, will be discussed. The significance of this for studies of the immune system, in which cell death is such a prominent feature at so many levels, will be apparent. Because this is a new area for most immunologists, the survey that follows is more broad than deep. A number of brief reviews of the general topic of programmed cell death and apoptosis are available (1–5,14–19). The present discussion will first focus on the mechanisms of programmed cell death, and then turn to a consideration of examples of cell death in the immune systems that appear operationally or on a mechanistic basis to be programmed.

II. Mechanisms of Programmed Cell Death

A. THE MORPHOLOGY OF APOPTOSIS

In 1972, Kerr, Wyllie, and Currie (15) proposed the term *apoptosis* to refer to the peculiar morphology of programmed cell death, as opposed to *necrosis*, the appearance of accidental cell death. Apoptosis is a word from the Hippocratic corpus and refers to the falling of leaves from trees in the autumn, a lovely example of programmed, physiologically appropriate death that also implies renewal. It is correctly pronounced by English speakers "ap-o-TO-sis," with a short *a* as in *apoplexy*, the second *p* silent and the stress on the third syllable (to pronounce it "a-pop-tosis" is to invoke a Modern Greek usage referring to hair loss). In necrosis, the histological changes are primarily cytoplasmic, and the mitochondria are the major target of damage. These organelles swell and may show precipitates of calcium phosphate; at this stage the process becomes irreversible, the cell loses the ability to make sufficient ATP and regulate its osmotic pressure, and it swells and lyses. Throughout the process, the nuclear structure, except for swelling, remains relatively intact. Apoptosis, on the other hand, involves loss of cell volume, membrane blebbing, and chromatin margination along the nuclear envelope followed by collapse of the nucleus into spherical "beads" of very dense chromatin, still surrounded by membrane (14). The cytoplasmic organelles are much less affected during this process; curiously, this seems to be true even in leaves falling off trees in autumn (20). The loss of cell volume is striking: in rat thymocytes, average volume drops from 75 to 50 fl (21). An important difference between necrosis and apoptosis is that the rupture of the cell in necrosis releases chemotactic cell contents, which attract an inflammatory reaction to clear the debris. In apoptosis, the cell remains intact or breaks up into membrane-bound apoptotic bodies, some of which contain chromatin. These are phagocytized by the nearest cell capable of it, usually as soon as the process of apoptosis has begun and usually before the cell dies (takes up vital dyes). An inflammatory reaction is not elicited. It is a very good design for the removal of cells undergoing physiological death in an otherwise normal tissue (see Section II,D).

B. DNA Damage

In 1980, Wyllie demonstrated that the death of thymocytes *in vitro*, when exposed to glucocorticoid, had the morphological appearance of apoptosis (22). He further demonstrated that the cell's chromatin became cleaved into a series of fragments whose average size was about 180 bp and integral multiples thereof. This "ladder" pattern was reminiscent of the one described earlier by Williamson (23) and by Hewish and Burgoyne (24–26) as due to the action of a nuclease found endogenously in the nuclei of certain normal cells, especially hepatocytes. This endogenous endonuclease could be activated in isolated nuclei in the presence of Ca^{2+} and Mg^{2+} ions, both in the millimolar range. We subsequently showed that thymocyte nuclei contained considerable quantities of this (or a similar) Ca^{2+}/Mg^{2+}-dependent endonuclease, and that the pattern of cleavage seen with isolated nuclei incubated with these ions was identical to that seen when intact cells were incubated with dexamethasone or corticosterone (27). Thus it seemed reasonable to assume that the cleavage in intact cells was in fact due to the activation of this endogenous endonuclease, although to this date it has not been directly established.

DNA fragmentation is induced in murine thymocytes *in vitro* by concentrations (0.01–1 μM) of glucocorticoids that are in the *in vivo* physiological range (at the peak of the circadian cycle) (28). Depending on techniques used to demonstrate fragmentation, it begins to be detectable in 1 or 2 hours, and plateaus at 70–90% of the total DNA in an unfractionated population of thymocytes by 24 hours (27). Cell death, as measured by uptake of vital dye, lags behind DNA cleavage by about 5 hours; DNA damage is definitely premortem, and this is helpful in distinguishing apoptosis from necrosis, in which, if there is DNA damage, it occurs after the cell is dead and its lysosomal enzymes are released. A definition of cell death is of course difficult; it is obvious that any cell sustaining the degree of DNA damage that is seen in apoptosis will never again divide, so if clonogenic potential is equated with death, DNA fragmentation and death must be essentially the same event. In this review, a looser definition is employed, one that is suitable for the study of cells that, like the bulk of thymocytes, are not readily able to divide further: death is an inability to regulate the composition of the cytoplasm in order to maintain homeostasis, and can be measured, for example, by the uptake of vital dyes or the release of large intracellular molecules as observed in a chromium release assay.

In apoptosing thymocytes *in vitro*, there is a very good correlation between the fraction of cells scored as apoptotic and the fraction of DNA that no longer sediments in a moderate centrifugal field (22,29). In addition, when apoptotic glucocorticoid-treated rat thymocytes were separated from normal cells by density gradient centrifugation, essentially all the fragmented DNA was found in the apoptotic (dense) cell fraction (21). In this study, no intermediate sizes were observed among normal thymocytes (modal diameter 5.2 μm) and apoptotic ones (modal diameter 4.6 μm). These data indicate that the process of DNA fragmentation is "all or nothing" in a single cell; cells that are half-apoptotic are not seen. They further suggest that the transition from normal to apoptotic takes place with extreme rapidity, although the lag time before the process begins is variable within a cell population and may be lengthy.

The pattern of DNA fragments observed results from DNA undergoing double-stranded cleavage in the linker regions between nucleosomes. This cutting is not sequence specific, but rather relates to the relative accessibility of DNA in the linker, where it is rather loosely associated with histone H1; DNA in the nucleosome is tightly complexed to the H2a/H2b/H3/H4 core. DNase I, in contrast, can readily degrade chromatin down to oligonucleotides on the order of 10 bp; it is perhaps a smaller enzyme than the endogenous endonuclease (30). There appear to be two classes of DNA cleavage in glucocorticoid-treated rat thymocytes; one generates long stretches of DNA, still attached to the nuclear matrix, and another generates the mono- and oligonucleosomes readily seen on agarose gels (18). Whether this means that there are two different enzymes involved, or whether two different classes of chromatin are being cleaved differentially (for example, condensed and transcriptionally active chromatin), remains to be determined.

After these initial observations in rodent thymocytes, it became apparent that nucleosomal cleavage of chromatin was a frequent characteristic of cell death that might be considered programmed, or observed as apoptotic. It was shown in interleukin-2 (IL-2)-dependent cells when their required growth factor was removed (31); this was true of fresh T lymphoblasts as well as of long-term cell lines. The same pattern of cleavage was found when resting small lymphocytes from thymus or spleen were subjected to low-dose γ-irradiation (32). Subsequently, this pattern of DNA fragmentation has been recorded in other cells deprived of trophic or growth factors (33–44); in thymocytes exposed to toxins (45); in the targets of cytotoxic T lymphocytes (CTLs) (46–47); and in the activation-induced death of thymocytes, T cell hybridomas, and pre-B cell lines (48–54). These and other examples are discussed below.

It is still not clearly established that the morphological event, apoptosis, is exactly equivalent to the biochemical event, DNA fragmentation. To date, a study comparing the two has not been carried out with a series of diverse cell types. There is a temptation to conclude that the nucleosomal "ladder pattern" is the *sine qua non* of programmed cell death. A controversy arose when the ladder was not seen in certain human-origin cell targets of cytotoxic T lymphocytes (CTLs) (55), although the cells were efficiently lysed. Did this mean that human cells were fundamentally different from mouse cells, and that the DNA fragmentation seen in mouse cells was a curious artifact devoid of human relevance? Probably not, as further work has shown that the DNA in many human cell types, for example, thymocytes and leukemia cell lines, is in fact fragmented into nucleosomes after appropriate stimuli (56–58). Furthermore, in cells that do not apparently fragment their DNA into nucleosomes, there can be extensive DNA damage, which can consist of either frequent single-stranded nicks or rarer double-stranded cuts (59,60). In several systems, the topoisomerase II inhibitor novobiocin has been found to inhibit DNA fragmentation (61,62). Although the inhibitor has pleiotropic effects, this result may indicate that part of the process of apoptosis is the unwinding of chromatin, perhaps to permit the endonuclease greater access to the DNA.

A word here about methods. When nucleosomal DNA fragmentation occurs, it can be observed by extracting the cells' DNA and running it on a 1% neutral agarose gel. This method allows the pattern to be recognized but is poorly adapted to quantitation, and an impressive gel pattern may represent only 2–3% of the total DNA. Because DNA fragmentation goes rapidly to completion in any cell that begins the process, this represents fragmentation in only a tiny proportion of the cells; if greater than 2–3% of the cells died in such an experiment, their death may have been by some other mechanism. It is essential to quantify the extent of DNA fragmentation, either by the method originally described by Wyllie (22), by our simpler modifications of his method (27,32,47), or by some other method. The morphological identification of apoptosis was defined in electron microscopic terms, and this is still the best way to be sure one is dealing with apoptosis, especially in tissue sections. With experience, however, it is quite easy to tell apoptotic from normal cells in the light microscope. We stain cells in suspension with acridine orange at 4 μg/ml final concentration, and examine them under a fluorescence microscope using fluorescein filters and dichroic mirrors. Normal nuclei fluoresce in a structured pattern (different for each cell type), whereas in apoptosis the DNA is collapsed into crescents around the nuclear envelope or appears as featureless bright beads. Observed at the right time, most apoptotic cells will still be phase bright. The characteristic shrinkage of apoptotic cells may be studied in a flow cytometer, or more readily in a particle size analysis system such as the Coulter Channelyzer.

C. BIOCHEMICAL PATHWAYS

Remarkably little is known about the biochemical events that take place in cells undergoing apoptosis, and even less about the events that are directly causal of the process. It is crucial to distinguish between pathways that are unique to a particular inducer of PCD, and those that are common to all inducers. Clearly, some form of chromatin damage is a common characteristic, and this implies the action of an endonuclease. The intracellar location of the putative endonuclease is still uncertain. Thymocyte nuclei contain large quantities of a Ca^{2+}/Mg^{2+}-dependent endonculease, and it is logical to believe that this enzyme is the one that is activated in glucocorticoid-treated cells. However, other cells that undergo apoptosis have no detectable endonuclease in their nuclei; a good example is the thymoma line S49, which shows DNA fragmentation on incubation with dexamethasone (63). In fact, we have never found a line of cells in tissue culture whose nuclei contain an endonuclease that can be activated *in situ* by Ca^{2+} and Mg^{2+}. Furthermore, if mouse thymocytes are made into single-cell suspensions and incubated for several hours, no enzyme is then detectable in their nuclei; but if they are then incubated with dexamethasone, the kinetics of DNA fragmentation are identical to those seen with fresh cells (29). This seems to imply that the readily detected endonuclease in thymocyte nuclei is not actually necessary for apoptosis. It may be that thymocytes, the vast majority of which are soon to die, overproduce the enzyme in preparation for their suicide. The only other population of cells in the mouse in which we detect large amounts of Ca^{2+}/Mg^{2+}-dependent endonuclease are splenic B cells (29). Curiously, the same fate—death within 24 hours—may await most of these cells as well (64).

In some PCD systems, there is an absolute requirement for new RNA and protein synthesis for apoptosis to take place; this will be discussed in Section IV. Is the endonuclease one of proteins whose synthesis is required? This is a controversial area. It has been claimed that a variety of new endonucleases can be observed in nuclear extracts from thymocytes treated with glucocorticoid (65). These extracts were active in inducing DNA fragmentation in nuclei without this activity, whereas similar extracts from control thymocytes were virtually inactive [but, in contrast, we find control thymocyte nuclear extracts to contain large quantities of readily extracted endonuclease (66)]. The nucleases were reported to range from 12 to 32 kDa. Subsequent to this publication, however, to other groups indicated that the observed nucleases may well have been histones, which could mimic nucleases in the assay that was used (67,68). Thus the question of induced endonucleases remains moot. The existence of systems in which new macromolecular synthesis is not required for DNA fragmentation (Section IV,C) suggests that the enzyme(s) may already be in place in most cell types, even if activity is not detectable in the nucleus upon addition of Ca^{2+} and Mg^{2+}. Until a reliable molecular probe for the endonuclease is available this question will be unanswered.

If there is resident endonuclease in the nuclei of most cell types, how does it become activated? It does not appear that the cell can mobilize the concentrations of Ca^{2+} necessary to activate the enzyme *in vitro* (1 mM or greater), as physiological levels of cytoplasmic Ca^{2+} rarely exceed 1 μM, even after cell activation, and the nuclear concentration appears to be even lower. We examined a number of cations but none was as effective as calcium; still, it is possible that the physiological activator of this enzyme is another mineral or organic cation. On the other hand, there is precedence for an enzyme changing its requirement for calcium upon binding to certain lipid membranes; for example, the protease calpain, which in isolation requires millimolar Ca^{2+}, is activated by micromolar Ca^{2+} in the presence of phosphatidylinositol (69).

The actual role of calcium in apoptosis is also controversial. There are reports on both sides of the question of the need for external calcium in thymocyte apoptosis (70,71). We have found that the calcium ionophore A23187 in Ca^{2+}-containing medium will induce some DNA fragmentation in thymocytes (27), but the same is not seen in any tissue culture line (72,73). The thymocyte story is complicated by the observation that A23187-induced fragmentation is prevented by the protein synthesis inhibitor cycloheximide, which should not be the case if a simple activation of an endonuclease by Ca^{2+} were all that was involved (73,74). There are examples of CTL-mediated DNA fragmentation in which external Ca^{2+} is clearly not involved. On the other hand, calcium mobilization from internal stores probably *is* required for apoptosis.

D. SURFACE CHANGES RESULTING IN PHAGOCYTOSIS

There are two extraordinarily interesting biological facets of apoptosis. The first is that it represents a physiological way of controlling the turnover of cells in an organ. The other is the way in which apoptotic cells are removed from the system. In necrosis, massive, unscheduled tissue damage requires the immediate attention of "professional" phagocytes. Vertebrates have evolved a number of effective mechanisms for attracting such cells to the site of tissue damage, and the immune system is not the least elegant of these. These heroic measures often result in damage to normal structures, which are "innocent bystanders," and are acceptable in emergencies but hardly so for the maintenance of homeostasis on a day-to-day basis. Apoptosis is different; the changes that take place seem to preserve cytoplasmic integrity for a considerable part of the process, and even when the cell is "dead" in terms of its reproductive future (the degree of DNA damage that characterizes apoptosis is beyond repair), it does not lyse, release major cytoplasmic contents, or admit normally excluded extracellular macromolecules. What does happen is that it is phagocytosed, either by a "professional" phagocyte or by a nearby amateur; this phenomenon is striking in epithelia, but also in the thymus, where up to a third of all lymphocytes may die each day, but a histological section is amazingly devoid of corpses. Undesirable thymocytes, beginning apoptosis, seem to be rapidly phagocytosed by macrophages (75). The phagocytosis of apoptotic cells does not seem to activate the macrophage's proinflammatory activity.

It seems clear that apoptotic cells have membrane differences that make them susceptible to phagocytosis. This has been well demonstrated with human neutrophils, cells of short half-life that die rapidly by apoptosis *in vivo* and in culture (see also Section III,D). Apoptotic neutrophils, viable by the dye exclusion test, are phagocytosed by macrophages, but viable nonapoptotic neutrophils are not (76). The means by which the phagocyte recognizes an apoptotic cell have been investigated by a number of groups, and the conclusions are all different. Duvall *et al.* (77) found that mouse thymocytes, apoptotic from treatment with glucocorticoid, were phagocytosed by macrophages, and this process was inhibited by N,N'-diacetylchitobiose or N-acetylglucosamine. They suggested that there is a lectinlike activity on macrophages, which recognizes newly expressed sugar moieties on apoptotic thymocytes. Savill *et al.* (78) have suggested that apoptotic neutrophils are recognized by macrophages by means of a "charge-sensitive" recognition system, more responsive to net charge than to a particular molecular structure, and different from known macrophage adhesion mechanisms such as that involved in the recognition of aged erythrocytes. Further work from this group (79) implicated a vitronectin receptor on macrophages in the recognition of apoptotic human lymphocytes and neutrophils. Finally, Fadok *et al.* (80) have evidence that the recognition of apoptotic thymocytes by murine macrophages involves phosphatidylserine expression on the thymocyte; this phospholipid is normally situated on the inner leaflet of the plasma membrane, and seems to "flip out" during apoptosis.

It is evident from these studies that there may be more than one way in which an apoptotic cell is recognized by a phagocyte. Why this should be so is not established, but it could provide a margin of safety and specificity in a biologically important system. It will also be interesting to see whether there is regional specialization in this recognition process; for example, will bone marrow macrophages preferentially phagocytose apoptotic pre-B cells? Parenthetically, if the recognition and phagocytosis of an apoptotic cell is the rule *in vivo*, then lysis must be a rare event, and *in vitro* studies that depend on lysis may be unphysiological or even irrelevant.

III. Programmed Cell Death in Immunologic Systems

A. PROGRAMMED CELL DEATH IN B CELLS

There are a surprisingly large number of stages in a B cell's life when it risks undergoing a programmed death. By "programmed" here is meant a death that any B cell might meet, and many do, as part of the design of the immune system. It can result from certain requirements of the immune system not being met at the right time or, conversely, from other requirements being met appropriately.

1. Faulty Recombination

Death is probably the fate of the pre-B cell that fails to rearrange its heavy- or light-chain genes in such a way as to produce a complete transcript. Although the signal sequences found 3' of V regions, 5' and 3' of D regions, and 5' of J regions are designed to associate so that the coding sequences are precisely apposed, mechanisms exist to remove bases at the junction and add others, so that somatic variation plays an important role (81). This imprecision of joining of coding sequences during DNA recombination in pre-B cells generates a great deal of desirable diversity, but also leads to frequent out-of-frame ligations, due to removal and/or replacement of nucleotides that yield an incomplete codon. The exact frequency of such erroneous recombinations is difficult to state, but it seems reasonable that fewer than a tenth of all pre-B cells that begin D-to-J rearrangements of the heavy-chain locus end up as antibody-producing B cells (82,83). The fate of cells with nonsense rearrangements has not been clearly delineated. Three are conceivable (D. Nemazee, personal communication): first, the cell may attempt and successfully achieve further rearrangements; second, the cell might differentiate into another cell type, for example, a macrophage; and third, such a cell may be programmed to die. The third possibility seems to be currently favored by the weight of experimental evidence. The severe combined immune-deficient (SCID) mouse is defective in the recombinase machinery and does not join coding sequences properly (84). Such mice have cells bearing pre-B cell markers but are virtually lacking in B cells. Transgenic mouse technology should allow an unequivocal decision between these models.

2. Clonal Abortion

If a pre-B cell successfully rearranges its heavy- and light-chain loci to produce functional immunoglobulin, it will eventually express surface IgM (sIgM) before expressing surface IgD (sIgD) (85). There is good evidence that, if it encounters antigen at this $sIgM^+$, $sIgD^-$ stage, it will not be stimulated but rather risks deletion. Support for this model was obtained many years ago, when it was shown that immature B cells cap their immunoglobulin very rapidly in response to cross-linking and, unlike mature B cells, reexpress it slowly if at all (86,87). This and other observations led Nossal to propose the concept of clonal abortion (88). B cells remain at this abortable stage for a brief time, being susceptible to abortion by antigen for its duration, and if they survive it, they are then stimulated rather than deleted by antigen. The antigens most likely to be present in the bone marrow during the abortable state are, of course, self molecules, so this is one way of deleting autoreactive B cell clones. Transgenic mouse models have recently demonstrated the correctness of this model (89,90). Assuming that this is a good model of PCD in developing B cells, it is a very interesting one, in that the same signals that result in proliferation and differentiation in a mature B cell cause death in an immature one. The study of how cells at different stages of maturation interpret the same signals differently will be a rewarding endeavor.

3. Growth Arrest of the WEHI-231 B Cell Line

WEHI-231 is a cell line derived from a (BALB/c × NZB)F_1 mouse injected with mineral oil (91). It bears a large amount of IgM on its surface but little or no IgD, and it does not secrete immunoglobulin.

Thus it resembles an early B cell, and it is not surprising that, when it is exposed to antibodies to its heavy or light chains, a condition that mimics the normal interaction of such a cell with its appropriate antigen, it undergoes growth arrest and may die (54). When treated with lipopolysaccharide (LPS), this cell line expresses sIgD and seems to become resistant to receptor-mediated growth arrest (92), although not all reports support this claim (93). If so, though, then WEHI-231 behaves as if it has a maturation block at the abortable stage, and can perhaps be driven past this block by LPS. This may represent an interesting model system to study some of the mechanisms of programmed cell death as well as clonal abortion; it has been shown that growth arrest induced by exposure to antiimmunoglobulin reagents induces nucleosomal DNA fragmentation (54) and is apoptotic. The essential difference between a WEHI-231 and a mature B cell, i.e., the former responds to stimulation via its surface immunoglobulin with death and the latter responds with proliferation, is under intensive investigation (94,95).

4. Splenic B Cells

The exact fate of splenic B cells is unclear. Freitas *et al.* (64) showed that about 10–20% of these cells are long-lived, and 80–90% die within 18–24 hours. The calculations are supported by the work of Osmond's group (96). Why do these cells die? From their surface phenotype (predominantly $sIgM^+$, $sIgD^+$), they should be past the clonal abortion stage. There is some evidence that these B cells may be rescued from death by antigenic stimulation (96,97), and then leave the spleen and recirculate to the lymph nodes. Teleologically, the rapid turnover of splenic B cells would allow a fresh set of receptors to be available constantly; the splenic B cell mass would be replaced every 3 days or so, with only those that were stimulated (by antigen? by LPS?) surviving to become relatively long-lived, and the rest dying a programmed death. The nuclei of splenic B cells, but not T cells, contain large amounts of endogenous endonuclease, which can be activated by incubating these nuclei with Ca^{2+} and Mg^{2+} (60). This is a striking parallel to cortical thymocytes, which also contain the endonuclease and which also have a very short average life expectancy. In these cells it may be that the program is primed for rapid activation.

5. Somatic Mutation and Terminal Differentiation

A feature of B cell differentiation is hypermutation of immunoglobulin variable region genes after antigen stimulation. This process generates variant antibodies and is probably the cause of affinity maturation during an immune response. There is evidence that the hypermutation process is activated in centroblasts within germinal centers, and that their progeny, centrocytes, die by apoptosis unless they are rescued by continued interaction with antigen (97).

It is generally acknowledged that many or most plasma cells, which devote an enormous part of their energy to secreting immunoglobulin molecules, do not divide and in fact die after several days of this feverish activity. The mechanism of cell death—whether inherently programmed, perhaps to limit the extent of somatic mutation, or due to metabolic exhaustion in the most active clones—is not known.

B. PROGRAMMED CELL DEATH IN T CELLS

1. Negative Selection in the Thymus

There is reasonable evidence that most pre-T cells rearrange their receptor genes within the thymus (98), and one must suppose that, as in the case of B cells, most rearrangements are unsuccessful and probably lead to cell death. Once a complete receptor is displayed on the cell surface, there are three possible fates awaiting the thymocyte: negative selection, positive selection, or no selection. If the receptor binds to self with high affinity (that is, affinity that would result in the activation of a mature T cell), the cell must be removed from the repertoire, as it represents a risk of autoreactivity. Evidence from transgenic mice indicates that for the most part such cells are deleted rather than inactivated, and so negative selection is *de facto* an example of programmed cell death (99,100). This argument does not specifically take into account the role of accessory binding molecules such as CD4 and CD8 and endogenous peptides in the selection process, but their involvement does not change the concept. It has also been demonstrated that at least some of the cells in the thymus die when stimulated *in vivo* by antibodies to their receptors or associated molecules such as CD3 (50,101). This death probably mimics the deletion of immature self-reactive T cells, and the mechanism has been shown to be similar to that seen in other PCD systems. There is nucleosomal DNA cleavage, and it seems to depend upon protein synthesis. The process is inhibited in mice treated with cyclosporin A (101). This interesting observation suggests a mechanism by which cyclosporin A, through preventing the activation-induced death of autoreactive thymocytes, may allow the survival of forbidden clones and their export to the periphery (102).

2. Activation-Induced Death of Hybridomas

Helper-type T cell hybridomas usually secrete lymphokines when their receptors are stimulated, either by a correct antigen–MHC combination or by antibodies to their antigen receptor, or to CD3, Ly-6, or Thy-1. This treatment may also cause growth arrest and cell death (103,104). In this respect the hybridomas behave like the immature thymocytes their thymoma parent was probably derived from; it is curious that this behavior is dominant over the mature phenotype in most hybridomas. The activation-induced death of hybridomas, like the similar death of WEHI-231 pre-B cells, may be a model for the study of negative selection or clonal abortion in the developing immune system. Death of hybridomas upon activation via their antigen receptor is morphologically apoptotic, and involves nucleosomal DNA fragmentation (53). It is blocked by cycloheximide and by actinomycin D, indicating that new protein and RNA synthesis are necessary for activation-induced death (105,106). In support of the idea that this is activation-induced death, cyclosporin A was found to inhibit death of hybridomas stimulated with anti-CD3 (101).

3. Death of Nonselected Thymocytes

If receptor expression is truly random, then, of the spectrum of T cell antigen receptor configurations available to an animal, a small number will be anti-self and have to be eliminated. A probably somewhat larger proportion will have the sort of affinity for self structures (that is, real but too low to activate the cell) that might in the periphery translate into high (activating) affinity for self plus some novel processed peptide, which is likely to be foreign (107). Such desirable cells are positively selected to undergo further maturation, upregulating receptor expression and shutting off either CD4 or CD8, depending on whether the MHC element recognized is class I or class II, respectively. Other markers may be expressed, including ones that permit the export of the cell into the periphery. What then is the fate of the large bulk of thymocytes that is neither negatively nor positively selected? Early studies showed that the majority—perhaps 99%—of thymocytes in young rodents die *in situ* (108). This death is clearly programmed. What is not clear is whether there is an external signal to these cells to die; this signal could be the presence of a factor, or the removal of one. Death could also be internally programmed, in which case the survival of positively selected thymocytes would represent rescue from such a program. One interesting possibility is that the death of the bulk of nonselected thymocytes is induced by endogenous glucocorticoids (28). This suggestion is based on the well-known sensitivity of most cortical thymocytes to peak physiological glucocorticoid concentrations, whereas mature T cells are resistant to these concentrations [reviewed by Cohen (28)]. The situation would be that developing thymocytes, when they have rearranged their receptor genes and expressed receptor on their membranes, have a period of time to be positively selected. What the molecular mechanism of positive selection is has not been established, but along with the acquisition of markers for export and so on, the program for glucocorticoid-induced death would be shut off. Then, when the daily peak of glucocorticoid concentration arrives at the thymus, those cells that have *not* been selected stand a high chance of having the death program induced, whereas the selected cells, with the phenotype of mature T cells, are not affected. The elimination of a large proportion of nonselected cells makes "room" for a new wave of candidates in the positive-selection sweepstakes. Although this hypothesis is entirely speculative, it is interesting that the thymus in adrenalectomized mice becomes greatly enlarged, as might be expected if a lethal influence were removed (109).

4. Death of T Cells Deprived of Growth Factors

It is clear that some T cells live a very long time, perhaps even as long as the individual who carries them. Others seem to die when the immediate need for them has passed. It probably suffices that helper T cells survive, because they can stimulate virgin B cells with secondary-response kinetics. A similar story may be true for cytotoxic T cells: if helpers are present it may not be necessary that CTLs also persist after antigen has been disposed of. It has been known for some time that certain proliferating T cells, usually of the CTL phenotype, are dependent on helper T cell-derived growth factors, including IL-2, not only for proliferation but for their very survival. In fact, use is made of this phenomenon in the common bioassay for IL-2, in which cell lines that are IL-2 dependent (for example, HT-2 or CTLL-2) are incubated in varying concentrations of authentic or putative IL-2 and their survival is measured after about 24 hours. Mitogen-stimulated normal T cell blasts also die when IL-2 is removed. It is probable that this is a true example of programmed cell death; the body is spared the expense of keeping large numbers of lymphocytes alive after they have become redundant. Death (rather than stasis) following growth factor removal is turning out to be a common theme in metazoan biology.

In 1986 it was shown (31) that death of T cell lines and T lymphoblasts following IL-2 removal is a fact apoptotic, and that their DNA is fragmented into nucleosomes. Curiously, death on removal of IL-2 could be prevented by cycloheximide or actinomycin D, indicating the activation of a death program.

5. Death of CTL Targets

Although the phenomenon of target cell killing by cytotoxic T cells has been known for over two decades, the mechanism remains confused. One model has it that the T cell engages the target, becomes activated, and releases the contents of "lytic granules" that diffuse across the narrow space separating the CTL's membrane from that of the target. Among the contents of the granules are a number of esterases, or "granzymes," and the precursor of a pore-forming protein called perforin or cytolysin (110–112). Perforin, which has been cloned, is certainly capable of causing rapid lysis of many erythrocytes and nucleated cells. This membrane attack model suggests that the CTL's role is to deliver the lytic molecules, and the target cell's role is essentially that of a passive victim. Although this is an attractive model and there is considerable support for some aspects of it, there remains the fact that there are target cell:CTL combinations with which normal cytolysis is observed, under conditions wherein exocytosis of granule contents is unlikely to occur (113). Either these experiments have some hidden flaw, or CTLs have available another mechanism of inflicting lethal damage on their targets.

The suggestion that there was another mechanism came from the work of Russell's group (46,114). Target cells exposed to CTLs were shown to undergo internal damage even before lysis; it would be difficult to explain this early damage just on the basis of pore formation in the plasma membrane. The nature of this damage, and its relationship to programmed cell death, is considered in Section IV,C.

C. INTERPHASE DEATH OF T AND B LYMPHOCYTES, AND NONSPECIFIC DAMAGE

One of the many curiosities of the immune system is the extraordinary sensitivity of resting lymphocytes to low doses of ionizing radiation; it has been reported that as little as 5 R will kill a human lymphocyte (115). Unlike other cells, lymphocytes are more sensitive to irradiation when in G_0 or G_1 than in S phase; thus the phenomenon has been known for a long time as "interphase death." Lymphocytes (murine thymocytes, splenic T, or B cells) exposed to low doses of ionizing radiation fragment their DNA in the characteristic nucleosomal ladder pattern (32,116–122) and undergo apoptosis. This process requires new RNA and protein synthesis (32); it is not the direct result of ionizing particle interaction with DNA. Cells irradiated at low doses appear healthy if kept several hours in cycloheximide, but die by apoptosis if the inhibitor is washed out. Thus the transient production of free radicals leads to long-term changes in the irradiated cell.

However similar the biochemical events in irradiated thymocytes may be to those in more physiological systems of apoptosis, one could hardly claim that death by irradiation is "programmed." One would certainly suppose that exposure to ionizing radiation at the required dose level is a rare event in an individual's or even a species' lifetime. However, radiation is damaging to cells. Sellins and Cohen (32) have suggested that lymphocytes, the most useful of cells, are also potentially the most dangerous; few other cells are allowed, indeed poised, to divide rapidly upon stimulation. If such a cell were damaged, it might be safer for it to commit suicide rather than to attempt repair with the attendant risks of error. For the good of the community of cells as a whole, it is preferable to lose a damaged lymphocyte. This is a very significant insight, because it suggests that certain cells, primarily lymphoid, might commit suicide by activating their death program *whenever* they detect a significant deviation from homeostasis. As an example, thymocytes, when heated to 43°C for an hour and then returned to 37°C, undergo apoptosis and die; fibroblasts similarly treated would repair the damage (123). It is important, then, to distinguish between a cell activating its suicide program because it senses damage, and one in which a specific physiological death pathway (one is tempted to say receptor-mediated pathway) has been triggered. If a toxin induces apoptosis in a thymocyte, it may not mean that there is a specific toxin-responsive death pathway, but rather that the toxin has injured the cell, and the cell has responded to injury by suicide. This is conceptually different from the death of the same cell induced by glucocorticoid, which may be a normal regulatory mechanism.

It is unknown at present how radiation or other injury could activate the death program. Clearly, changes are produced in the damaged target that result in gene induction and expression of lethal products (32,124). One possible scheme (125) might be that the injury (heat; reactive oxygen species) denatures intracellular proteins, and they are then bound by ubiquitin, which has an affinity for such proteins. This would lower free ubiquitin in the cell, and thus by mass action "pull" ubiquitin off histones; this could result in altered expression of certain genes, including heat-shock genes. In some cells the outcome would be the activation of repair mechanisms; in others, the outcome would be suicide.

IV. Triggering and Regulation of Programmed Cell Death

A. INTRODUCTION

The agents that induce programmed cell death in different cell types constitute a bewildering list (126). In most cases *in vivo,* it is not clear whether an external inducer is required or whether death is genetically programmed intrinsically. In some cases it may be that death is prepared for, being a highly probable event, but something external to the cell makes the decision as to whether the cell should proceed with its suicide or be rescued from it. In Puccini's version of Belasco's *Madama Butterfly* (127), the emperor has sent a sword to Butterfly's father, inscribed "Death with honor, for those who cannot live with honor." Both father and daughter know when and how they must respond to this message, and cells evidently do, too.

In spite of the apparent complexity of the induction of apoptosis, there may be just a few basic routes to a final common pathway. Until we know all the biochemical events in programmed cell death, this will remain speculation, but it may turn out that there are only a few different mechanisms, each of which can be triggered in multiple ways. We have already discussed how apoptosis may be triggered in lymphocytes by a variety of agents, which may all have in common the ability to denature proteins; thus denatured protein becomes the nexus for that pathway. In the next three sections we will discuss apoptosis-triggering mechanisms that can be categorized on the basis of their susceptibility to inhibition by cycloheximide and other protein synthesis inhibitors. This may not be the ultimate categorization, but at our state of knowledge it serves a useful purpose: that of simplifying a complicated situation.

B. INDUCTION: ACTIVATION OF DEATH GENES

When the induction of apoptosis requires protein (and usually mRNA) synthesis, and is therefore blocked by cycloheximide and related agents, I call that mechanism *induction*. The death of thymocytes induced by glucocorticoids or by irradiation, activation-induced death of B cells and T cell hybridomas, and death of interleukin-dependent cells are all examples of induction. In these situations, new proteins must be made for the cell to die—dramatic evidence for the concept of cellular suicide. What these new proteins are is not established; even though thymocytes contain readily detectable endogenous endonuclease, other cells that require induction do not, and the enzyme may have to be made *de novo*. However, other evidence (68) (and see Section IV,C) suggests that the enzyme must be present in virtually all cells, so it is more likely that the new proteins required act, among other things, to activate the endogenous endonuclease. The answer to these questions will have to wait until the death genes are cloned and characterized.

The identification of death genes is at a very early stage. If a gene is expressed in cells that are undergoing apoptosis, it is not automatic that it is a death gene. Here we define death genes as genes that are part of the final common pathway of apoptosis; their messages or protein products should be expressed in different cell types following exposure to different inducers of apoptosis. Other genes may be expressed as part of the triggering pathway. For example, in glucocorticoid-mediated thymocyte death, the glucocorticoid receptor is necessary for the induction of apoptosis; receptor-negative mutants do not die in the presence of steroid. The receptor gene is part of the triggering mechanism for this cell type and with this inducer, but it is not part of the final common pathway.

Investigators working in invertebrate systems have made progress in identifying death genes, especially in simple organisms such as the nematode *Caenorhabditis elegans*, in which the fate of every cell has been mapped. Certain mutations of the gene *deg-1* cause the accumulation of a toxic protein late in the development of certain neurons, which die (128). Although this gene may provide an interesting model for human neurodegenerative disease, the death process differs morphologically from that seen in programmed cell death in the same organism, and may not represent apoptosis; *deg-1* mutants may have an abnormal ion channel. If this mechanism also occurs in normal cell death in the nervous system, it would imply that there are two or more different mechanisms for programmed cell death, but evidence for this is currently minimal, except perhaps in the nervous system (129).

The genes *ced-3* and *ced-4*, on the other hand, are clearly involved in programmed cell death of a type that resembles apoptosis (130–132). Mutations in either of these genes in *C. elegans* result in animals with cells that do not die as they should, and which in fact may differentiate and take over the function of neighboring cells. These genes are thought to be involved in the initiation of programmed cell death. Other genes, *ced-1* and *ced-2*, seem to be necessary for the next step, which is the phagocytosis of the doomed cell. Its DNA is then degraded and this step is regulated by another gene, *nuc-1*. There is the suggestion that the *nuc-1*-encoded nuclease is supplied by the phagocytizing cell (133), in which case the *C. elegans* story diverges from that known in vertebrates.

In human follicular B cell lymphoma, the common t(14;18) translocation brings the *Bcl-2* gene into the Ig H region and activates expression of *Bcl-2*, which therefore may be a cellular protooncogene (134,135). Certain cell lines infected with a recombinant retroviral vector containing *Bcl-2* did not become independent of growth factors, but their death upon removal of required growth factors was delayed; they seemed to go into G_0 rather than undergo apoptosis (39,136). This effect was cell-type specific, and was seen in a pre-B cell line and a mastocytoma but not in a plasmocytoma or an IL-2-dependent T cell line. Therefore, the mechanism of *Bcl-2* action is unknown, but it seems to work at a stage prior to the final common pathway. Recent evidence is that *Bcl-2* encodes a mitochondrion-associated protein (137).

A curious observation has been made in a rat chloroleukemia cell line (138): when these cells become crowded in culture, they suddenly activate transcription of the repeat element *L1Rn;* growth stops and the cells undergo apoptosis. Death results from the incorporation of up to 300,000 copies of *L1Rn*, apparently by retroposition via an RNA intermediate, into random locations in the genome, which must produce many lethal mutations. Although other cell lines will undergo apoptosis when crowded [WEHI-231, human B-CLL cells (139), and many plasmacytomas, for example], we do not find increased transcription of *L1Rn* or related repeat elements in them (124), and so the situation in the chloroleukemia line, though fascinating, may be unique.

In recent years, the *TRPM-2* gene has been shown to be activated in a number of cell types, mostly of the urogenital tract, during involution following hormone removal or after injury (140–143). There is substantial evidence that the involution is apoptotic, so the transcriptional activation of *TRPM-2* is of considerable interest. It does not seem to be expressed in other cell types, for example, lymphocytes, so a role in the final pathway is unlikely. Furthermore, it seems to be expressed too late in the apoptotic process to be involved in initiation (144), and it may be part of a late response to injury or perhaps involved in repair. *TRPM-2* is apparently identical to the previously described genes *CL-1* (clusterin) and *Sgp-2* (sulfated glycoprotein-2) (145), and its protein may be the same as serum protein 40,40 (146).

C. Transduction

The definitive example of a *transduction* mechanism is the destruction of target cells by cytotoxic T lymphocytes. In this context we refer only to the "internal disintegration" events produced within the target by its interaction with the CTL, and not to any membrane attack. DNA in many target cells is fragmented within 5 minutes of CTL contact (47,147), well before lysis. It appears that the CTL has acquired the means to activate a self-destruction pathway that closely resembles apoptosis in virtually any nucleated target. Why? Perhaps this relates to the CTL's primary role in immune surveillance against cells infected by viruses or other internal parasites. If the idea is that the infected cell must be killed before virus has had a chance to assemble complete infectious particles within it, it may be that the process of lysis is too slow: if the cell is not lysed until after virus has replicated, lysis would make matters worse. Thus it was interesting to find (148) that virus DNA was fragmented *parri passu* with cellular DNA in the targets of CTLs. Perhaps it is exactly for this purpose—to destroy viral nucleic acid—that the CTLs as well as natural killer (NK) cells and antibody-dependent cellular cytoxic (ADCC) killer cells (149,150) have learned to activate the death program (126).

A key characteristic of the transduction mechanism—a corollary of its rapidity—is that there seems to be no need for mRNA or protein synthesis in either killer or target, because neither cycloheximide nor actinomycin D blocks cytotoxicity (47,113). This could mean that the CTL secretes or injects the substances necessary for target cell apoptosis. We do not subscribe to this view, as there are CTL–target systems described in which secretion is blocked but DNA fragmentation proceeds normally (113), and cytoplasmic connections between killer and target have not been convincingly described. The suggestion that the lytic granules of CTLs contain factors that induce DNA fragmentation in targets has not been supported by the evidence (151–153). An alternative model involves transmembrane signaling, in which the activated CTL engages a structure on the surface of the target—it need not be the structure recognized by the killer's receptor for antigen—and thereby sends a signal into the target's interior, the ultimate result of which is apoptosis. All of this must be accomplished without the synthesis of new proteins, so all the pieces must be in place. We have suggested (154) that there may be a membrane-associated molecule that is activated by the CTL, and which then triggers the apoptotic cascade. In the induction mechanism, in contrast, this molecule cannot be activated directly and so a cytoplasmic equivalent of it must be synthesized *de novo*. The rest of the required molecules could already be in place.

Allbritton *et al.* (72) showed that the potassium ionophore valinomycin induces DNA fragmentation in a variety of cell types. We found that protein synthesis is not necessary for this effect of valinomycin (155), which is clearly apoptosis by morphology as well as by the pattern of DNA fragmentation. The kinetics resemble those of CTL-mediated killing, and it is intriguing to speculate that valinomycin may do to a tumor cell essentially what a CTL does. If this is true, it would facilitate the study of CTL mechanisms and could put the transduction mechanism on a sound molecular basis. It also establishes that all the required molecules for apoptosis may be present in most cells.

D. Release of the Death Program

Some years ago, as a control in another experiment, we found that certain cells in mouse bone marrow underwent DNA fragmentation when treated with cycloheximide; this is different from the induction or transduction mechanisms, in which cycloheximide inhibits or has no effect upon apoptosis, respectively. We have found that the cells affected are metamyelocytes (156). These cells give rise to granulocytes, predominantly neutrophils, whose life expectancy in the circulation is 24 hours or less. When metamyelocytes are placed in cell culture they survive at least 24 hours, but all are dead by apoptosis in that time if exposed to cycloheximide. Mature neutrophils die in culture in 24 hours, and cycloheximide only accelerates the process by about an hour. From these data we conclude that in anticipation of the shortness of their useful life spans, metamyelocytes activate the death program and synthesize the suicide proteins; to avoid premature death they also synthesize suicide inhibitor proteins. These inhibitors have a relatively short half-life; when protein synthesis is stopped by cycloheximide, the inhibitors decay first and the suicide proteins are released from inhibition. When neutrophils mature and leave the marrow they lose most of their endoplasmic reticulum and make very little protein; as soon as the inhibitor protein cannot be replenished as fast as it is decaying, the death program would be released. Thus in the mature neutrophil cycloheximide cannot accelerate a program that has already begun physiologically.

It is probable that the release mechanism will be found in other cell systems, especially those in which the cell's life expectancy is limited and imminent for physiological reasons. We have found it in basal keratinocytes from mouse skin (61); these cells give rise to granular keratinocytes whose lifetime is short (in isolated granular keratinocytes, the DNA is already beginning to fragment). We have also found it in several cell lines, as have Martin *et al.* (157) in the human promyelocytic cell line HL-60. It is an elegant way to regulate cell survival, and significantly it is a "fail-safe" mechanism; if a cell with a release mechanism is accidentally injured, even so badly that its ability to make macromolecules is impaired, its last act will be to commit apoptotic suicide and to spare the environment the release of its (possibly toxic) contents.

V. Conclusions

Programmed cell death is an old phenomenon that has awakened new interest. Wyllie deserves credit for first showing that the mechanism is susceptible to biochemical understanding (22). What is amazing is the number of different systems in which it can be demonstrated. They all seem to have in common nuclear disintegration, the morphology of apoptosis, DNA damage, and early recognition by phagocytic cells. It is likely that there is a death program that all cells follow when they embark on this pathway, even if the specific (or nonspecific) triggers were very different. It is also likely that most examples of physiological cell death are apoptotic, and the mere cataloging of examples is no longer an intellectually stimulating exercise, but understanding the mechanisms—signaling, biochemistry, and genetics—definitely *is*. If we fully understand how a cell makes the decision to live or to die, will we be able to prevent the abnormally programmed death of cells in diseases such as Huntington's (158), Parkinson's, or Alzheimer's (159)? Will we be able to induce cells that would rather live, such as cancer cells, to commit suicide?

Acknowledgments

I thank my colleagues Rick Duke, Pam Smith, Greg Owens, Peg Taylor, Julie Lang, Cathy McCall, Peter Henson, and Mary Schleicher for their enormous contribution to the work described here, and for their many valuable insights and arguments. The author's work has been supported by NIH Grant AI-11661.

References

1. Walker, N. I., Harmon, B. V., Gobe, G. C., and Kerr, J. F. (1988). Patterns of cell death. *Methods Achiev. Exp. Pathol.* **13**, 18.
2. Searle, J., Kerr, J. F. R., and Bishop, C. J. (1982). Necrosis and apoptosis: Distinct modes of cell death with fundamentally different significance. *Pathol. Annu.* **17**, 229.
3. Wyllie, A. H. (1981). Cell death: A new classification separating apoptosis from necrosis. *In* "Cell Death in Biology and Pathology" (I. D. Bowen and R. A. Lockshin, eds.), n. p. Chapman & Hall, London.
4. Wyllie, A. H. (1987). Cell death. *Int. Rev. Cytol. (Suppl.)* **17**, 755.
5. Duvall, E., and Wyllie, A. H. (1986). Death and the cell. *Immunol. Today* **7**, 115.
6. Glucksmann, A. (1951). Cell deaths in normal vertebrate ontogeny. *Biol. Rev.* **26**, 59.
7. Saunders, J. W. Jr. (1966). Death in embryonic systems. *Science* **154**, 604.

8. Hinchliffe, J. R. (1981). Cell death in embryogenesis. In "Cell Death in Biology and Pathology" (I. D. Bowen and R. A. Lockshin, eds.), p. 35. Champan & Hall, London.
9. Antalikova, L., Kren, V., Kasparek, R., and Bila, V. (1989). Patterns of physiological cell death and mitoses in the apical ectodermal ridge in normodactylous and polydactylous rat limb buds. A quantitative evaluation. *Folia Biol. (Prague)* **35**, 339.
10. Gage, F. H., Bjorklund, A., and Stenevi, U. (1984). Denervation releases a neuronal survival factor in adult rat hippocampus. *Nature (London)* **308**, 637.
11. Oppenheim, R. W. (1985). Naturally occurring cell death during neural development. *Trends NeuroSci* **8**, 487.
12. Hamburger, V., and Levi-Montalcini, R. (1949). Proliferation, differentiation and degeneration in the spinal ganglion of the chick under normal and experimental conditions. *J. Exp. Zool.* **111**, 457.
13. Kyprianou, N., English, H. F., and Isaacs, J. T. (1988). Activation of a $Ca^{2+}-Mg^{2+}$-dependent endonuclease as an early event in castration-induced prostatic cell death. *Prostate* **13**, 103.
14. Wyllie, A. H. (1988). Apoptosis. "ISI Atlas Science: Immunology," p. 192. Institute for Scientific Information, Philadelphia, PA.
15. Kerr, J. F. R., Wyllie, A. H., and Currie, A. R. (1972). Apoptosis: A basic biological phenomenon with wide-ranging implications in tissue kinetics. *Brit. J. Cancer* **26**, 239.
16. Wyllie, A. H., Kerr, J. F. R., and Currie, A. R. (1980). Cell death: The significance of apoptosis. *Int. Rev. Cytol.* **68**, 251.
17. Bowen, I. D., and Lockshin, R. A. (1981). "Cell Death in Biology and Pathology" (I. D. Bowen and R. A. Lockshin, eds.). Chapman & Hall, London.
18. Arends, M. J., Morris, R. G., and Wyllie, A. H. (1990). Apoptosis. The role of the endonuclease. *Am. J. Pathol.* **136**, 593.
19. McConkey, D. J., Orrenius, S., and Jondal, M. (1990). Cellular signalling in programmed cell death (apoptosis). *Immunol. Today.* **11**, 120.
20. Zeiger, E., and Schwartz, A. (1982). Longevity of guard cell chloroplasts in falling leaves: Implication for stomatal function and cellular aging. *Science* **218**, 680.
21. Thomas, N., and Bell, P. A. (1981). Glucocorticoid-induced cell-size changes and nuclear fragility in rat thymocytes. *Mol. Cell. Endocrinol.* **22**, 71.
22. Wyllie, A. H. (1980). Glucocorticoid-induced thymocyte apoptosis is associated with endogenous nuclease activation. *Nature (London)* **284**, 555.
23. Williamson, R. (1970). Properties of rapidly labelled DNA fragments isolated from the cytoplasm of primary cultures of embryonic mouse liver cells. *J. Mol. Biol.* **51**, 157.
24. Hewish, D. R., and Burgoyne, L. A. (1973). Chromatin substructure. The digestion of chromatin at regularly spaced sites by a nuclear deoxyribonuclease. *Biochem. Biophys. Res. Commun.* **52**, 504.
25. Hewish, D. R., and Burgoyne, L. A. (1973). The calcium dependent endonuclease activity in isolated nuclear preparations. Relationships between its occurrence and the occurrence of other classes of enzymes found on nuclear preparation. *Biochem. Biophys. Res. Commun.* **52**, 475.
26. Burgoyne, L. A., and Mobbs, J. (1975). The reaction of the calcium–magnesium endonuclease with the A-sites of rat nucleoprotein. *Nuc. Acids Res.* **2**, 1551.
27. Cohen, J. J., and Duke, R. C. (1984). Glucocorticoid activation of a calcium-dependent endonuclease in thymocyte nuclei leads to cell death. *J. Immunol.* **132**, 38.

28. Cohen, J. J. (1989). Lymphocyte death induced by glucocorticoids. In "Antiinflammatory Steroid Action: Basic and Clinical Aspects" (R. P. Schleimer, H. N. Claman, and A. Oronsky, eds.), p. 110. Academic Press, San Diego, California.
29. Cohen, J. J. Unpublished results.
30. Burgoyne, L. A., Hewish, D. R., and Mobbs, J. (1974). Mammalian cromatin substructure studies with the calcium–magnesium endonuclease and two dimensional polyacrylamide-gel electrophoresis. *Biochem. J.* **143**, 67.
31. Duke, R. C., and Cohen, J. J. (1986). IL-2 addiction: Withdrawal of growth factor activates a suicide program in dependent T cells. *Lymphokine Res.* **5**, 289.
32. Sellins, K. S., and Cohen, J. J. (1987). Gene induction by gamma-irradiation leads to DNA fragmentation in lymphocytes. *J. Immunol.* **139**, 3199.
33. Koury, M. J., and Bondurant, M. C. (1990). Erythropoietin retards DNA breakdown and prevents programmed death in erythroid progenitor cells. *Science* **248**, 378.
34. Kyprianou, N., English, H. F., and Isaacs, J. T. (1990). Programmed cell death during regression of PC-82 human prostate cancer following androgen ablation. *Cancer Res.* **50**, 3748.
35. Araki, S., Shimada, Y., Kaji, K., and Hayashi, H. (1990). Apoptosis of vascular endothelial cells by fibroblast growth factor deprivation. *Biochem. Biophys. Res. Commun.* **168**, 1194.
36. Sendtner, M., Kreutzberg, G. W., and Thoenen, H. (1990). Ciliary neurotrophic factor prevents the degeneration of motor neurons after axotomy. *Nature (London)* **345**, 440.
37. Schubert, D., Kimura, H., LaCorbiere, M., Vaughan, J., Karr, D., and Fischer, W. H. (1990). Activin is a nerve cell survival molecule. *Nature (London)* **344**, 868.
38. Walker, N. I., Bennett, R. E., and Kerr, J. F. (1989). Cell death by apoptosis during involution of the lactating breast in mice and rats. *Am. J. Anat.* **185**, 19.
39. Nunez, G., London, L., Hockenbery, D., Alexander, M., McKearn, J. P., and Korsmeyer, S. J. (1990). Deregulated *Bcl-2* gene expression selectively prolongs survival of growth factor-deprived hemopoietic cell lines. *J. Immunol.* **144**, 3602.
40. Szende, B., Srkalovic, G., Groot, K., Lapis, K., and Schally, A. V. (1990). Growth inhibition of mouse MXT mammary tumor by the luteinizing hormone-releasing hormone antagonist SB-75. *J. Natl. Cancer Inst.* **83**, 513.
41. Sloviter, R. S., Valiquette, G., Abrams, G. M., Ronk, E. C., Sollas, A. L., Paul, L. A., and Neubort, S. (1989). Selective loss of hippocampal granule cells in the mature rat brain after adrenalectomy. *Science* **243**, 535.
42. Pincus, D. W., DiCicco-Bloom, E. M., and Black, I. B. (1990). Vasoactive intestinal peptide regulates mitosis, differentiation and survival of cultured sympathetic neuroblasts. *Nature (London)* **343**, 564.
43. Williams, G. T., Smith, C. A., Spooncer, E., Dexter, T. M., and Taylor, D. R. (1990). Haemopoietic colony stimulating factors promote cell survival by suppressing apoptosis. *Nature (London)* **343**, 76.
44. English, H. F., Kyprianou, N., and Isaacs, J. T. (1989). Relationship between DNA fragmentation and apoptosis in the programmed cell death in the rat prostate following castration. *Prostate* **15**, 233.
45. McConkey, D. J., Hartzell, P., Duddy, S. K., Hakansson, H., and Orrenius, S. (1988). 2,3,7,8-Tetrachlorodibenzo-*p*-dioxin kills immature thymocytes by Ca^{++}-mediated endonuclease activation. *Science* **242**, 256.
46. Russell, J. H. (1983). Internal disintegration model of cytotoxic lymphocyte-induced target damage. *Immunol. Res.* **72**, 97.
47. Duke, R. C., Chervenak, R., and Cohen, J. J. (1983). Endogenous endonuclease-

induced DNA fragmentation: An early event in cell-mediated cytolysis. *Proc. Natl. Acad. Sci. U.S.A.* **80**, 6361.
48. Smith, C. A., Williams, G. T., Kingston, R., Jenkinson, E. J., and Owen, J. J. T. (1989). Antibodies to CD3/T-cell receptor complex induce death by apoptosis in immature T cells in thymic cultures. *Nature (London)* **337**, 181.
49. Takahashi, S., Maecker, H. T., and Levy, R. (1989). DNA fragmentation and cell death mediated by T cell antigen receptor/CD3 complex on a leukemia T cell line. *Eur. J. Immunol.* **19**, 1911.
50. Jenkinson, E. J., Kingston, R., Smith, C. A., Williams, G. T., and Owen, J. J. (1989). Antigen-induced apoptosis in developing T cells: A mechanism for negative selection of the T cell receptor repertoire. *Eur. J. Immunol.* **19**, 2175.
51. McConkey, D. J., Hartzell, P., Amador-Perez, J. F., Orrenius, S., and Jondal, M. (1989). Calcium-dependent killing of immature thymocytes by stimulation via the CD3/T cell receptor complex. *J. Immunol.* **143**, 1801.
52. MacDonald, H. R., and Lees, R. K. (1990). Programmed death of autoreactive thymocytes. *Nature (London)* **343**, 642.
53. Shi, Y. F., Szalay, M. G., Paskar, L., Boyer, M., Singh, B., and Green, D. R. (1990). Activation-induced cell death in T cell hybridomas is due to apoptosis. Morphologic aspects and DNA fragmentation. *J. Immunol.* **144**, 3326.
54. Benhamou, L. E., Cazenave, P. A., and Sarthou, P. (1990). Anti-immunoglobulins induce death by apoptosis in WEHI-231 B lymphoma cells. *Eur. J. Immunol.* **20**, 1405.
55. Christiaansen, J. E., and Sears, D. W. (1985). Lack of lymphocyte-induced DNA fragmentation in human targets during lysis represents a species-specific difference between human and murine cells. *Proc. Natl. Acad. Sci. U.S.A.* **82**, 4482.
56. Bell, P. A., and Jones, C. N. (1982). Cytotoxic effects of butyrate and other "differentiation inducers" on immature lymphoid cells. *Biochem. Biophys. Res. Commun.* **104**, 1202.
57. Kumar, S., Baxter, G. D., Smith, P. J., Pemble, L., Collins, R. J., Prentice, R. L., and Lavin, M. F. (1987). DNA fragmentation in childhood T-cell acute lymphoblastic leukemia. *Mol. Biol. Med.* **4**, 111.
58. Baxter, G. D., Collins, R. J., Harmon, B. V., Kumar, S., Prentice, R. L., Smith, P. J., and Lavin, M. F. (1989). Cell death by apoptosis in acute leukaemia. *J. Pathol.* **158**, 123.
59. Gromkowski, S. H., Brown, T. C., Cerutti, P. A., and Cerottini, J. C. (1986). DNA of human Raji target cells is damaged upon lymphocyte-mediated lysis. *J. Immunol.* **136**, 752.
60. Sellins, K. S., and Cohen, J. J. Unpublished results.
61. McCall, C. A., and Cohen, J. J. (1991). Programmed cell death in terminally differentiating keratinocytes: Role of an endogenous endonuclease. *J. Invest. Dermatol.* **98**, in press.
62. Duke, R. C., Sellins, K. S. (1989). Target cell nuclear damage in addition to DNA fragmentation during cytotoxic T lymphocyte-mediated cytolysis. *Prog. Leukocyte Biol.* **9**, 311.
63. Ucker, D. S. (1987). Cytotoxic T lymphocytes and glucocorticoids activate an endogenous suicide process in target cells. *Nature (London)* **327**, 62.
64. Freitas, A. A., Rocha, B., and Coutinho, A. A. (1986). Two major classes of mitogen-reactive B lymphocytes defined by life span. *J. Immunol.* **136**, 466.
65. Compton, M. M., and Cidlowski, J. A. (1987). Identification of a glucocorticoid-induced nuclease in thymocytes. *J. Biol. Chem.* **262**, 8288.

66. Cohen, J. J., Duke, R. C., Chervenak, R., Sellins, K. S., and Olson, L. K. (1985). DNA fragmentation in targets of CTL: An example of programmed cell death in the immune system. In "Mechanisms of Cell-Mediated Cytotoxicity II" (P. Henkart and E. Martz, eds.), p. 493. Plenum, New York.
67. Baxter, G. D., Smith, P. J., and Lavin, M. F. (1989). Molecular changes associated with induction of cell death in a human T-cell leukaemia line: Putative nucleases identified as histones. *Biochem. Biophys. Res. Commun.* **162**, 30.
68. Alnemri, E. S., and Litwack, G. (1989). Glucocorticoid-induced lymphocytolysis is not mediated by an induced endonuclease. *J. Biol. Chem.* **264**, 4104.
69. Suzuki, K. (1987). Calcium activated neutral protease: Domain structure and activity regulation. *Trends Biosci.* **12**, 103.
70. Kaiser, N., and Edelman, I. S. (1977). Calcium dependence of glucocordicoid-induced lymphocytolysis. *Proc. Natl. Acad. Sci. U.S.A.* **74**, 638.
71. Bansal, N., Houle, A. G., and Melnykovych, G. (1990). Dexamethasone-induced killing of neoplastic cells of lymphoid derivation: Lack of early calcium involvement. *J. Cell Physiol.* **143**, 105.
72. Allbritton, N., Verret, C. R., Wolley, R. C., and Eisen, H. N. (1988). Calcium ion concentrations and DNA fragmentation in target cell destruction by murine cloned cytotoxic T lymphocytes. *J. Exp. Med.* **167**, 514.
73. Duke, R. C., and Zulauf, R. Unpublished results.
74. Wyllie, A. H., Morris, R. G., Smith, A. L., and Dunlop, D. (1984). Chromatin cleavage in apoptosis: Association with condensed chromatin morphology and dependence on macromolecular synthesis. *J. Pathol.* **142**, 67.
75. Inaba, K., Inaba, M., Kinashi, T., Tashiro, K., Witmer-Pack, M., Crowley, M., Kaplan, G., Valinsky, J., Romani, N., Ikehara, S., Muramatsu, S., Honjo, T., and Steinman, R. M. (1988). Macrophages phagocytose thymic lymphocytes with productively rearranged T cell receptor alpha and beta genes. *J. Exp. Med.* **168**, 2279.
76. Savill, J. S., Wyllie, A. H., Henson, J. E., Walport, M. J., Henson, P. M., and Haslett, C. (1989). Macrophage phagocytosis of aging neutrophils in inflammation programmed cell death in the neutrophil leads to its recognition by macrophages. *J. Clin. Invest.* **83**, 865.
77. Duvall, E., Wyllie, A. H., and Morris, R. G. (1985). Macrophage recognition of cells undergoing programmed cell death (apoptosis). *Immunology* **56**, 351.
78. Savill, J. S., Henson, P. M., and Haslett, C. (1989). Phagocytosis of aged human neutrophils by macrophages is mediated by a novel "charge-sensitive" recognition mechanism. *J. Clin. Invest.* **84**, 1518.
79. Savill, J. S., Dransfield, I., Hogg, N., and Haslett, C. (1990). Vitronectin receptor-mediated phagocytosis of cells undergoing apoptosis. *Nature (London)* **343**, 170.
80. Fadok, V. A., Voelker, D. R., Campbell, P. A., Cohen, J. J., Bratton, D. L., and Henson, P. M. (1991). Specific recognition by macrophages of phosphatidylserine on apoptotic lymphocytes. Submitted for publication.
81. Lewis, S., and Gellert, M. (1989). The mechanism of antigen receptor gene assembly. *Cell* **59**, 585.
82. Yancopoulos, G. D., and Alt, F. W. (1986). Regulation of the assembly and expression of variable-region genes. *Annu. Rev. Immunol.* **4**, 339.
83. Tonegawa, S. (1983). Somatic generation of antibody diversity. *Nature (London)* **302**, 575.
84. Malynn, B. A., Blackwell, T. K., Fulop, G. M., Rathbun, G. A., Furley, A. J. W., Ferrier, P., Heinke, L. B., Phillips, R. A., Yancopoulos, G. D., and Alt, F. W. (1988).

The *scid* defect affects the final step of the immunoglobulin VDJ recombinase mechanism. *Cell* **54**, 453.
85. Cooper, M. D., and Burrows, P. D. (1989). B cell differentiation. *In* "Immunoglobulin Genes" (T. Honjo, F. W. Alt, and T. H. Rabbits, eds.), p. 1. Academic Press, London.
86. Taylor, R. B., Duffus, W. P. H., Raff, M. C., and dePetris, S. (1971). Redistribution and pinocytosis of lymphocyte surface immunoglobulin molecules induced by anti-immunoglobulin antibody. *Nature (New Biol.)* **233**, 225.
87. Ashman, R. F. (1984). Lymphocyte activation. *In* "Fundamental Immunology" (W. E. Paul, ed.), p. 267. Raven Press, New York.
88. Nossal, G. J. V. (1983). Cellular mechanisms of immunologic tolerance. *Annu. Rev. Immunol.* **1**, 33.
89. Nemazee, D. A., and Buerki, K. (1989). Clonal deletion of B lymphocytes in a transgenic mouse bearing anti-MHC class I antibody genes. *Nature (London)* **337**, 562.
90. Nemazee, D., and Buerki, K. (1989). Clonal deletion of autoreactive B lymphocytes in bone marrow chimeras. *Proc. Natl. Acad. Sci. U.S.A.* **86**, 8039.
91. Boyd, A. W., Goding, J. W., and Schrader, J. W. (1981). The regulation of growth and differentiation of a murine B cell lymphoma. I. Lipopolysaccharide-induced differentiation. *J. Immunol.* **126**, 2461.
92. Jakway, J. P., Usinger, W. R., Gold, M. R., Mishell, R. I., and DeFranco, A. L. (1986). Growth regulation of the B lymphoma cell line WEHI-231 by anti-immunoglobulin, lipopolysaccharide, and other bacterial products. *J. Immunol.* **137**, 2225.
93. Boyd, A. W., and Schrader, J. W. (1981). The regulation of growth and differentiation of a murine B cell lymphoma. II. The inhibition of WEHI 231 by anti-immunoglobulin antibodies. *J. Immunol.* **126**, 2466.
94. Seyfert, V. L., Sukhatme, V. P., and Monroe, J. G. (1989). Differential expression of a zinc finger-encoding gene in response to positive versus negative signaling through receptor immunoglobulin in murine B lymphocytes. *Mol. Cell Biol.* **9**, 2083.
95. Monroe, J. G., Seyfert, V. L., Owen, C. S., and Sykes, N. (1989). Isolation and characterization of a B lymphocyte mutant with altered signal transduction through its antigen receptor. *J. Exp. Med.* **169**, 1059.
96. Osmond, D. G. (1986). Population dynamics of bone marrow B lymphocytes. *Immunol. Rev.* **93**, 103.
97. Liu, Y. J., Joshua, D. E., Williams, G. T., Smith, C. A., Gordon, J., and MacLennan, I. C. M. (1989). Mechanism of antigen-driven selection in germinal centres. *Nature (London)* **342**, 929.
98. Raulet, D. H., Garman, R. D., Saito, H., and Tonegawa, S. (1985). Developmental regulation of T-cell receptor gene expression. *Nature (London)* **314**, 103.
99. Kappler, J. W., Roehm, N., and Marrack, P. (1987). T cell tolerance by clonal elimination in the thymus. *Cell* **49**, 273.
100. von Boehmer, H. (1988). The developmental biology of T lymphocytes. *Annu. Rev. Immunol.* **6**, 309.
101. Shi, Y. F., Sahai, B. M., and Green, D. R. (1989). Cyclosporin A inhibits activation-induced cell death in T-cell hybridomas and thymocytes. *Nature (London)* **339**, 625.
102. Gao, E.-K., Lo, D., Cheney, R., Kanagawa, O., and Sprent, J. (1988). Abnormal

differentiation of thymocytes in mice treated with cyclosporin A. *Nature (London)* **336**, 176.
103. Ashwell, J. D., Cunningham, R. E., Noguchi, P. D., and Hernandez, D. (1987). Cell growth cycle block of T cell hybridomas upon activation with antigen. *J. Exp. Med.* **165**, 173.
104. Mercep, M., Bluestone, J. A., Noguchi, P. D., and Ashwell, J. D. (1988). Inhibition of transformed T cell growth *in vitro* by monoclonal antibodies directed against distinct activating molecules. *J. Immunol.* **140**, 324.
105. Odaka, C., Kizaki, H., and Tadakuma, T. (1990). T cell receptor-mediated DNA fragmentation and cell death in T cell hybridomas. *J. Immunol.* **144**, 2096.
106. Ucker, D. S., Ashwell, J. D., and Nickas, G. (1989). Activation-driven T cell death. 1. Requirements for *de novo* transcription and translation and association with genome fragmentation. *J. Immunol.* **143**, 3461.
107. Duke, R. C. (1989). Self-recognition by T cells. Bystander killing of target cells bearing syngeneic MHC antigens. *J. Exp. Med.* **170**, 59.
108. McPhee, D., Pye, J., and Shortman, K. (1979). The differentiation of T lymphocytes. V. Evidence for intrathymic death of most thymocytes. *Thymus* **1**, 151.
109. Shortman, K., and Jackson, H. (1974). The differentiation of T lymphocytes. I. Proliferation kinetics and the interrelationships of subpopulations of mouse thymus cells. *Cell. Immunol.* **12**, 230.
110. Podack, E. R. (1985). The molecular mechanism of lymphocyte-mediated tumor cell lysis. *Immunol. Today* **6**, 21.
111. Young, J. D.-E., and Cohn, Z. A. (1986). Cell-mediated killing: A common mechanism? *Cell* **46**, 641.
112. Henkart, P. A. (1985). Mechanism of lymphocyte-mediated cytotoxicity. *Annu. Rev. Immunol.* **3**, 31.
113. Ostergaard, H. L., and Clark, W. R. (1989). Evidence for multiple lytic pathways used by cytolytic T lymphocytes. *J. Immunol.* **143**, 2120.
114. Russell, J. H. (1981). Internal disintegration model of cytotoxic lymphocyte-mediated cytotoxicity. *Biol. Rev.* **56**, 153.
115. Anderson, R. E., and Warner, N. L. (1976). Ionizing radiation and the immune response. *Adv. Immunol.* **24**, 215.
116. Filippovich, I. V., Sorokina, N. I., Soldatenkov, V. A., Alfyerova, T. M., and Trebenok, Z. A. (1988). Effect of the inducers of cellular differentiation and of ionizing radiation of thymus lymphocytes: Chromatin degradation and programmed cell death. *Int. J. Radiat. Biol.* **53**, 617.
117. Kondo, S. (1988). Altruistic cell suicide in relation to radiation hormesis. *Int. J. Radiat. Biol.* **53**, 95.
118. Yamada, T., and Ohyama, H. (1988). Radiation-induced interphase death of rat thymocytes is internally programmed (apoptosis). *Int. J. Radiat. Biol.* **53**, 65.
119. Afanas'ev, V. N., Korol', B. A., Matylevich, N. P., Pechatnikov, V. A., and Sorokina, N. I. (1988). General similarities and differences in the postradiation death of various animal species and of man. *Radiobiologiia.* **28**, 731.
120. Soldatenkov, V. A., Denisenko, M. F., Khodarev, N. N., Votrin, I. I., and Filippovich, I. V (1989). Early postirradiation chromatin degradation in thymocytes. *Int. J. Radiat. Biol.* **55**, 943.
121. Skalka, M., Matyasova, J., and Cejkova, M. (1976). DNA in chromatin of irradiated lymphoid tissues degrades *in vivo* into regular fragments. *FEBS Lett.* **72**, 271.
122. Yamada, T., Ohyama, H., Kinjo, Y., and Watanabe, M. (1981). Evidence for the

internucleosomal breakage of chromatin in rat thymocytes irradiated *in vitro.* *Radiat. Res.* **85,** 544.
123. Sellins, K. S., and Cohen, J. J. (1991). Hyperthermia induces apoptosis in thymocytes. *Radiat. Res.* **126,** 88.
124. Owens, G., Hahn, W., and Cohen, J. J. (1991). Identification of mRNAs associated with programmed cell death in immature thymocytes. Submitted for publication.
125. Parag, H. A., Raboy, B., and Kulka, R. G. (1987). Effect of heat shock on protein degradation in mammalian cells: Involvement of the ubiquitin system. *EMBO J.* **6,** 55.
126. Martz, E., and Howell, D. M. (1989). CTL: Virus control cells first and cytolytic cells second? DNA fragmentation, apoptosis and the prelytic halt hypothesis. *Immunol. Today* **10,** 79.
127. Puccini, G. (1983). *"Madama Butterfly,"* p. 70. Dover Publications, New York.
128. Chalfie, M., and Wolinsky, E. (1990). The identification and suppression of inherited neurodegeneration in *Caenorhabditis elegans. Nature (London)* **345,** 410.
129. Clarke, P. G. H. (1990). Developmental cell death: Morphological diversity and multiple mechanisms. *Anat. Embryol.* **181,** 195.
130. Ellis, H. M., and Horvitz, H. R. (1986). Genetic control of programmed cell death in the nematode *C. elegans Cell* **44,** 817.
131. Yuan, J. Y., and Horvitz, H. R. (1990). The *Caenorhabditis elegans* genes *ced-3* and *ced-4* act cell autonomously to cause programmed cell death. *Dev. Biol.* **138,** 33.
132. Horvitz, H. R., Ellis, H. M., and Sternberg, P. W. (1982). Programmed cell death in nematode development. *Neurosci Commentaries* **1,** 56.
133. Hedgecock, E., Sulston, J. E., and Thomson, N. (1983). Mutations affecting programmed cell deaths in the nematode *Caenorhabditis elegans. Science* **220,** 1277.
134. Cleary, M. L., and Sklar, J. (1985). Nucleotide sequence of a t[14;18] chromosomal breakpoint in follicular lymphoma and demonstration of a breakpoint cluster region near a transcriptionally active locus on chromosome 18. *Proc. Natl. Acad. Sci. U.S.A.* **82,** 7439.
135. Bakhshi, A., Wright, J. J., Graninger, W., Seto, M., Owens, J., Cossman, J., Jenson, J. P., Goldman, P., and Korsmeyer, S. J. (1987). Mechanism of the t[14;18] chromosomal translocation: Structural analysis of both derivative 14 and 18 reciprocal partners. *Proc. Natl. Acad. Sci. U.S.A.* **84,** 2396.
136. Vaux, D. L., Cory, S., and Adams, J. M. (1988). Bcl-2 gene promotes haemopoietic cell survival and cooperates with c-*myc* to immortalize pre-B cells. *Nature (London)* **335,** 440.
137. Hockenberry, D., Nunez, G., Milliman, C., Schreiber, R. D., and Korsmeyer, S. J. (1990). Bcl-2 is an inner mitochondrial membrane protein that blocks programmed cell death. *Nature (London)* **348,** 334.
138. Servomaa, K., and Rytomaa, T. (1988). Suicidal death of rat chloroleukaemia cells by activation of the long interspersed repetitive DNA element (L1Rn). *Cell Tissue Kinet.* **21,** 33.
139. Collins, R. J., Verschuer, L. A., Harmon, B. V., Prentice, R. L., Pope, J. H., and Kerr, J. F. (1989). Spontaneous programmed death (apoptosis) of B-chronic lymphocytic leukaemia cells following their culture *in vitro. Br. J. Haematol.* **71,** 343.
140. Buttyan, R., Olsson, C. A., Pintar, J., Chang, C., Bandyk, M., Ng, P. Y., and Sawczuk, I. S. (1989). Induction of the *TRPM-2* gene in cells undergoing programmed death. *Mol. Cell. Biol.* **9,** 3473.
141. Sawczuk, I. S., Hoke, G., Olsson, C. A., Connor, J., and Buttyan, R. (1989). Gene

expression in response to acute unilateral ureteral obstruction. *Kidney Int.* **35**, 1315.
142. Bandyk, M. G., Sawczuk, I. S., Olsson, C. A., Katz, A. E., and Buttyan, R. (1990). Characterization of the products of a gene expressed during androgen-programmed cell death and their potential use as a marker of urogenital injury. *J. Urol.* **143**, 407.
143. Connor, J., Sawczuk, I. S., Benson, M. C., Tomashefsky, P., O'Toole, K. M., Olsson, C. A., and Buttyan, R. (1988). Calcium channel antagonists delay regression of androgen-dependent tissues and suppress gene activity associated with cell death. *Prostate*, **13**, 119.
144. Rennie, P. S., Bruchovsky, N., Buttyan, R., Benson, M., and Cheng, H. (1988). Gene expression during the early phases of regression of the androgen-dependent Shionogi mouse mammary carcinoma. *Cancer Res.* **48**, 6309.
145. Cheng, C. Y., Chen, C. L., Feng, Z. M., Marshall, A., and Bardin, C. W. (1988). Rat clusterin isolated from primary Sertoli cell-enriched culture medium is sulfated glycoprotein-2 (SGP-2). *Biochem. Biophys. Res. Comm.* **155**, 398.
146. Tsuruta, J. K., Wong, K., Fritz, I. B., and Griswold, M. D. (1990). Structural analysis of sulfated glycoprotein-2 from amino acid sequence. Relationship to clusterin and serum protein 40,40. *Biochem. J.* **268**, 571.
147. Duke, R. C., Cohen, J. J., and Chervenak, R. (1986). Differences in target cell DNA fragmentation induced by mouse cytotoxic T lymphocytes and natural killer cells. *J. Immunol.* **137**, 1442.
148. Sellins, K. S., and Cohen, J. J. (1989). Polyoma viral DNA is damaged in target cells during cytotoxic T lymphocyte-mediated killing. *J. Virol.* **63**, 572.
149. Stacey, N. H., Bishop, C. J., Halliday, J. W., Halliday, W. J., Cooksley, W. G. E., Powell, L. W., and Kerr, J. F. R. (1985). Apoptosis as the mode of cell death in antibody-dependent lymphocytotoxicity. *J. Cell Sci.* **74**, 169.
150. Liu, Y., Mullbacher, A., and Waring, P. (1989). Natural killer cells and cytotoxic T cells induce DNA fragmentation in both human and murine target cells *in vitro*. *Scand. J. Immunol.* **30**, 31.
151. Gromkowski, S. H., Brown, T. C., Masson, D., and Tschopp, J. (1988). Lack of DNA degradation in target cells lysed by granules derived from cytolytic T lymphocytes. *J. Immunol.* **141**, 774.
152. Duke, R. C., Sellins, K. S., and Cohen, J. J. (1988). Cytotoxic lymphocyte-derived lytic granules do not induce DNA fragmentation in target cells. *J. Immunol.* **141**, 2191.
153. Duke, R. C., Persechini, P. M., Chang, S., Liu, C.-C., Cohen, J. J., and Young, D.-E. (1989). Purified perforin induces target cell lysis but not DNA fragmentation. *J. Exp. Med.* **170**, 1451.
154. Duke, R. C., and Cohen, J. J. (1988). The role of nuclear damage in lysis of target cells by cytotoxic T lymphocytes. *In* "Cytolytic Lymphocytes and Complement: Effectors of the Immune System" (E. R. Podack, ed.), p. 35. CRC Press, Boca Raton, Florida.
155. Cohen, J. J., and Smith, P. A. (1990). Apoptosis induced by the potassium ionophore valinomycin: Potassium flux is not involved. *FASCEB J.* **4**, A1707.
156. Cohen, J. J., Lang, J. A., and Henson, P. M. (1991). Suicide program released by the loss of inhibitors during polymorphonuclear cell development. Submitted for publication.
157. Martin, S. J., Lennon, S. V., Bonham, A. M., and Cotter, T. G. (1990). Induction of apoptosis (programmed cell death) in human leukemic HL-60 cells by inhibition of RNA or protein synthesis. *J. Immunol.* **145**, 1859.

158. Martin, J. B. (1982). Huntington's disease: Genetically programmed cell death in the human central nervous system. *Nature (London)* **299,** 205.
159. Whitehouse, P. J., Price, D. L., Struble, R. G., Clark, A. W., Coyle, J. T., and DeLong, M. R. (1982). Alzheimer's disease and senile dementia: Loss of neurons in the basal forebrain. *Science* **215,** 1237.

This article was accepted for publication on 4 March 1991.

Avian T Cell Ontogeny

MAX D. COOPER,*† CHEN-LO H. CHEN,* R. PAT BUCY,* AND CRAIG B. THOMPSON†‡

* *Division of Developmental and Clinical Immunology, Departments of Pediatrics, Medicine, Microbiology and Pathology, and the Comprehensive Cancer Center, University of Alabama at Birmingham, Birmingham, Alabama 35294*
† *Howard Hughes Medical Institute, Birmingham, Alabama 35294*
‡ *Departments of Internal Medicine and Microbiology/Immunology, University of Michigan Medical Center, Ann Arbor, Michigan 48109*

I. Introduction

A renewed interest in avian T cell development has emerged with the use of monoclonal and functional antibodies to elucidate T cell differentiation antigens and molecular and functional definitions of mammalian T cell receptors (TCRs). Birds provide a valuable model system for the study of early developmental events because of experimental accessibility to the embryo in the egg. Studies using chickens played a key role in enabling recognition of the separate developmental pathways of T and B cells and their hematopoietic stem cell precursors (Cooper *et al.*, 1965; Moore and Owen, 1965). Gene conversion, originally identified as an important mechanism for somatic diversification of the immunoglobulin variable regions in the chicken (Weill and Reynaud, 1987; Thompson and Neiman, 1987), may also contribute to development of the antibody repertoire in mammals (Knight and Becker, 1990; J.-C. Weill, personal communication). Comparative studies of mammalian and avian immune systems may thus help to define central features of the vertebrate immune system and provide insight into the evolution of the immune system.

Birds and mammals evolved from their reptilian ancestors over 200 million years ago. Although thousands of avian species have been identified, the domestic chicken (*Gallus gallus*) has served as the avian representative in most studies of the immune system. Chick–quail chimeras (Le Douarin, 1978) have also proved to be an especially informative model. Accordingly, this review of avian T cell development will focus on information gained in studies of these model systems.

II. Avian T Cell Differentiation Antigens

Monoclonal antibodies have been produced against a variety of functionally important molecules expressed on the surface of chicken T cells (Table I), and most of these have well-defined mammalian counterparts. The relative ease with which the mouse monoclonal antibodies have been produced probably reflects the evolutionary divergence of these conserved molecules.

The chicken T cell receptors can be divided into three groups, each of which is recognized by a different monoclonal antibody. One anti-TCR antibody recognizes receptors of the γ/δ type (TCR1) (Sowder et al., 1988) and two others identify subtypes of α/β receptors called

TABLE I
CELL SURFACE MOLECULES IDENTIFIED ON CHICKEN T CELLS BY MONOCLONAL ANTIBODIES

Antigen	Molecular weight ($\times 10^{-3}$)	Antibody	Reference
γ/δTCR	90 (subunits, 40 and 50)	TCR1	Sowder et al. (1988)
α/βTCR	90 (α and β subunits, 40 and 50)	TCR2	Cihak et al. (1988); Chen et al. (1988)
α/βTCR	88 (α and β subunits, 40 and 48)	TCR3	Char et al. (1990)
CD3	17, 19, and 20	CT3	Chen et al. (1986)
CD4	64	CT4	Chan et al. (1988)
		CTLA1 and 6	Lillehoj et al. (1988)
		2-6 and 2-35	Vainio and Lassila (1989)
CD8	63 (subunits, 34)	CT8	Chan et al. (1988)
		CTLA3, 4, and 9	Lillehoj et al. (1988)
CT1	65	CT1 and CT1a, $T_{10}A_6$	Chen et al. (1984), Houssaint et al. (1985)
CD2	40	2-4, 2-102	Vainio and Lassila (1989)
CD5	56	3-58	Vainio and Lassila (1989)
	45 and 46	CTLA 5 and 8	Lillehoj et al. (1988)
CD45	180 and 210	L-17	Pink and Rijnbeek (1983)
	185, 200, and 215	CL1	Houssaint et al. (1987)
MHC class I	45	F21-2	Crone et al. (1985)
MHC class II	65 (α and β subunits, 32 and 27)	Cla-1	Ewert et al. (1984)
		TAP1, TAC	Guillemot et al. (1984)
IL-2 receptor (CD25)	50 (subunit)	Inn-CH-16	Hála et al. (1986)

TCR2 (Cihak *et al.*, 1988; Chen *et al.*, 1988) and TCR3 (Chen *et al.*, 1989; Char *et al.*, 1990). All of these molecules are disulfide-linked heterodimers that are noncovalently associated with a CD3 complex of proteins to form a signal transduction unit (Chen *et al.*, 1986; Char *et al.*, 1990) similar to that previously defined in mammals (Clevers *et al.*, 1988). Avian CD4 and CD8 homologues with molecular weights and tissue distribution similar to their mammalian counterparts (Littman, 1987) have also been identified with monoclonal antibodies (Chan *et al.*, 1988), as have CD2 (Vainio and Lassila, 1989), CD5 (Lillehoj *et al.*, 1988; Vainio and Lassila, 1989), and CD45 homologues (Pink and Rijnbeek, 1983; Houssaint *et al.*, 1987). The CT1 antigen (Chen *et al.*, 1984; Houssaint *et al.*, 1985), for which a mammalian homologue is not evident, is prominently expressed on cortical thymocytes. This antigen is also present on a small subpopulation of peripheral T cells that may represent recent emigrants from the thymus. Activated chicken T cells express MHC class II antigen (Ewert *et al.*, 1984), resembling human T cells in this way more than mouse T cells (Frelinger, 1982). An IL-2 receptor homologue has also been identified by an antibody that identifies the β chain of this receptor (Hála *et al.*, 1986).

Most of these murine antibodies react specifically with chicken T cell antigens. Their failure to react with related molecules on quail T cells is a useful feature, because this specificity allows the precise identification of chick T cells in chick–quail chimeras (Coltey *et al.*, 1989; Bucy *et al.*, 1989). Interestingly, however, the TCR1, TCR2, and TCR3 antibodies may cross-react with homologous T cell epitopes present in other species classified in the same order as chicken (Galliformes). The TCR1 antibody identifies a subpopulation of T cells in the chukar, anti-TCR2 identifies a subset of T cells in the peacock, turkey, and guinea fowl, and anti-TCR3 recognizes some of the T cells in the chukar and turkey, indicating an interesting conservation pattern of these TCR epitopes in gallinaceous species (Char *et al.*, 1990).

III. Thymocyte Precursors: Origin and Thymic Attraction

Before their hematopoietic stem cell (HSC) origin was recognized, thymocytes were thought to be derived from epithelial precursors in the thymus that could be converted into lymphocytes under the inductive influence of surrounding mesenchymal tissue (Auerbach, 1961). Evidence obtained later indicated that the embryonic yolk sac contains the HSC precursors of lymphoid cells (Moore and Owen, 1967). Yolk sac HSCs were then shown to be derived from the embryo itself (Martin and Dieterlen-Lievre, 1978). Recent studies indicate the presence of pluripotent HSCs in the region of the thoracic aorta as early as the fourth day in embryonation (E4), a few days before these can be found in the spleen (F. Dieterlen-Lievre, J. M. Pickel, C. H. Chen, R. P. Bucy, and M. D. Cooper, unpublished). These studies, conducted in chick–quail chimeras, indicate an aortic or paraaortic origin of the HSC precursors of thymocytes, B cells, and myeloid cells. Embryonic HSCs native to the aortic region then migrate via the circulation to colonize the spleen, yolk sac, and, finally, the bone marrow. This principle is likely to hold for mammals with minor variation. A key difference is that the fetal liver in mammals plays an important transient role in hematopoiesis and in B cell production, whereas these functions are not evident in the avian liver during embryonic development.

The chick–quail model was originally used to define the positive attraction of circulating thymocyte precursors into the epithelial thymus derived from the third pharyngeal pouch (Le Douarin, 1978). Thymocyte precursors enter the epithelial thymus in waves (Le Douarin *et al.*, 1984), the first of which begins on E6.5, the second on E12, and the third around E18 (Fig. 1a). Each wave of thymic influx lasts for 1 or 2 days and is followed by the transient production of thymocyte progeny (Coltey *et al.*, 1987, 1989). This same pattern of thymic colonization has been demonstrated in quails (Jotereau and Le Douarin, 1982), except that each colonization wave occurs a day or two earlier. These observations lead to the hypothesis that the thymic epithelial cells produce chemoattractants in a periodic fashion, generating the waves of receptivity to the blood-borne stem cells (Le Douarin, 1978).

Direct evidence for thymus-directed chemotaxis of hematopoietic precursors has been obtained by observing migration of cells toward the thymic epithelium in an *in vitro* model. Free β_2-microglobulin (β_2m) produced by the embryonic thymus has been shown to serve as a chemoattractant factor in the chick (Dunon *et al.*, 1990). However, mice lacking the β_2m gene are still capable of thymocyte development (Zÿlstra *et al.*, 1990; Koller *et al.*, 1990), suggesting that chicks and mice either must use different sets of chemoattractants or alternate modes of thymocyte precursor homing.

IV. Diversification in the Thymus

During their complex differentiation in the thymus, the progeny of the thymocyte precursors undergo proliferation, diversification, and maturation. Both positive and negative clonal selection occurs during this period of T cell development (reviewed by Blackman *et al.*, 1990). The features that mark these intrathymic events in birds appear to resemble those that have been defined in the mammalian thymus.

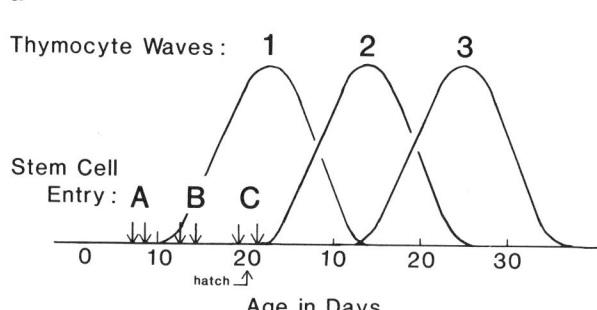

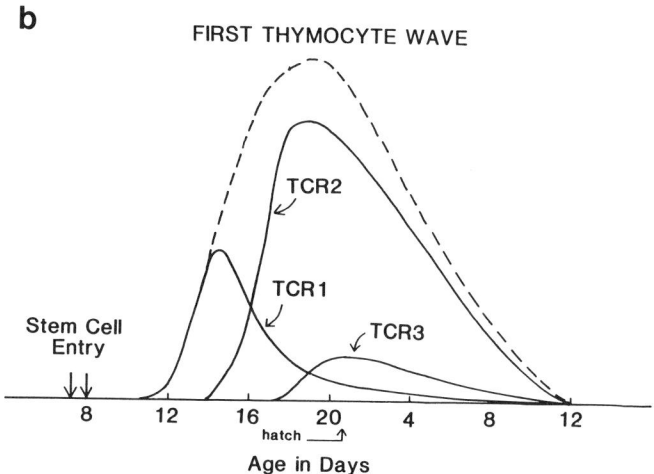

Fig. 1. Chick thymocyte development. (a) Schematic model illustrating the initial waves of thymocyte precursor influx (A, B, C) and subsequent thymocyte production (1, 2, 3). (b) TCR diversity generated in first thymocyte wave (after Coltey et al., 1989).

A. Ontogeny of TCR1, TCR2, and TCR3

The ontogeny of three separate sublines of T cells has been traced in the chicken with the TCR1, TCR2, and TCR3 monoclonal antibodies. These antibodies were originally named in accordance with the ontogenetic order observed for the generation of thymocytes expressing the three types of receptor molecules (Fig. 2). On E12, about 5 days after the initial influx of thymocyte precursors, a few chick thymocytes begin to express the TCR1/CD3 complex on their surface (Sowder et al., 1988), and the receptors are expressed in relatively high levels

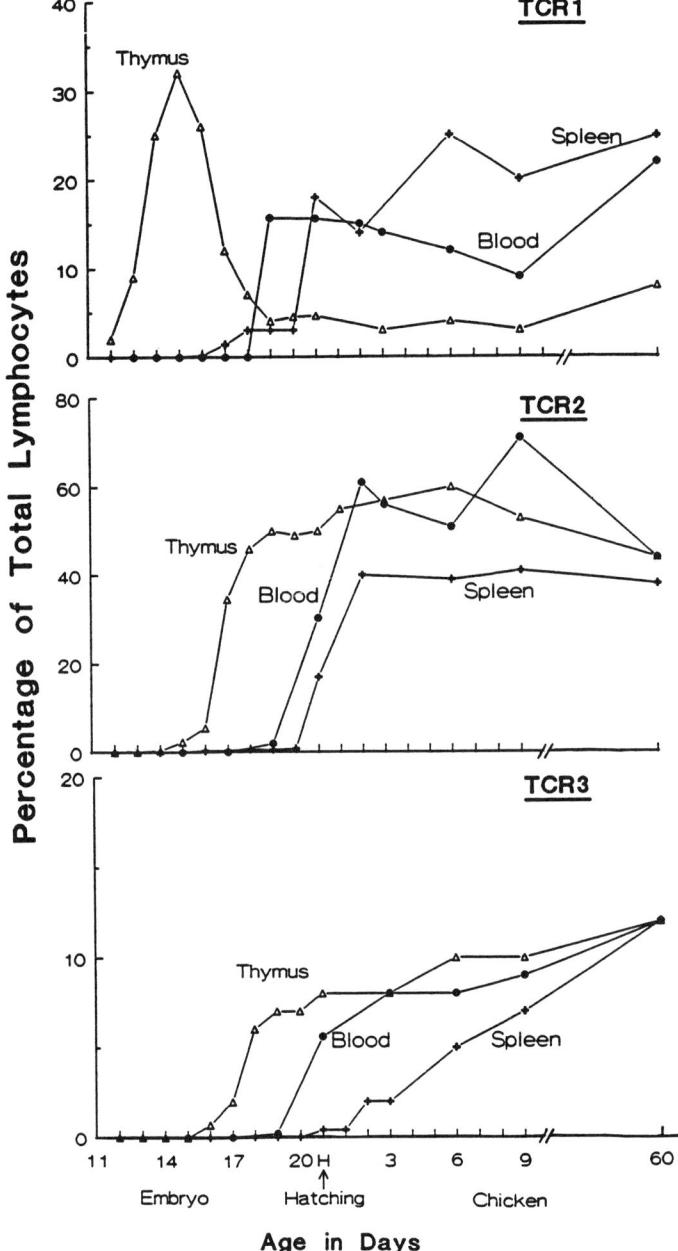

FIG. 2. Ontogeny of T cell subpopulations in the chicken. Cell frequencies were determined by staining with fluorochrome-labeled monoclonal antibodies or by indirect immunofluorescence and cytofluorometric analysis.

($\sim 10^4$/cell) (George and Cooper, 1990). The TCR1/CD3$^+$ cells increase numerically to reach peak frequency levels by E15, when more than 30% of the thymocytes express TCR1. These constitute virtually all of the surface CD3$^+$ thymocytes until around E15. During this period, the TCR1 cells begin to enter the medulla and then to exit via the bloodstream to reach the spleen (Bucy et al., 1990a). Within the thymus, the cells that express the TCR1/CD3 receptor complex characteristically fail to express the CD4 accessory molecule in detectable amounts and relatively few of them express the CD8 molecule (Sowder et al., 1988; Bucy et al., 1990a).

Thymocytes expressing the TCR2 and TCR3 receptor complexes exhibit very similar developmental patterns, which differ strikingly from the TCR1 developmental pattern (Chen et al., 1988; Char et al., 1990). Circumstantial evidence suggests that the anti-TCR2 antibody recognizes the β chain of the α/β TCR2 heterodimer, and the cytoplasmic expression of this epitope can be seen as early as E12 in a subpopulation of thymocytes (Bucy et al., 1990a). However, thymocytes that express surface TCR2/CD3 receptors do not appear until around E15 (Chen et al., 1988). The TCR2/CD3$^+$ subpopulation then expands to become the predominant thymocyte subpopulation by E17–18.

The gradual maturation of the TCR2 subpopulation is manifested in several ways. These cells linger in the thymic cortex for approximately 3–4 days before they enter the medulla; they begin to migrate to the spleen around E19 (Bucy et al., 1990a). Expression of the cell surface TCR2/CD3 complex is a gradual process in which the level of expression is relatively low on embryonic cortical thymocytes ($<10^3$ receptors/cell) and increases as a function of developmental age and stage of intrathymic maturation (George and Cooper, 1990).

The pattern of accessory molecule expression by avian TCR2 cells is the same as that described for α/β T cells in mammals (Chen et al., 1988). CD8 expression at moderately high levels precedes the dual expression of CD4 and CD8 on the surface of cortical TCR2$^+$ thymocytes (Bucy et al., 1990a). As this population of "double-positive" cells undergoes clonal selection and maturation, selected TCR2 cells apparently cease to express either the CD4 or the CD8 accessory molecule in order to become "single positives."

Apart from the fact that the intrathymic development of the TCR3$^+$ thymocytes begins later (i.e., around E18), the developmental pattern of the TCR3$^+$ thymocytes is very similar to that of the TCR2$^+$ thymocytes (Char et al., 1990). This subpopulation of cells also undergoes expansion in the cortex. During this developmental phase, the cells express both CD4 and CD8 as they begin to express gradually increasing levels of the TCR3/CD3 receptor complex before maturing to become either $CD4^+$ or $CD8^+$ T cells. Interestingly, the final CD4/CD8 ratio among mature TCR3 cells is higher than that observed for the TCR2 subpopulation (4/1 versus 2/1) (Chen et al., 1988; Char et al., 1990).

B. Intrathymic Clonal Selection

This process has not been examined extensively in birds. However, the remarkable developmental parallels among the avian TCR2 and TCR3 subpopulations and mammalian α/β T cells suggest that birds may employ the same positive and negative clonal selection mechanisms that have been defined in mammals (Kisielow and von Boehmer, 1990; Marrack and Kappler, 1990). Expression of the MHC class II molecules by cortical epithelial cells and cells of hematopoietic origin that are located near the thymic corticomedullary junction has been well defined through use of the chick–quail chimeric model (Oliver and Le Douarin, 1984). Interruption of the TCR/CD3 signal transduction pathway by cyclosporin A (CsA) treatment blocks maturation of the cortical TCR2 and TCR3 subpopulations in the chicken (Bucy et al., 1990b) as efficiently as this treatment blocks α/β thymocyte development in the mouse (Jenkins et al., 1988; Gao et al., 1988; Kosugi et al., 1989).

Several observations indicate that $TCR1^+$ cells may not be influenced by the same selection pressures encountered by α/β T cells within the thymus. As noted above, TCR1 cells express the $\gamma/\delta/CD3$ complex in relatively high levels on their surface as soon as these cells appear in embryonic development (George and Cooper, 1990). The newly formed TCR1 cells are relatively resistant to receptor modulation via receptor cross-linkage (George and Cooper, 1990), a finding that contrasts with the striking susceptibility of TCR2 thymocytes to receptor modulation and death by apoptosis (Smith et al., 1989). Moreover, in situ expansion of the cortical TCR1 thymocyte subpopulation appears to be minimal. Instead, the TCR1 cells are dispersed throughout the thymic cortical region, which they rapidly traverse to enter the medulla and to make their way out to the peripheral lymphoid tissues (Bucy et al., 1990a). Finally, CsA treatment has no demonstrable effect on the intrathymic development and peripheral migration of $TCR1^+$ cells (Bucy et al., 1990b). These observations suggest that the TCR1 subpopulation does not undergo the intrathymic selection that has been defined for the α/β T cell subpopulations (Kisielow and von Boehmer, 1990; Marrack and Kappler, 1990).

C. Embryonic Waves of Thymocyte Development

From the time of the initial influx of thymocyte precursors during the sixth day of chick embryonic life, approximately 2 weeks are required to generate peak numbers of thymocyte progeny (Fig. 1). Thymocyte attrition and emigration of maturing T cells then contribute to a sharp decline of this intrathymic population (Jotereau et al., 1980; Jotereau and Le Douarin, 1982). The duration of the initial thymic wave is thus approximately 3 weeks (Coltey et al., 1987, 1989). Because successive waves of thymocyte precursors enter the thymus after rest intervals of approximately 4 days, there is overlap in the waves of thymocyte production that follow. Similar waves of thymocyte production have also been shown in the mouse (Jotereau et al., 1987). Persistence of lymphopoietic activity in the adult depends on continued influx of thymocyte precursors, because sustained thymic reconstitution can be achieved with bone marrow cells, but not with thymocytes, in irradiated mice (Fowlkes et al., 1985; Shortman et al., 1990). In birds, as in mammals, the decline in size of the thymus with advancing age reflects a reduction in thymopoietic activity in adult life.

Experimental analysis of thymocyte development in chick–quail chimeras indicates that the TCR1, TCR2, and TCR3 subpopulations can all be derived from each wave of thymocyte precursors (schematically illustrated in Fig. 1b) (Coltey et al., 1989). When limiting numbers of thymocyte precursors are supplied from chick E4 aortic transplants, all three T cell subpopulations can be found clustered within thymic lobules of quail embryo recipients (F. Dieterlen-Lievre, J. M. Pickel, C. H. Chen, M. D. Cooper, and R. P. Bucy, unpublished observations), implying that these are derived from an individual stem cell precursor. This would suggest interplay of environmental factors with the genetic program of thymocyte precursors to determine the final differentiation pathways to be followed by different thymocyte progeny.

The first waves of thymocyte precursors and their progeny may encounter a relatively unique thymic microenvironment in terms of the antigens presented, the nature of the antigen-presenting cells, and perhaps the diversity of cell types in the embryo relative to these variables in older animals. For example, peripheral T cells have the capacity to return to the thymus medulla (Michie et al., 1988; Bucy et al., 1989), although an influence of these experienced voyagers on naive thymocytes has not been demonstrated.

A question that has received much attention in mammals concerns whether T cells can be generated in nonthymic locations. Analysis of this issue in chick–quail chimeras indicates that during embryonic life the generation of T cells is restricted to the thymic microenvironment (Bucy et al., 1989; Coltey et al., 1989). Considering the continued importance of the thymus after hatching, it would perhaps be surprising if other tissues could assume its inductive role for T cells later in life. However, this issue is difficult to resolve experimentally once even a few postthymic cells exist elsewhere in the body, because of the considerable proliferative capacity of peripheral T cells.

V. T Cell Migration to the Periphery

Monoclonal antibodies against chicken T cell antigens have been used to trace their T cell migration patterns during ontogeny. Complementary information on tissue origins and migration routes has been obtained in chick–quail chimeras. In these studies, different lymphoid organs were obtained from chick embryos of various ages and implanted into quail embryos. Serial examination of the quail recipients has thus allowed the migration routes of T cells to be mapped.

A. Migration of the TCR1, TCR2, and TCR3 Subpopulations

The exit of T cells from the thymus begins relatively early in the chick embryo, especially for the TCR1 cells. These can be found in the spleen as early as E15, and in the intestine and bursa of Fabricius on the next day. The T cells are sparsely distributed in peripheral tissues of the embryo, and immunohistological techniques are required to identify the microenvironments to which they home. Immunohistologic analysis reveals that the TCR2 cells do not reach the spleen until around E19 and the TCR3 cells are not found until the second day after hatching, which takes place at 20–21 days of incubation. A rapid increase in the peripheral T cell pool begins just prior to hatching. Thus the seeding of the three T cell subpopulations occurs in the order of their generation in the thymus (Fig. 2).

B. Homing Preferences

The proportion of TCR1 cells reaches levels 20–50% of the circulating T cells in adult chickens (see Fig. 2) (J. Cihak and U. Lösch, personal communication). This relatively high proportion of TCR1 cells makes the chicken a practical model for study of the γ/δ T cell population. Many of these cells home to the spleen and the intestine, where they characteristically populate specific microenvironments (Fig. 3) (Bucy et al., 1988). In the spleen, the TCR1 cells are located predominantly in the red pulp. In the intestine, the TCR1 cells are preferentially enriched in the epithelium, although some of them

(20–30%) can be found in the lamina propria, which they must traverse in order to reach, or to leave, the intestinal epithelium. The TCR1 cells are rarely found within germinal centers in spleen or intestine, although TCR2 cells are common in this microenvironment (Fig. 4).

Two additional features of the peripheral TCR1 cells are notable. The first is that the TCR1 cells are relatively dispersed even in peripheral tissues. This is true both in the spleen and the intestine, where the TCR1 cells rarely form lymphoid nodules or aggregates. Second, although the TCR1 cells seldom express accessory molecules before leaving the thymus, approximately two-thirds of TCR1 cells in the spleen and the intestine may express CD8 on their surface (Chen et al., 1988; Bucy et al., 1988). Because $CD8^+$ TCR1 cells are relatively infrequent in the circulation, this may suggest that functional activities of TCR1 cells involving the CD8 accessory molecule are conducted primarily in the peripheral lymphoid tissues, namely, the spleen and intestines because birds lack well-formed lymph nodes.

The homing pattern of both TCR2 and TCR3 cells within the spleen is very different from that of the TCR1 cells. Both of these subpopulations home to the periarteriolar lymphatic sheaths in the spleen, where they tend to form dense aggregates. Fewer TCR2 and TCR3 cells are present within the red pulp region. Interestingly, this region appears to be the site of preference for $CD8^+$ T cells regardless of their TCR type. In the intestines, the TCR2 cells are located largely within the lamina propria, although a subpopulation of these (~30%), which are mainly $CD8^+$, can be found in the epithelial layer. TCR3 cells appear to constitute the least abundant subpopulation of T cells in the intestines. In fact, it is often impossible to find TCR3 cells in intestinal tissue samples (Char et al., 1990).

C. INTESTINAL T CELL SURFACE GLYCOPROTEIN

An interesting cell surface molecule that is specifically expressed on intestinal T cells has been recently identified (M. Haury, S. Schaal, R. P. Bucy, and M. D. Cooper, unpublished observations). This molecule, identified by the A 19 monoclonal antibody, is a heterodimer of approximately 200,000 Da, with subunits of approximately 120,000 Da. Most of the T cells in the intestinal epithelium (80–90%) express high levels of this cell surface antigen, whereas, even after activation, lymphocytes from other peripheral tissues rarely express the antigen in detectable levels. Intestinal epithelial T cells of all TCR isotypes may express high levels of the A19 antigen, but B cells and other types of intestinal cells do not. Interestingly, a subpopulation of mature thymocytes may express relatively low levels of the A19 antigen. The unique

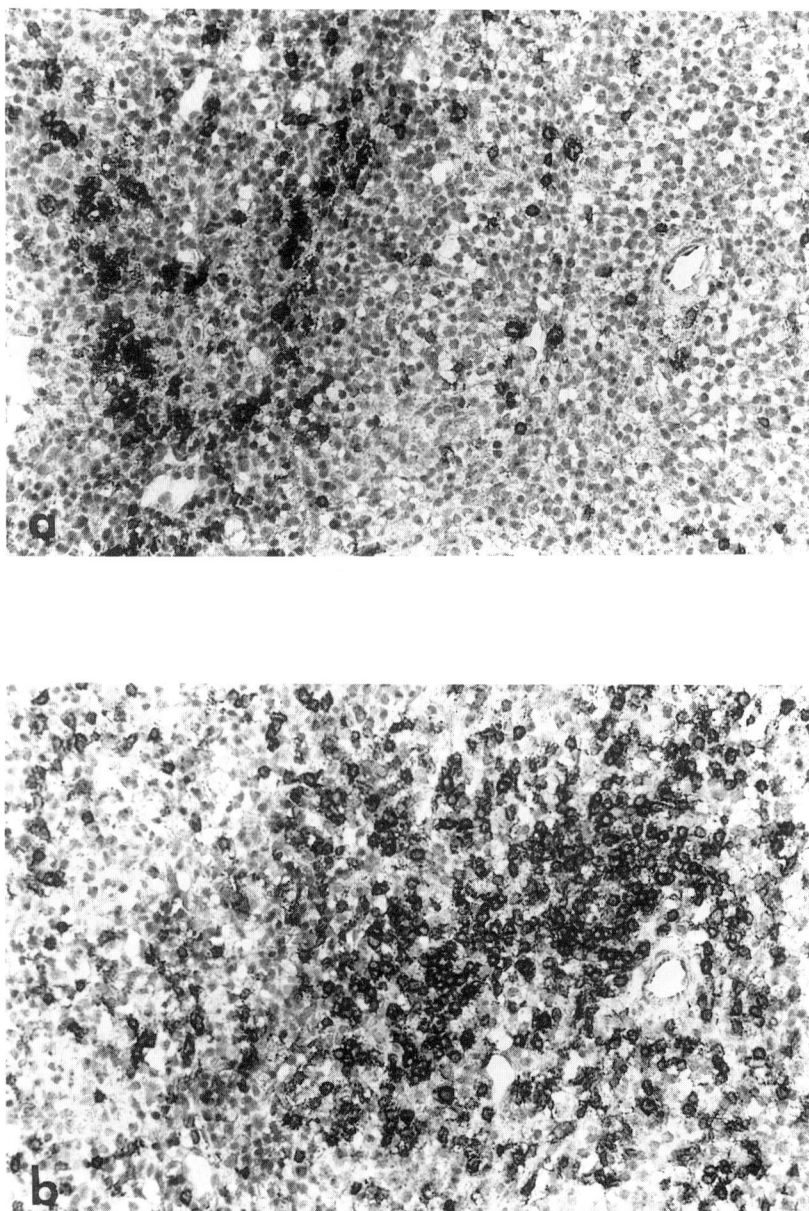

FIG. 3. Localization of TCR1 (A and C) and TCR2 (B and D) cells in the spleen (A and B) and intestinal villi (C and D) illustrated by immunoperoxidase staining of frozen sections (Bucy *et al.*, 1988).

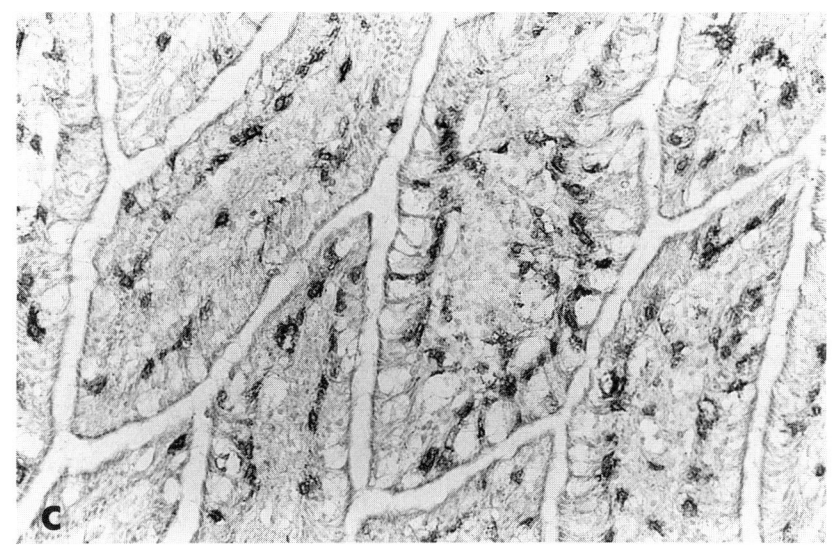

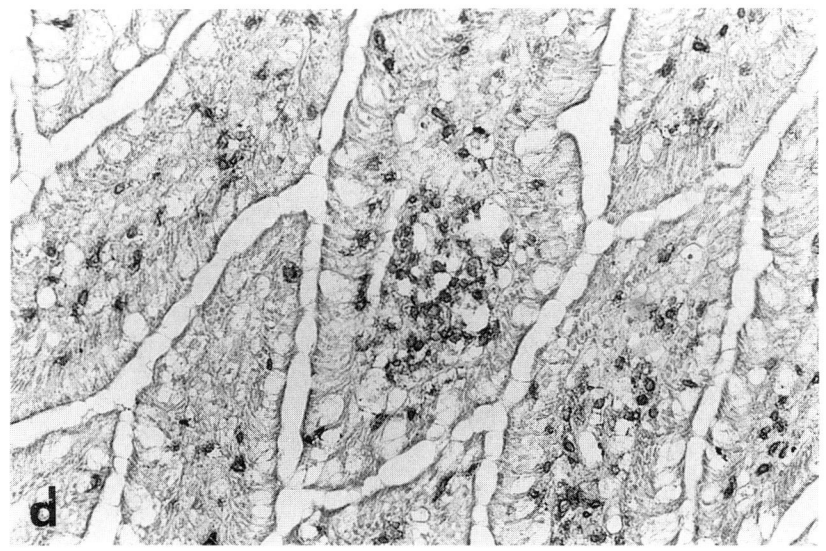

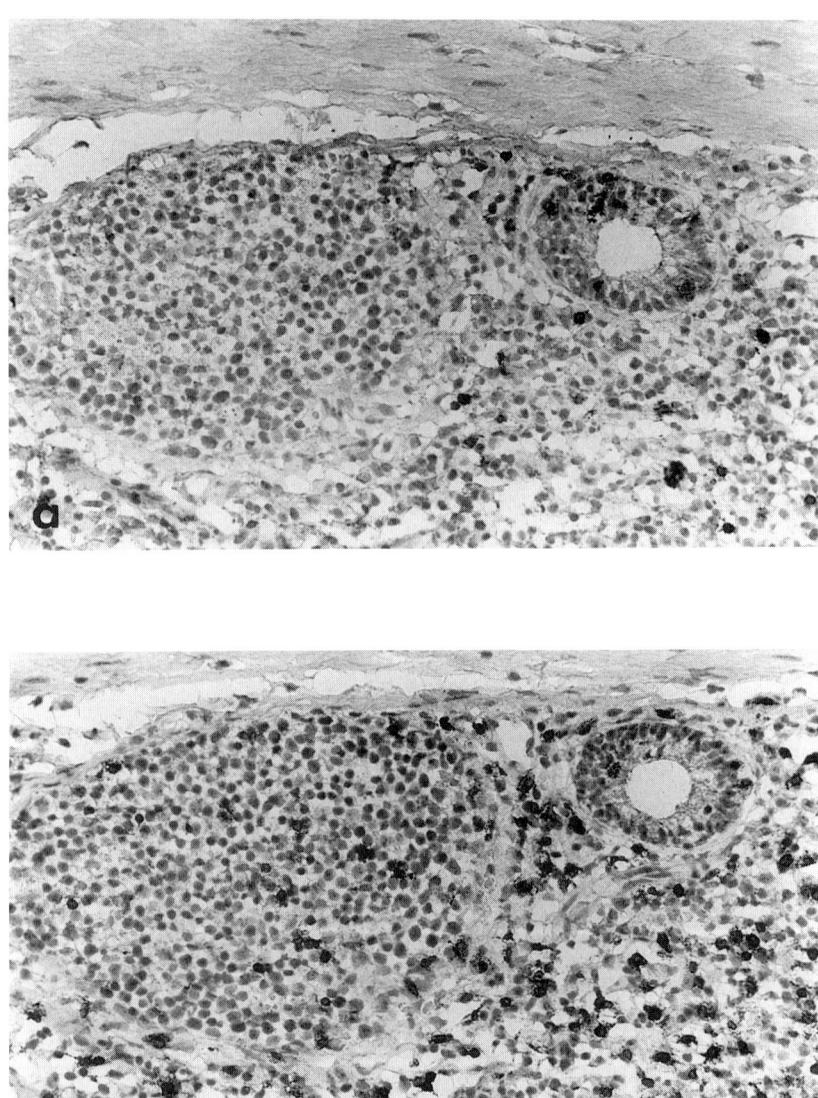

FIG. 4. Localization of TCR1[+] (A) and TCR2[+] (B) T cells in the germinal centers of the lymphoid nodule in the cecal tonsil of an adult chicken. Note the absence of TCR1[+] cells from the germinal center, whereas TCR2[+] cells are present within this microenvironment.

distribution of this cell surface molecule thus makes it an attractive candidate as an adhesion molecule for intestinal T cells. Alternatively, the molecule could be involved in another functional activity of intestinal T cells.

VI. Functional Capabilities of TCR1, TCR2, and TCR3 Cells

Mature T cells belonging to each of the three subpopulations can be triggered by cross-linkage of their TCR/CD3 receptor units by TCR type-specific antibodies coupled to sepharose beads (Chen et al., 1986). Although TCR1, TCR2, and TCR3 cells can be induced to proliferate in this way, the proliferative response of the TCR1 cells is relatively low (Sowder et al., 1988). Similarly, although all three types of T cells respond with proliferation to phytohemagglutinin, pokeweed mitogen, and concanavalin A, the TCR1 cells consistently respond less well (our unpublished observations). Avian $CD4^+$ T cells, like their mammalian counterparts, appear to be especially efficient in producing soluble T cell growth factors (Chan et al., 1988). However, avian lymphokines and their cellular production have not yet been well defined.

The potential for graft-versus-host (GVH) reactions has been analyzed for purified T cell subpopulations in the Simonsen assay (Char et al., 1991). In this analysis, TCR1, TCR2, or TCR3 cells, isolated by negative selection via fluorescence-activated cell sorting, were injected into the chorioallantoic vessels of allogeneic embryos, which were then examined several days later for splenic enlargement. Histologic examination of the spleens revealed focal necrotic lesions with surrounding lymphoid proliferation and edema. Perhaps the most striking observation in these experiments was that the TCR1 cells appeared to lack GVH potential, whereas both the TCR2 and TCR3 cells were capable of GVH alloreactivity. In some strain combinations differing in MHC class II alleles, the frequency of TCR2 cells capable of inducing splenic foci was much higher than that for TCR3 cells, whereas in other allogeneic combinations the reverse was true, suggesting basic differences in the TCR2 and TCR3 repertoires.

When the T cells were separated into $CD4^+$ and $CD8^+$ cells, the $CD8^+$ cells were incapable of inducing GVH lesions regardless of their TCR type. This implies that MHC class II recognition is the key element in this GVH reaction. A failure to express the CD4 accessory molecule may thus account for the inability of the TCR1 cell to function by itself in this assay. TCR1 cells are nevertheless capable of cytotoxic activity. This was shown by incubating chick TCR1 cells with mouse hybridoma cells expressing antiavian CD3 antibodies on their surface (Chan *et al.*, 1988). The hybridoma cells were specifically lysed by the TCR1 cells, thus activated via their receptors.

Although avian T cells have been shown to mediate MHC class II-restricted antigen recognition (Vainio *et al.*, 1988) and B cell help (Vainio *et al.*, 1987), graft rejection (Cooper *et al.*, 1966), tumor cell lysis (Maccubbin and Schierman, 1986), and experimental allergic encephalomyelitis (Blaw *et al.*, 1967), the T cell subpopulations involved in these reactions have not been well characterized. Cells involved in down-regulation of immune responses, conventionally termed suppressor cells, may include a variety of cell types, one of which belongs to the TCR1 $CD8^+$ subpopulation (Quere *et al.*, 1990).

VII. Experimental Manipulation of T Cell Development

The avian thymus differs from the mammalian thymus in that there are approximately 14 individual lobes located bilaterally along the jugular veins in the neck of birds. The architecture of the individual thymic lobes, however, is similar in birds and mammals.

A. Thymectomy

T cell migration from the thymus and the population of peripheral tissues are well underway by the time of hatching. Probably because of this and the extrathymic growth potential of T cells, removal of the thymus alone after hatching does not eliminate the thymus-dependent T cell system or most cell-mediated immune functions. In earlier studies, thymectomy plus whole-body irradiation was required to delineate the thymus-dependent system of T cells and their functions in the chicken (Cooper *et al.*, 1965, 1966).

The analysis of the peripheral T cell subpopulations in birds thymectomized at hatching reveals an interesting pattern. Although development of the TCR2 and TCR3 subpopulations in the circulation is only moderately compromised, a striking deficit of TCR1 cells is consistently observed in the thymectomized birds (Chen *et al.*, 1989). This suggests that sustained thymic seeding of the TCR1 cells is required for the development of the peripheral TCR1 population, at least for the circulating pool of TCR1 cells. Preliminary data suggest that the intestinal TCR1 population is not as severely compromised in the birds thymectomized at hatching. Current experiments are designed to determine whether these intestinal TCR1 cells are derived from the embryonic thymus.

B. Embryonic Treatment with Anti-TCR1, -TCR2, or -TCR3 Antibodies

Cells expressing more than one type of TCR could not be detected during the sequential development of the TCR1, TCR2, and TCR3 subpopulations of cells in the thymus (Chen *et al.*, 1988), suggesting that they are not generated by TCR isotype switching. This possibility was examined further in suppression experiments that involved treatment with TCR-specific antibodies (Chen *et al.*, 1989; Cihak *et al.*, 1989; Char *et al.*, 1990; our unpublished observations). The development of T cells could be inhibited by either the administration of anti-CD3 or anti-TCR antibodies. In the latter case, each of the T cell subsets could be selectively inhibited by the cross-linking, TCR-specific antibodies. Embryonic treatment with anti-TCR1 antibodies thus inhibited the development of TCR1 cells, both in the thymus and periphery, but had not apparent effect on the TCR2 and TCR3 cells. The suppressive effects of embryonic treatment with the anti-TCR2 or anti-TCR3 monoclonal antibodies were equally selective.

The deficiency in specific T cell sublineages induced by anti-TCR treatment could be extended by surgically removing the thymus. The circulating pool of TCR1 cells could be virtually eliminated by combining thymectomy at hatching with embryonic anti-TCR1 treatment. When anti-TCR2 treatment was combined with thymectomy, the suppressive effect on the TCR2 subpopulation was prolonged and a compensatory increase in the percentage of TCR3 cells was observed. The extent of TCR2 depletion depended on the anti-TCR2 dosage (Cihak *et al.*, 1989). In thymectomized birds receiving the highest dose of anti-TCR2 as embryos, the TCR3 cells accounted for up to 80% of the circulating T cell pool, whereas they normally constitute about 15% of the T cell pool. Conversely, a compensatory increase in the percentage of $TCR2^+$ cells was observed in birds treated as embryos with anti-TCR3 antibodies and thymectomy (Char *et al.*, 1990).

The long-term physiological effects of selective depletion of the TCR1, TCR2, or TCR3 subpopulations should prove to be informative. It may be particularly interesting to determine the effect of combined anti-TCR1 treatment and thymectomy on development of the intestinal TCR1 population in birds, because studies in mice suggest a possible extrathymic origin of these cells (L. LeFrancois, personal communication). One particularly interesting result is an inhibitory effect of anti-TCR2 and thymectomy on IgA production (J. Cihak, G. Hoffmann-Fezer, H. W. L. Ziegler-Heitbrock, H. Stein, B. Caspers, C. H. Chen, M. D. Cooper, and U. Lösch, unpublished).

C. CYCLOSPORIN TREATMENT

Another experimental approach to examine mechanisms of T cell development involves treatment with cyclosporin A, a drug that blocks TCR-initiated signal transduction (Shevach, 1985; Hess et al., 1988; Crabtree, 1989; June et al., 1987; Thompson et al., 1989; Emmel et al., 1989; Shi et al., 1989). This biochemical effect is the basis for the immunosuppressive effect of CsA because TCR-initiated activation is essential for the induction of immune responses. In the developing thymus, blockade of TCR-mediated signals prevents clonal selection by self antigens. In both mice (Jenkins et al., 1988; Kosugi et al., 1989) and chickens (Bucy et al., 1990b), CsA treatment arrests α/β T cell development at the cortical thymocyte stage (CD4$^+$ CD8$^+$; low TCR density), at which α/β repertoire selection occurs (Fig. 5). Development of the TCR3 subpopulation is inhibited even more drastically than that of the TCR2 subpopulation. The TCR3 thymocytes fail to complete their initial cortical development following CsA treatment. Despite this drastic effect on the α/β T cells, no apparent effect of CsA can be found on the thymic development of γ/δ T cells in either species. Because mature γ/δ T cells are fully sensitive to CsA inhibition of activation (Bucy et al., 1990b), the lack of a CsA effect on γ/δ T cell thymic maturation strongly implies that there is no requirement

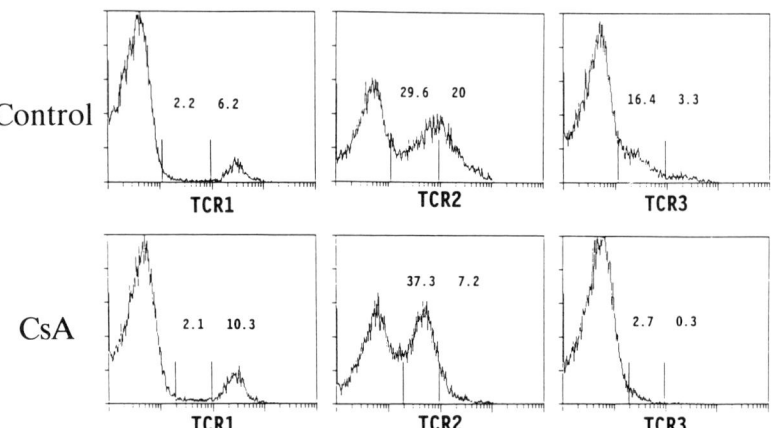

FIG. 5. Effect of CsA treatment (20 mg/kg daily for the first 3 weeks of life) on the development of the thymocyte sublineages. Cells were stained by indirect immunofluorescence on day 21 and analyzed by flow cytometry. The frequencies of cells between the indicated markers are shown. The CsA treated thymi had approximately one-third fewer total cells (redrawn from Bucy et al., 1990b).

for positive intrathymic selection of this sublineage. However, studies of mice that are transgenic for a γ/δTCR may suggest a role for negative intrathymic selection of the γ/δTCR repertoire (Dent et al., 1990; Bonneville et al., 1990).

We have exploited the CsA disruption of TCR selection to develop a model of autoimmune disease in the chicken, similar to that previously described in the mouse (S. Sakaguchi and N. Sakaguchi, 1989). Chickens treated with CsA the first 14 days of life, and then thymectomized, often (five out of seven) developed widespread inflammatory lesions 4–6 months later. Although the TCR2 cells were decreased in frequency immediately after discontinuation of the CsA treatment (due to compromised positive selection), by the time lesions developed the frequencies of the T cell sublineages were similar to thymectomized controls not given CsA treatment. The tissue lesions contained TCR2 T cells and monocytic cells, but few TCR1 or TCR3 cells. Thus, the CsA-induced deficit in intrathymic TCR2 repertoire selection is associated with autoimmune lesions in peripheral organs.

VIII. T Cell Tumors

Avian models have been valuable in the analysis of virus-induced B cell tumorigenesis. Following infection at hatching with the avian leukosis virus, widespread lymphoid tumors of B lineage occur 5 to 9 months later (Burmester et al., 1946; Peterson et al., 1966; Cooper et al., 1974). The multistage transformation process begins in the first few weeks after retroviral infection with the development of preneoplastic bursal foci as a consequence of proviral insertion near the c-*myc* gene. The resultant upregulated c-*myc* expression appears to be a necessary event but is insufficient to cause the final malignant transformation (Cooper et al., 1968; Hayward and Neal, 1981; Thompson et al., 1987). Although less is known about the pathogenesis of avian T cell tumors, two models of T cell tumors in the chicken promise to be especially informative. The most extensively analyzed avian T cell tumor, found in Marek's disease, is associated with infection by a DNA herpesvirus. Another type of avian T cell tumor occurs after infection with the reticuloendotheliosis retrovirus. The target cells appear to differ in the two model systems.

A. MAREK'S DISEASE VIRUS-INDUCED TUMORS

Marek's disease in the chicken is a well-characterized herpesvirus infection. Following exposure to the Marek's disease virus (MDV), B cells are the primary lymphocytic targets of the early phase of the infection (Shek *et al.*, 1983). The necrotizing effects of this early cytolytic B cell infection provoke an acute lymphoreticulitis and splenic enlargement (Payne *et al.*, 1976). T cells become activated, express MHC class II (Ia$^+$) antigen, and some enter a latent phase of the infection (Calnek *et al.*, 1984). These latently infected lymphocytes are thought to be responsible for dissemination of the MDV to other tissues. Reactivation of the virus in these cells can lead to the reappearance of cytolytic infection in lymphoid organs. Infiltrates of lymphocytes and monocytes can also be found in neural tissues of infected birds (reviewed by Calnek, 1986).

The final response to infection with MDV is often a neoplastic transformation of T lymphocytes. These Ia$^+$ lymphoma cells carry multiple copies of the MDV genome (Nazerian *et al.*, 1973). Although both integrated and episomal forms of viral DNA have been reported in the tumor cells (Tanaka *et al.*, 1978; Kaschka-Dierich *et al.*, 1979), integration may not be required for tumor transformation. Tumors from different birds vary in the number of viral DNA copies, but tumors from the same chicken usually have the same number of copies (Lee *et al.*, 1975), suggesting a clonal origin. However, because the exact numbers of viral copies per tumor are difficult to determine when fresh tumors are examined (Witter *et al.*, 1975), the issue of tumor clonality has not been formally resolved.

T cell lines can be established from MD tumors (Akiyama and Kato, 1974; Powell *et al.*, 1974; Calnek *et al.*, 1978) and a large number of these have been typed according to their differentiation antigen profiles. The tumor cell lines uniformly express Ia (Brogren *et al.*, 1981; Schat *et al.*, 1982) and all express the α/βTCR/CD3 surface complex; about 80% express the TCR2 form, and the rest use TCR3. Within both subsets the CD4$^+$ phenotype is the more common, although CD8$^+$ lines have been identified (Schat *et al.*, 1988; our unpublished observations). Lymphoblastoid T cell lines have also been derived from acute lesions induced by the intramuscular inoculation of MDV-infected allogeneic cells of renal origin (Calnek *et al.*, 1989). These cell lines are also α/βTCR$^+$ Ia$^+$, but the proportion of lines expressing TCR3 is greater and a few CD4$^-$ CD8$^-$ cell lines have been identified (K. A. Schat *et al.*, unpublished observations).

Of 100 cell lines examined, no TCR1 lines were found (Schat *et al.*, 1991), suggesting that TCR1 cells are rarely the targets for MDV infection and transformation. It is possible, however, that TCR1 cells are transformed but are not easily propagated. In fact, *in vitro* transformation of spleen cells by MDV may result in the transient growth of MDV$^+$ Ia$^+$ TCR1$^+$ T cells (B. Calnek and T. Schat, personal communication), but cells with this phenotype are difficult to maintain in culture.

The target cells for MDV-induced tumorigenesis thus appear to be mature T cells and cell activation may be a key element in the transformation process. This view is supported by the observation that neonatal thymectomy does not prevent development of MDV-induced lymphomas (Sharma et al., 1975). Moreover, when either thymic or splenic lymphocytes from MD-susceptible chickens (line 7) are transplanted into MD-resistant chickens (line 6), the recipients become susceptible to tumor formation (Gallatin and Longenecker, 1979; Powell et al., 1982). These results indicate an extrathymic origin of the MDV-induced T cell tumors, although a preneoplastic intrathymic event could still explain the clonal susceptibility to MDV-induced transformation.

B. Reticuloendotheliosis Virus-Induced Tumors

Reticuloendotheliosis viruses (REVs) are nondefective avian retroviruses with multipotent pathogenic potential. Most REV strains induce chronic neoplastic lesions that include bursal involvement in the development of lymphoid tumors after a relatively long latent period (Witter et al., 1981). The transformed cells are of the B cell lineage and proviral insertion adjacent to the c-*myc* gene is essential to the transformation process (Swift et al., 1985). In contrast, the splenic necrosis (SN) REV strain has been reported to induce T-lineage malignancies after a relatively short latency period in certain inbred lines of chickens (Witter et al., 1986). These lymphomas frequently involve the thymus, rarely affect the bursa, and often infiltrate peripheral nerves.

Each T cell tumor contains an average of two to four proviruses (Isfort et al., 1987), one of which is near the c-*myc* gene. The correlation between c-*myc* activation and tumor induction (100%) is even higher than that found in B lymphomas (~70%) (Hayward and Neel, 1981). In these T cell tumors, c-*myc* mRNA levels are increased 6- to 18-fold over normal, whereas 50- to 100-fold increases are characteristically found in B lymphomas (Hayward and Neel, 1981). Similarly, transformations of both murine and feline T cells have been found to be more dependent on c-*myc* deregulation than on the actual level of transcriptional amplification (Reicin et al., 1986; Forrest et al., 1987).

Phenotypic analysis of the REV SN-induced lymphoma cells (Chen et al., 1991) indicates that all of these tumors are $CD3^+$ and Ia^-. Most express TCR2 and CD8. Occasional tumors express TCR3 and some are $CD4^+$ or $CD4^+ CD8^+$. As with MDV, no $TCR1^+$ tumors have been identified. The consistent thymic involvement and occurrence of immature T cell phenotypes suggest a thymic origin for these retrovirus-induced T cells tumors. The proviral insertion can serve as a precise clonal marker for these tumors, further study of which could provide valuable information on T cell tumorigenesis.

IX. TCR Genes

The antigen specificity of a T cell is determined by the TCR. In mammals, genes that encode the TCR polypeptides, α, β, γ, and δ, are composed of immunoglobulin-like gene segments that undergo rearrangement to create a distinct α/β or γ/δ heterodimer in each T cell during development (Davis and Bjorkman, 1988). During this process, TCR diversity is generated by the apparent joining of individual members of the variable (V), diversity (D), joining (J), and constant (C) gene segment families. Additional diversity is created at the junctions of the V, D, and J elements by junctional variability and N nucleotide addition. Combinatorial diversity and junctional variability are also primary determinants of the mammalian immunoglobulin (Ig) repertoire (Tonegawa, 1983). In contrast, avian species have evolved a distinct mechanism for the generation of an Ig repertoire. The chicken heavy and light chain Ig repertoires are each created during B cell development by gene conversion of a single rearranged V gene segment using V segment pseudogenes as donors (Reynaud *et al.*, 1987, 1989; Thompson and Neiman, 1987). Therefore, it was of interest to determine whether birds also utilize gene conversion to create a T cell repertoire.

The high degree of structural similarity among mammalian and chicken TCR polypeptides permitted a novel strategy to be utilized for the cloning of the chicken TCRβ chain gene (Tjoelker *et al.*, 1990). A large number of mammalian TCRβ cDNAs and genomic clones were pooled and labeled by nick translation using a high concentration of DNase I. This resulted in a pool of randomly generated probes approximately 100 bp in length, which together generated enough TCRβ specificity to identify several TCRβ cDNAs in a thymocyte cDNA library. These cDNAs have been used to characterize the genomic organization, diversity, and structural conservation of the chicken TCRβ chain locus.

A. Genomic Organization

The chicken TCRβ chain locus undergoes rearrangement during T cell development in the thymus (Fig. 6). Multiple distinct recombination events are observed in developing thymocytes and T cell lines, demonstrating that at least some combinatorial diversity is generated

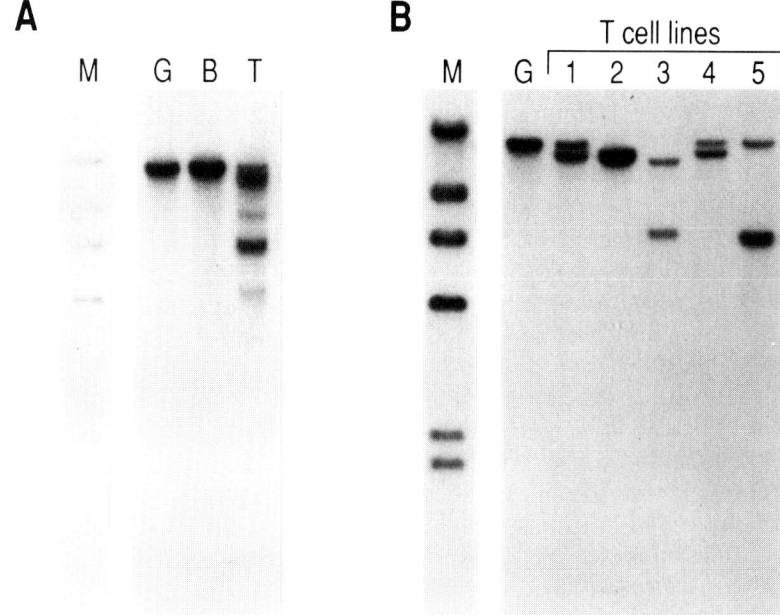

FIG. 6. TCRβ rearrangements. (A) DNA from germ-line (G), bursal (B), and thymic (T) tissue isolated from a 4-week-old chicken. (B) Germ-line DNA and DNA from five TCR cell lines. DNA was digested with SACI and probed with a TCRβ constant-region probe.

during TCRβ recombination. Analysis of the genomic organization of the chicken TCRβ has revealed that it is composed of two families of functional V segments, one D element, four J segments, and one C region. To date, no V segment pseudogenes have been identified. Germ-line V, D, and J gene segments are flanked by typical recombination signal sequences. Within the thymus a functional receptor is created by ordered recombination first of D to J and then of V to DJ. Unlike mammals, the chicken does not contain a duplication of the J and C regions. However, the exon structure of the constant region is identical to that of mammals, suggesting that the basic structure of the TCRβ gene evolved prior to the evolutionary separation of the mammalian and avian orders.

B. TCRβ DIVERSITY

The consensus sequences of the two chicken Vβ families bear little similarity to each other. However, within each Vβ family, surprisingly little sequence variation is observed. Most chicken strains contain six members of the Vβ1 family and three to five members of the Vβ2 family (Fig. 7). Of the six Vβ1 alleles that have been sequenced, none differs by more than five amino acids. The Vβ2 family is only slightly more diverse. Similarly, the four J segments were also found to be strikingly similar to each other. Together these data suggest that because of the lack of sequence heterogeneity among germ-line elements, combinatorial diversity appears limited in its ability to generate a TCRβ repertoire. Instead, recombination appears to generate two distinct TCRβ polypeptides based on the usage of either a Vβ1 or a Vβ2 element. The sequence diversity that was observed within the two Vβ families was not clustered in the complementarity-determining regions (CDRs) thought to be important to MHC–TCR interactions.

To determine if a significant TCRβ repertoire is generated, a series of recombinations between a single Vβ1 and a single J segment were analyzed. No sequence modifications were observed within the V, J, or C regions. Instead, every clone was found to have a distinct sequence at the sites of V–D and D–J recombinations (Fig. 8). These sequences are the result of variable exonuclease digestion and N nucleotide addition at the coding joints. Therefore, the chicken T cell receptor repertoire is created almost exclusively during the process of TCR rearrangement. As noted previously for mammalian TCRs, this concentration of diversity to one small region of the receptor is consistent with the role of the TCR in recognizing a small antigenic peptide in the context of an MHC molecule (Davis and Bjorkman, 1988). To examine the relationship between Vβ usage and the TCR2 and TCR3 sub-

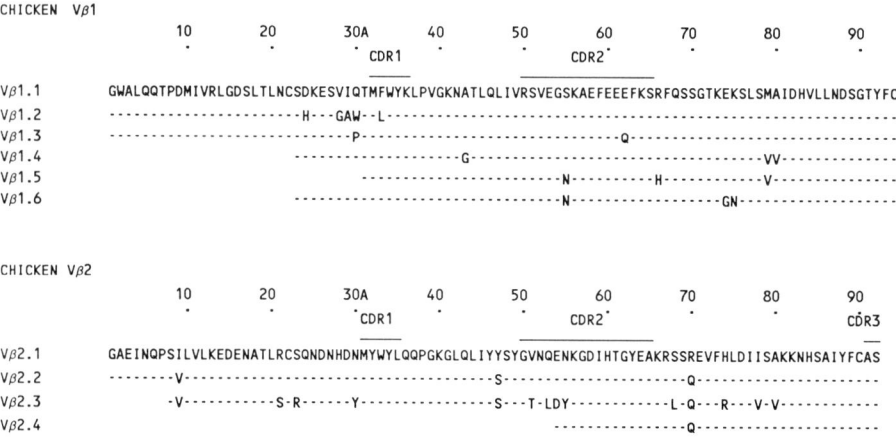

FIG. 7. Amino acid coding sequences of chicken Vβ1 and Vβ2 gene segments.

	← Vβ1 →	N	← Dβ →	N	← Jβ3 →	
germline	GCTAAGCAAGATA		GGGACAGGGGGATC		CAAACACACCACT	
clone 1	----------	CGA	-------	AGGC	----------	(+)
clone 2	----------	CTTCCC	----------	ACACC	----------	(+)
clone 3	----------	GGA	----------	ATCGTTCT	-------	(+)
clone 4	------	AAGTTAGTAC	--------	ACGG	----------	(+)
clone 5	-----	AGTATCA	----------	TCGGGAAGG	----------	(+)

Fig. 8. Diversity of VDJ joints isolated from the thymus of a 4-week-old chicken.

lineages defined by murine mAbs, we (Lahti et al., 1991) used defined TCR2 and TCR3 cell lines. A perfect correlation between Vβ usage and TCR subtype was found. All TCR2 cells were found to have a TCRβ chain containing a V region from the Vβ1 family. TCR3 cells, in turn, exclusively express V regions of the Vβ2 family.

C. CONSERVED STRUCTURAL FEATURES

At the amino acid level the human and chicken TCRβ polypeptides are only 31% identical. This confirms that considerable evolutionary modification of the ancestral TCRβ gene has occurred in both species. Despite this divergence, both genes encode the β chain of a functional MHC-restricted, antigen-specific receptor that assembles with CD3. By comparing the amino acid sequences of the chicken and mammalian genes, several conserved structural features of TCRβ can be identified, which probably result from evolutionary selection at the protein level.

The chicken Vβ1 family displays remarkable similarity to the mammalian Vβ1 subgroup consensus sequence defined by Schiffer et al. (1986). In particular, the chicken Vβ1 family has the conserved arginine at position 64 required to form a salt bridge with the aspartic acid at position 86. The chicken Vβ2 family has a tyrosine at position 65 that is found in all members of the mammalian Vβ2 subgroup. Like the mammalian Vβ2 subgroup, the chicken Vβ2 family lacks the ability to form a salt bridge between amino acids 64 and 86.

Only limited amino acid conservation is observed between the mammalian and chicken TCRβ transmembrane domains (Tjoelker et al., 1990). However, the conservation that is observed suggests a model by which the TCRβ chain assembles with the other conserved component of the TCR/CD3 complex. If the transmembrane domain were to form a helix through the membrane, the three conserved leucines and one conserved valine could be positioned on the same face, generating a structure that resembles the leucine zipper motif used by transcription factors to form heterodimers. By analogy, these residues might be used to associate TCRβ with another chain of the TCR/CD3 complex. In such a helical domain, the conserved lysine would be on the opposite face, making it available to interact with another chain of the TCR/CD3 complex through charge interactions. The lack of conservation of the intracellular domain of TCRβ is consistent with the hypothesis that intracellular signal transduction from TCRβ occurs through its membrane association with CD3.

X. Concluding Remarks

Comparative analysis of avian T cell development and function reveals striking conservation of the major features defined for the mammalian T cell system. In both mammals and birds the T cells utilize either γ/δTCR or α/βTCR together with a CD3 protein complex as signal-transducing receptors for antigen presented in the context of MHC restriction elements. CD4 and CD8 molecules appear to be used in common as assessory receptors for class II or class I (or class I-like) molecules, respectively, on antigen-presenting cells.

The γ/δTCR or TCR1 cells, which are always the first to appear in development, are present in relatively large numbers in birds. This feature and the early availability of an anti-TCR1 antibody have made the chicken an informative model for the study of this conserved T cell subpopulation. The avian TCR1 cells represent a separate sublineage of T cells (see Fig. 8) that express relatively high receptor levels from the onset of their appearance in the thymic cortex, initially lack detectable CD4 and CD8 assessory receptors, and show no obvious dependence on receptor-mediated signals for their intrathymic development and early migration, homing preferentially to the splenic red pulp and intestinal epithelial areas. On reaching these peripheral tissues, the TCR1 cells frequently express CD8 molecules, which may be essential recognition elements in their immune responses. The TCR1$^+$ CD8$^+$ cells also appear capable of down-regulating α/β T cell help of B cells, at least in an experimental model system. Interestingly, the avian TCR1 cells alone appear to lack the capacity to initiate a graft-versus-host attack. Most of these features appear to be conserved in mammalian TCR1 cells, which are much better defined on molecular grounds.

Two additional sublineages of avian T cells have been defined as α/β T cells, each of which uses a different Vβ gene family. The chicken

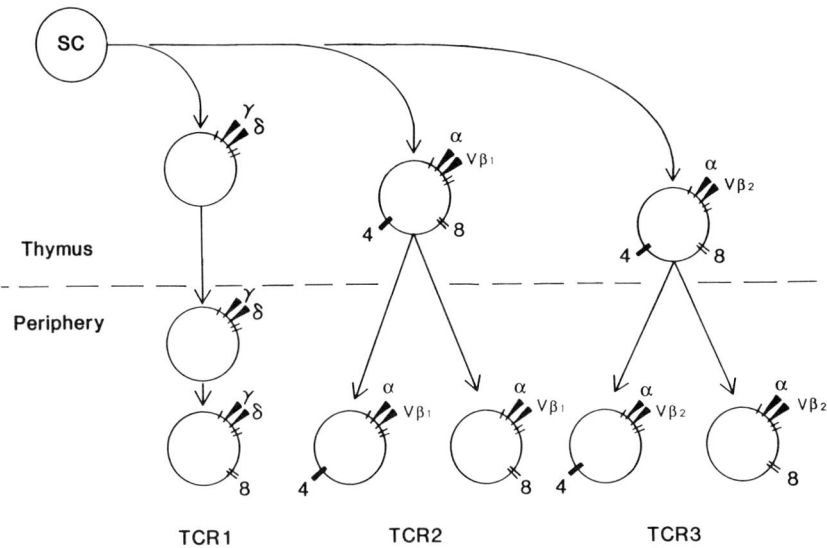

FIG. 9. Hypothetical model of avian T cell development featuring three sublineages of cells: the γ/δTCR1 sublineage and the two α/β sublineages, TCR2 and TCR3, which utilize Vβ1 and Vβ2 genes, respectively.

TCRβ locus, which is the only nonmammalian TCR locus yet identified, is composed of single Cβ and Dβ genes, four Js, and the Vβ1 and Vβ2 gene families used, respectively, by the TCR2 and TCR3 sublineages. The six Vβ1 genes are all functional and are very similar to each other. The three to five Vβ2 genes are also functional and closely similar, but the Vβ1 and Vβ2 gene families show limited homology to each other. As in mammals, combinatorial diversification appears to be a major mechanism for generating the TCR2 and TCR3 repertoires. No evidence for the gene conversion mechanism, so prominent in generating the avian Ig repertoire, has been found in our studies of the avian TCRβ locus.

The definition of the two separate sublineages of α/β T cells, which utilize two different Vβ gene families, raises several interesting questions. The most obvious is the genetic mechanism by which these Vβ genes are sequentially rearranged and expressed during development. Although the entire locus has not yet been mapped, members of the two gene families appear to be intermixed, thus making gene location unlikely to be the prime factor. Conserved framework features used to assign mammalian Vβ genes to two major families are conserved in the two avian Vβ gene families. It will thus be of interest to determine if the Vβ genes in the two mammalian families are sequentially rearranged and expressed by two sublineages of mammalian T cells.

Perhaps the most interesting question concerns the differences in MHC interactions that must exist for the TCR2 and TCR3 receptor classes, as these may govern distinctive physiological roles for these two α/β T cell sublineages. Both TCR2 and TCR3 cells may recognize the same MHC class II alloantigens and initiate a graft-versus-host reaction, but the apparent differences in frequency of alloreactive cells indicate differences in their repertoires. The limited representation of the TCR3 cells in the intestine and the importance of the TCR2 cells for IgA antibody production also imply different physiological roles for these two sublineages of α/β T cells.

Acknowledgments

We thank our colleagues for sharing their ideas and unpublished data, and Mrs. E. A. Brookshire for secretarial assistance. This work has been supported by Grants CA 16673, CA 13148, AI 30879, and GM 42571, awarded by the National Institutes of Health. MDC and CT are HHMI investigators.

References

Akiyama, Y, and Kato, S. (1974). *Biken J.* **17**, 105.
Auerbach, R. (1961). *Dev. Biol.* **3**, 336.
Blackman, M., Kappler, J., and Marrack, P. (1990). *Science* **248**, 1335.
Blaw, M. E., Cooper, M. D., and Good, R. A. (1967). *Science* **158**, 1198.
Bonneville, M., Ishida, I., Itohara, S., Verbeek, S., Berns, A., Kanagawa, O., Haas, W., and Tonegawa, S. (1990). *Nature (London)* **344**, 163.
Brogren, C. H., Settnes, O. P., von Bulow, V., Bisati, S., Berjonneau, C., and Simonsen, M. (1981). *Immunobiology* **159**, 65.
Bucy, R. P., Chen, C. H., Cihak, J., Lösch, U., and Cooper, M. D. (1988). *J. Immunol.* **141**, 2200.
Bucy, R. P., Coltey, M., Chen, C. H., Char, D., Le Douarin, N. M., and Cooper, M. D. (1989). *Eur. J. Immunol.* **19**, 1449.
Bucy, R. P., Chen, C. H., and Cooper, M. D. (1990a). *J. Immunol.* **144**, 1161.
Bucy, R. P., Li, J., Xu, X. Y., Char, D., and Chen, C. H. (1990b). *J. Immunol.* **144**, 3257.
Burmester, B. R., Prickett, C. O., and Belding, T. C. (1946). *Cancer Res.* **6**, 189.
Calnek, B. W. (1986). *Crit. Rev. Microbiol.* **12**, 293.
Calnek, B. W., Murthy, K. K., and Schat, K. A. (1978). *Int. J. Cancer* **21**, 100.
Calnek, B. W., Schat, K. A., Ross, L. J. N., Shek, W. R., and Chen, C. H. (1984). *Int. J. Cancer* **33**, 389.
Calnek, B. W., Lucio, B., Schat, K. A., and Lillehoj, H. S. (1989). *Avian Dis.* **33**, 291.
Chan, M. M., Chen, C. H., Ager, L. L., and Cooper, M. D. (1988). *J. Immunol.* **140**, 2133.
Char, D., Sanchez, P., Chen, C. H., Bucy, R. P., and Cooper, M. D. (1990). *J. Immunol.* **145**, 3547.
Char, D., Chen, C. H., Bucy, R. P., Simonsen, M., and Cooper, M. D. (1991). *FASEB J.* **5**, A731.

Chen, C. H., Chanh, T. C., and Cooper, M. D. (1984). *Eur. J. Immunol.* **14**, 385.
Chen, C. H., Ager, L. L., Gartland, G. L., and Cooper, M. D. (1986). *J. Exp. Med.* **164**, 375.
Chen, C. H., Cihak, J., Lösch, U., and Cooper, M. D. (1988). *Eur. J. Immunol.* **18**, 538.
Chen, C. H., Sowder, J. T., Lahti, J. M., Cihak, J., Lösch, U., and Cooper, M. D. (1989). *Proc. Natl. Acad. Sci. U.S.A.* **88**, 2352.
Chen, C. H., Witter, R. L., Bucy, R. P., and Cooper, M. D. (1991). *FASEB J.* **5**, A706.
Cihak, J., Ziegler-Heitbrook, H. W. L., Trainer, H., Schranner, I., Merkenschlager, M., and Lösch, U. (1988). *Eur. J. Immunol.* **18**, 533.
Cihak, J., Ziegler-Heitbrook, H. W. L., Chen, C. H., and Cooper, M. D. (1989). *Int. Congr. Immunol.*, 7th, 256.
Clevers, H., Alarcon, B., Wileman, T., and Terhorst, C. (1988). *Ann. Rev. Immunol.* **6**, 629.
Coltey, M., Jotereau, F. V., and Le Douarin, N. M. (1987). *Cell Differentiation* **22**, 71.
Coltey, M., Bucy, R. P., Chen, C. H., Cihak, J., Lösch, U., Char, D., Le Douarin, N. M., and Cooper, M. D. (1989). *J. Exp. Med.* **170**, 543.
Cooper, M. D., Peterson, D. A., and Good, R. A. (1965). *Nature (London)* **205**, 143.
Cooper, M. D., Peterson, R. D. A., South, M. A., and Good, R. A. (1966). *J. Exp. Med.* **123**, 75.
Cooper, M. D., Payne, L. N., Dent, P. B., Burmester, B. R. and Good, R. A. (1968). *J. Natl. Cancer Inst.* **41**, 373.
Cooper, M. D., Purchase, H. G., Bockman, D. E., and Gathings, W. E. (1974). *J. Immunol.* **113**, 1210.
Crabtree, G. R. (1989). *Science* **243**, 355.
Crone, M., Simonsen, M., Skjodt, K., Linnet, K., and Olsson, R. (1985). *Immunogenetics* **21**, 181.
Davis, M. M., and Bjorkman, P. J. (1988). *Nature (London)* **334**, 395.
Dent, A. L., Matis, L. A., Hooshmand, F., Widacki, S. M., Bluestone, J. A., and Hedrick, S. M. (1990). *Nature (London)* **343**, 714.
Dunon, D., Kaufman, J., Salomonsen, J., Skjodt, K., Thiery, P., and Imhof, B. A. (1990). *EMBO J.* **9**, 3315.
Emmel, E. A., Verweij, C. L., Durand, D. B., Higgins, K. M., Lacy, E., and Crabtree, G. R. (1989). *Science* **246**, 1617.
Ewert, D. L., Munchus, M. S., Chen, C. H., and Cooper, M. D. (1984). *J. Immunol.* **132**, 2524.
Forrest, D., Onions, D., Lees, G., and Neil, J. C. (1987). *Virology* **158**, 194.
Fowlkes, B. J., Edison, L., Mathieson, B. J., and Chused, T. M. (1985). *J. Exp. Med.* **162**, 162, 802.
Frelinger, J. A. (1982). *In* "Ia Antigens" (S. Ferrone and C. S. David, eds.), p. 37. CRC Press, Boca Raton, Florida.
Gallatin, W. M., and Longenecker, B. M. (1979). *Nature (London)* **280**, 587.
Gao, E. K., Lo, D., Cheney, R., Kanagawa, O., and Sprent, J. (1988). *Nature (London)* **336**, 176.
George, J. F., and Cooper, M. D. (1990). *Eur. J. Immunol.* **20**, 2171.
Guillemot, F. P., Oliver, P. D., Peault, B. M., and Le Douarin, N. M. (1984). *J. Exp. Med.* **160**, 1803.
Hála, K., Schaunstein, K., Neu, N., Krömer, G., Wolf, H., Böck, G., and Wick, G. (1986). *Eur. J. Immunol.* **16**, 1331.
Hayward, W. S., and Neel, B. G. (1981). *Curr. Top. Microbiol. Immunol.* **91**, 219.
Hess, A. D., Esa, A. H., and Colombani, P. M. (1988). *Transplant. Proc.* **20**, 29.
Houssaint, E., Diez, E., and Jotereau, F. V. (1985). *Eur. J. Immunol.* **15**, 385.

Houssaint, E., Tobin, S., Cihak, J., and Lösch, U. (1987). *Eur. J. Immunol.* **17**, 287.
Isfort, R., Witter, R. L., and Kung, H. J. (1987). *Oncogene Res.* **2**, 81.
Jenkins, M. K., Schwartz, R. H., and Pardoll, D. M. (1988). *Science* **241**, 1655.
Jotereau, F. V., and Le Douarin, N. M. (1982). *J. Immunol.* **129**, 1869.
Jotereau, F. V., Houssaint, E., and Le Douarin, N. M. (1980). *Eur. J. Immunol.* **10**, 620.
Jotereau, F., Heuze, K. F., Samon-vie, V., and Gascan, H. (1987). *J. Immunol.* **138**, 1026.
June, C. H., Ledbetter, J. A., Gillespie, M. M., Lindsten, T., and Thompson, C. B. (1987). *Mol. Cell. Biol.* **7**, 4472.
Kaschka-Dierich, C., Nazerian, K., and Thomssen, R. (1979). *J. Gen. Virol.* **44**, 271.
Kisielow, P., and von Boehmer, H. (1990). *Sem. Immunol.* **2**, 35.
Knight, K. L., and Becker, R. S. (1990). *Cell* **60**, 963.
Koller, B. H., Marrack, P., Kappler, J. W., and Smithies, O. (1990). *Science* **248**, 1227.
Kosugi, A., Zuniga-Pfucker, J. C., Sharrow, S. O., Kruisbeek, A. M., and Shearer, G. M. (1989). *J. Immunol.* **143**, 3134.
Lahti, J. M., Chen, C. H., Tjoelker, L. W., Calnek, K. A., Schat, K. A., Thompson, C. B. and Cooper, M. D. (1991). *FASEB J.* **5**, A613.
Le Douarin, N. (1978). In "Differentiation of Normal and Neoplastic Hematopoietic Cells," p. 5. Cold Spring Harbor Laboratory, Cold Spring Harbor, New York.
Le Douarin, N. M., Dieterlen-Lievre, F., and Oliver, P. D. (1984). *Am. J. Anat.* **170**, 261.
Lee, L. F., Nazerian, K., and Boezi, J. A. (1975). In "Oncogenesis and Herpesviruses II" (G. de-Thé, M. A. Epstein, and H. Zur Hausen, eds.), p. 199. Int. Agency Res. Cancer, Lyon, France.
Lillehoj, H. S., Lillehoj, E. P., Weinstock, D., and Schat, K. A. (1988). *Eur. J. Immunol.* **18**, 2059.
Littman, D. R. (1987). *Annu. Rev. Immunol.* **5**, 561.
Maccubbin, D. L., and Schierman, L. W. (1986). *J. Immunol.* **136**, 12.
Marrack, P., and Kappler, J. (1990). *Sem. Immunol.* **2**, 45.
Martin, C., and Dieterlen-Lievre, F. (1978). *Nature (London)* **272**, 353.
Michie, S. A., Kirkpatrick, E. A., and Rouse, R. V. (1988). *J. Exp. Med.* **168**, 1929.
Moore, M. A. S., and Owen, J. J. (1965). *Lancet* **2**, 658.
Moore, M. A. S., and Owen, J. J. (1967). *Nature (London)* **215**, 1081.
Nazerian, K., Lindahl, T., Klein, G., and Lee, L. F. (1973). *J. Virol.* **12**, 841.
Oliver, P. D., and Le Douarin, N. M. (1984). *J. Immunol.* **132**, 1748.
Payne, L. N., Frazier, J. A., and Powell, P. C. (1976). *Int. Rev. Exp. Pathol.* **16**, 59.
Peterson, R. D. A., Purchase, H. G., Burmester, B. R., Cooper, M. D., and Good, R. A. (1966). *J. Natl. Cancer Inst.* **36**, 585.
Pink, J. R. L., and Rijnbeek, A. M. (1983). *Hybridoma* **2**, 287.
Powell, P. C., Payne, L. N., Frazier, J. A., and Rennie, M. (1974). *Nature (London)* **251**, 79.
Powell, P. C., Lee, L. F., Mustill, B. M., and Rennie, M. (1982). *Int. J. Cancer* **29**, 169.
Quere, P., Cooper, M. D. and Thorbecke, G. J. (1990). *Immunology* **71**, 517.
Reicin, A., Yang, J. Q., Marcu, K. B., Fleissmer, E., and Koehne, C. F. (1986). *Mol. Cell. Biol.* **6**, 2088.
Reynaud, C. A., Anquez, V., Grimal, H., and Weill, J.-C. (1987). *Cell* **48**, 379.
Reynaud, C. A., Dahan, A., Anquez, V., and Weill, J.-C. (1989). *Cell* **59**, 171.
Sakaguchi, S., and Sakaguchi, N. (1989). *J. Immunol.* **142**, 471.
Schat, K. A., Chen, C. H., Shek, W. R., and Calnek, B. W. (1982). *J. Natl. Cancer Inst.* **69**, 715.
Schat, K. A., Chen, C. H., Lillehoj, H., Calnek, B. W., and Weinstock, D. (1988). In "Advances in Marek's Disease Research" (S. Kato, T. Horiuchi, T. Mikami, and K. Hirai, eds.), p. 220. Japan. Assoc. in Marek's Dis., Tokyo, Japan.

Schat, K. A., Chen, C. H., Calnik, B. W., and Char, D. (1991). *J. Virol.* (in press).
Schiffer, M., Wu, T. T., and Kabat, E. A. (1986). *Proc. Natl. Acad. Sci. U.S.A.* **83**, 4461.
Sharma, J. M., Witter, R. L., and Purchase, H. G. (1975). *Nature (London)* **253**, 477.
Shek, W. R., Calnek, B. W., Schat, K. A., and Chen, C. H. (1983). *J. Natl. Cancer Inst.* **70**, 485.
Shevach, E. M. (1985). *Annu. Rev. Immunol.* **3**, 397.
Shi, Y., Sahai, B. M., and Green, D. R. (1989). *Nature (London)* **339**, 625.
Shortman, K., Egerton, M., Spangrude, G. J., and Scollay, R. G. (1990). *Sem. Immunol.* **2**, 3.
Smith, C. A., Williams, G. T., Kingston, R., Jenkinson, E. J., and Owen, J. J. (1989). *Nature (London)* **337**, 181.
Sowder, J. T., Chen, C. H., Ager, L. L., Chan, M. M., and Cooper, M. D. (1988). *J. Exp. Med.* **167**, 315.
Swift, R. A., Shaller, E., Witter, R. L., and Kung, H. J. (1985). *J. Virol.* **54**, 869.
Tanaka, A., Silver, S., and Nonoyama, M. (1978). *Virology* **88**, 19.
Thompson, C., and Neiman, P. (1987). *Cell* **48**, 369.
Thompson, C. B., Humphries, E. H., Carlson, L. M., Chen, C. H., and Neiman, P. E. (1987). *Cell* **51**, 371–381.
Thompson, C. B., Lindsten, T., Ledbetter, J. A., Kunkel, S. L., Young, H. A., Emerson, S. G., Leiden, J. M., and June, C. H. (1989). *Proc. Natl. Acad. Sci. U.S.A.* **86**, 1333.
Tjoelker, L. W., Carlson, L. M., Lee, K., Lahti, J. M., McCormack, W. T., Leiden, J. M., Chen, C. H., Cooper, M. D., and Thompson, C. B. (1990). *Proc. Natl. Acad. Sci. U.S.A.* **87**, 7856.
Tonegawa, S. (1983). *Nature (London)* **302**, 575.
Vainio, O., and Lassila, O. (1989). *Crit. Rev. Poultry Biol.* **2**, 97.
Vainio, O., Toivanen, P., and Toivanen, O. (1987). *Poultry Sci.* **66**, 795.
Vainio, O., Veromaa, T., Eerola, E., Toivanen, P., and Radcliffe, M. J. H. (1988). *J. Immunol.* **140**, 2864.
Weill, J.-C., and Reynaud, C.-A. (1987). *Science* **238**, 1054.
Witter, R. L., Stephens, E. A., Sharma, J. M., and Nazerian, K. (1975). *J. Immunol.* **115**, 177.
Witter, R. L., Smith, E. J., and Crittenden, L. B. (1981). *Avian Dis.* **25**, 374.
Witter, R. L., Sharma, J. M., and Fadly, A. M. (1986). *Avian Pathol.* **15**, 467.
Zÿlstra, M., Bix, M., Simester, N. E., Loring, J. M., Raulet, D. H., and Jaenish, R. (1990). *Nature (London)* **344**, 742.

This article was accepted for publication on 4 March 1991.

Structural and Functional Chimerism Results from Chromosomal Translocation in Lymphoid Tumors

T. H. RABBITTS AND T. BOEHM

MRC Laboratory of Molecular Biology, Cambridge CB2 2QH, England

I. Introduction

The presence of chromosomal abnormalities in lymphoid tumors has been established for many years, and the consistent nature of the abnormalities in specific tumor types and subtypes has consolidated the view that the tumor phenotype is established by the creation of the chromosomal abnormality. The obvious corollary of this view is that the new chromosomal environment alters a gene(s) located near the junction of the abnormal chromosome, and that this alteration is a key feature of lymphoid tumorigenesis. In the light of this, the affected genes are operationally defined as oncogenes. We review here molecular analyses of human lymphoid tumor chromosomal abnormalities (translocations and inversions) that have vindicated these ideas and point out the recurring features of oncogene activation illustrated by the different chromosomal abnormalities thus far studied.

As far as the practicality of molecular analyses is concerned, there are two types of chromosomal abnormalities. The first has been particularly amenable to study because one side of the aberrant chromosomal junction is bordered by a known rearranging gene [i.e., immunoglobulin (Ig) or T cell receptor (TCR)] (see Boehm and Rabbitts, 1989a,b), thus making cloning of the breakpoints a rather trivial exercise. The study of the second type, for which no known rearranging gene is present, has relied on the presence of a known gene near the chromosome junction facilitating cloning directly (e.g., see de Klein *et al.*, 1982) or after utilizing procedures involving larger chromosome segments (see von Lindern *et al.*, 1990).

The major lymphoid abnormalities that have been cloned so far are shown in Table I, together with relevant information about the location of the presumptive oncogenes involved. The oncogenes affected by the abnormalities range from previously known oncogenes (such as c-*myc* in Burkitt's lymphoma) to newly discovered ones [such as *bcl-2* in follicular lymphoma or *rhombotin* (*Ttg-1*) in T cell acute lymphoblastic leukemia (T-ALL)].

TABLE I
PROTOONCOGENES NEAR CHROMOSOMAL BREAKPOINTS[a]

Abnormality	Affected Gene	Type of Gene Product[b]	Type of Alteration
B cell			
t(8;14)(q24;q32)	c-myc	b-HLH/ZIP protein	Ig enhancer control/mutation
t(2;8)(p12;q24)			
t(8;22)(q24;q11)			
t(14;18)(q32;q21)[c]	bcl-2	Inner mitochondrial membrane protein	Not known (Ig enhancer control?)
t(1;19)(q23;p13.3)	prl-1 (1q23)	Homeodomain protein	Gene fusion
	E2A (19p13.3)	Ig enhancer binding protein (b-HLH/ZIP)	
t(14;19)(q32;q13.1)	bcl-3	CDC-10-like protein	Not known (Ig enhancer control; promoter truncation)
t(11;14)(q13;q32)	Not yet identified	—	—
T cell			
t(8;14)(q24;q11)[d]	c-myc	b-HLH/ZIP protein	Not known (TCRα enhancer control?)
t(11;14)(p15;q11)	rhombotin	LIM protein	Promoter truncation
t(11;14)(p13;q11)	rhom-2	LIM protein	Promoter truncation
t(7;11)(q35;p13)			
t(7;19)(q35;p13)	lyl-1	b-HLH protein	TCRβ enhancer control; promoter truncation
t(1;14)(p32;q11)	tal/SCL/TCL5	b-HLH protein	Not known; probably exon truncation
t(7;9)(q35;q34)	Not characterized	—	—
t(10;14)(q24;q11)	hox-t	Homeodomain protein	Promoter truncation
t(7;10)(q35;q24)			
inv14(q11;q32.1)[e]	Not yet identified	—	—
t(14;14)(q11;q32.1)			
t(7;14)(q35;q32.1)			

[a] See text for specific references to the various abnormalities.
[b] b-HLH, Basic helix-loop-helix motif; ZIP, leucine zipper motif; LIM, cysteine-rich motif.
[c] Other bcl-2 translocations have been seen with Ig κ and λ L chain genes (see text).
[d] Different c-myc translocations have been identified in T cell tumors (see text).
[e] t(14;14) also occurs in T cell tumors with breakpoints identical to those in inv14 except involving the homologous chromosomes 14 (see text). In addition, a translocation to 14q32.1 involving the TCRβ gene has been described (see text).

The mechanism of translocations in lymphoid tumors has been studied widely (for recent reviews see Boehm and Rabbitts, 1989a,b), and this is discussed briefly with particular reference to the timing of translocations in T cell tumorigenesis. The main part of this review is concerned with the consequences of chromosomal abnormalities: first, the consequences to the protooncogene near the site of the abnormality (a process that, in this context, we may loosely term "oncogene activation"), and second, the consequences to the cell of an "activated" oncogene (i.e., the process of "tumorigenesis"). It should be pointed out that in most cases the genes thought to be affected by the formation of the abnormal chromosome must be regarded as presumptive oncogenes, because often functional evidence of a role in tumorigenicity is lacking. Some examples, such as c-*myc*, have been clearly demonstrated to have oncogenic potential in transgenic mice (Adams *et al.*, 1985; Leder *et al.*, 1986), as expected from the existence of a retroviral counterpart (i.e., v-*myc*). However, as discussed in the conclusion in this review, the weight of evidence suggests that the majority of the translocation genes are "low-grade oncogenes," whose effect may be in the initiating stages of the prolonged process of leukemia development.

II. Mechanism of Translocation and Inversion

The mechanism by which chromosome abnormalities associated with the rearranging genes are created has been well studied, and the enzymes carrying out normal antigen receptor gene rearrangement (i.e., the so-called recombinase) are implicated in the creation of these abnormalities (Tsujimoto *et al.*, 1985a,b; Croce, 1987). Normal antigen receptor gene rearrangement involves joining of separate variable (V), diversity (D), and joining (J) segments by the recombinase (reviewed in Tonegawa, 1983; Davis and Bjorkman, 1988; Boehm and Rabbitts, 1989b). Further, the DNA sequences that are primarily recognized in this process are short stretches of seven (heptamer, or 7-mer) or nine (nonamer, or 9-mer) bases adjacent to the 3' side of V segments, on either side of D segments, and on the 5' side of J segments. Such sequences are referred to as recombinase signal sequences. The presence and involvement of signal-like sequences in the various chromosomal regions involved in chromosome abnormalities have been shown (see Croce, 1987), but this is certainly not universal (see Boehm and Rabbitts, 1989b). In addition, it has been suggested that accessibility to recombinase may be facilitated by stretches of alternating purine–pyrimidine residues (Boehm *et al.*, 1989), which might transiently adopt left-handed helical or Z-DNA conformations, disrupting chromatin structure. Such stretches have been located near 11p13, 10q24, 11q13 (Boehm et al., 1989), and 11p15 breakpoints (Boehm et al., 1988a), providing a good correlation between DNA sequence and propensity to the formation of abnormal chromosomes.

Studies on T cell abnormalities have been particularly informative about recombinase involvement, especially breakpoints within the chromosome band 11p13. Breaks at a small region of 11p13 occur in two types of translocation, t(11;14)(p13;q11) or t(7;11)(q35;p13), and involve either TCRδ, TCRα, TCRβ on 14q11 or 7q35 (Boehm et al., 1988b; Champagne et al., 1989; Harvey et al., 1989; Royer-Pokora et al., 1989; Yoffe et al., 1989; Cheng et al., 1990; Foroni et al., 1990; Sanchez-Garcia et al., 1991). Whereas some breakpoints at 11p13 occur at sequences with very good homology to the canonical 7-mer, other breakpoints do not (Boehm et al., 1988b; Sanchez-Garcia et al., 1991), even though they occur within a short distance of 7-mer-like sequences. For example, as illustrated in Fig. 1, within a region of 1108 bp of chromosome 11p13, three independent translocations occur. The translocation breakpoints of tumors 13 and 1 are at different heptamer-like (no nonamer-like homology) sequences. A third example, tumor 2, breaks in the middle, at a sequence lacking heptamer homology ("ignoring" both heptamer-like sequences). However, although signal-like sequences are not always involved, it seems that recombinase is responsible for the abnormal chromosomal unions, because often both reciprocal translocated chromosomes have N-region nucleotide addition (this is true for all three examples in Fig. 1), which is a hallmark of recombinase activity (Alt and Baltimore, 1982). Thus these translocations can be viewed as mutations of the normal joining process.

III. Timing of Chromosome Translocation and Inversion

The developmental timing of chromosomal translocation and inversion events is of importance to understanding etiology. Studies of T cell tumors have lent themselves to this problem because normal T cell developmental pathways are fairly well described. In outline, bone marrow stem cells give rise to the pre-T cells, which are primarily destined to differentiate in the thymus. Within this compartment several relevant features of T cell maturation occur, including T cell receptor gene rearrangement, surface phenotype changes, and stage-specific gene expression (see Strominger, 1989). TCR gene rearrangement is crucial to the formation of chromosomal abnormalities,

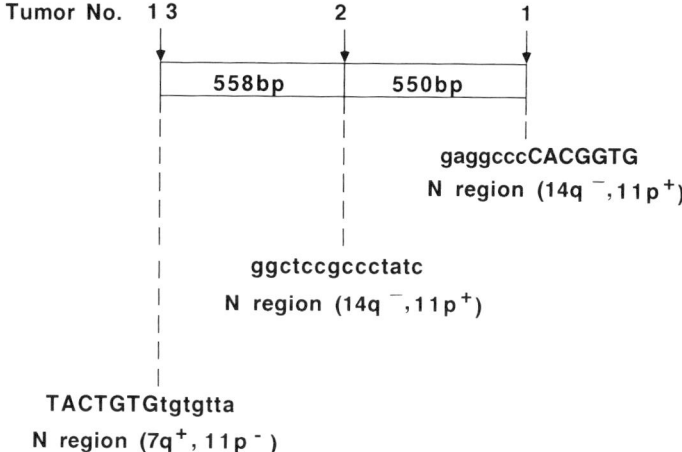

FIG. 1. Junctional sequences and location of translocation breakpoints in chromosome band 11p13. A common translocation in childhood T-ALL involves joining of TCRδ, TCRα, or TCRβ loci to chromosome band 11p13, at a region termed the T-ALL^{bcr} (Boehm et al., 1988b). The sequence and mechanistic features of the various individual translocations in the T-ALL^{bcr} illustrate the erroneous involvement of the VDJ recombinase. The figure shows three separate breakpoints [in tumors designated 1, 2, and 13 (Sanchez-Garcia et al., 1991)] that occur in a 1108-bp stretch. The reciprocal translocations in the three examples all show N-region diversity whereas only cases 1 and 13 break adjacent to heptamer-like sequences (i.e., CACAGGTG and TACTGTG, respectively). Thus translocations in this region are generally recombinase mediated but are not always sequence directed.

because obviously this is the time of recombinase activity. Indeed, studies of expression of the recombinase-activating genes (*RAG1* and *RAG2*) (Schatz et al., 1989; Oettinger et al., 1990) have shown that immature cortical thymocytes are the main site of expression of these genes (Boehm et al., 1991a). Thus recombinase-mediated chromosomal translocation/inversion must be early events in T cell maturation, resulting in the creation of the preleukemic cell.

Thus far tumor-associated translocations in T cells have been shown to involve TCRδ, TCRα, and TCRβ. No tumor-associated translocations with TCRγ have been cloned, although cytogenetic abnormalities of the same chromosomal band have been described (Kaneko et al., 1989). The only reported TCRγ involvement in a chromosome abnormality is an inversion chromosome 7 in a nonleukemic ataxia telangiectasia patient (Stern et al., 1989). This presents an interesting paradox, because TCR (or Ig) gene rearrangement *per se* seems to be the key feature leading to chromosome translocation. Perhaps the lack of TCRγ tumor-associated abnormalities reflects an inability of this locus to provide sustained oncogene activation after chromosome translocation. It has been established for some years that most or all human T cells have rearrangements of one TCRγ allele, and more usually both TCRγ alleles (Lefranc *et al.*, 1986), but TCRγ transcription seems to be switched off if the T cell differentiates to an α/β cell rather than a γ/δ cell (Tighe *et al.*, 1987). Thus although there is the means for TCRγ to be involved in chromosome translocations in most T cells (i.e., gene rearrangement), lack of sustained γ gene expression would effectively shut down the transcriptional accessibility of the locus, presumably including any translocated oncogene in the locality. Thus tumor outgrowth (i.e., selection for the transformed phenotype) would occur only rarely, for example, in those cells with translocations that survive thymus selection, which become γ/δ cells, or the rare cases wherein transient expression of the translocated oncogene is sufficient to allow the ultimate appearance of the transformed phenotype. If correct, a corollary of the argument is that most translocated oncogenes require continued expression to manifest their effect on the normal growth control of the afflicted cell.

IV. Consequences of the Formation of Chromosomal Abnormalities on Adjacent Oncogenes

The somatic formation of a chromosomal abnormality near a protooncogene has consequences in two ways. First, there are consequences to the protooncogene, resulting in its "activation," and second, there are consequences to the somatic cell with the abnormality. We deal with these two events separately.

The abnormalities and the protooncogenes discussed here are summarized in Table I. Interestingly, the only previously known oncogene found to be involved in chromosomal translocations is the c-*myc* protooncogene. All the other translocation genes are newly identified, and this may reflect the fact that the lymphoid tumor-associated abnormalities are often initiating events in tumorigenesis (see below).

A. c-*myc* IN BURKITT'S LYMPHOMA TRANSLOCATIONS: Ig ENHANCER PLACEMENT

The well-described translocations in Burkitt's lymphoma (BL) always involve a chromosomal region adjacent to the c-*myc* gene located at chromosome band 8q24 (reviewed in Boehm and Rabbitts, 1989a). This occurs upstream of c-*myc* in the major translocation t(8;14) or downstream in the variant translocations t(2;8) and t(8;22). In addition, the distance from the c-*myc* gene can be quite variable. In the t(8;14) translocations, the endemic BLs (Hamlyn and Rabbitts, 1983; Taub *et al.*, 1982; Bernard *et al.*, 1983; Haluska *et al.*, 1986; Pelicci *et al.*, 1986) have 8q24 breakpoints that tend to be a large distance upstream from the c-*myc* promoter region, whereas the t(8;14) sporadic translocations are close to or within the gene (Pelicci *et al.*, 1986). The variant translocations occur at variable distances from the 3' end of c-*myc* (Croce *et al.*, 1983; Davis *et al.*, 1984; Taub *et al.*, 1984), in a region called the *pvt*-like region (Graham and Adams, 1986; Mengle-Gaw and Rabbitts, 1987; Henglein *et al.*, 1989), because of its homology to the equivalent region in mouse plasmacytomas. The *pvt*-like breakpoints can be as much as 300 kb from the end of c-*myc*. This large variation in the position of chromosomal translocation breakpoints embodies the c-*myc* paradox (Rabbitts *et al.*, 1984a).

Somatic mutations are a frequent feature of the translocated c-*myc* gene (Rabbitts *et al.*, 1983a, 1984b; Leder *et al.*, 1983), generally affecting the first exon of the gene (Rabbitts *et al.*, 1983a; Leder *et al.*, 1983; Pelicci *et al.*, 1986; Morse *et al.*, 1989). Some coding region changes have also been found (Rabbitts *et al.*, 1983a; Murphy *et al.*, 1986; Showe *et al.*, 1985). However, BL translocated c-*myc* genes have been found with no structural alterations, so it is possible that the coding mutations enhance but do not initiate the transforming nature of the translocated c-*myc* in the cell. The exon 1 mutations are probably more important because they seem to affect a transcription block that normally operates on the normal c-*myc* gene (Cesarman *et al.*, 1987).

A common feature of BL translocations is, however, that the constant regions of the various Ig loci always lie adjacent to the translocated c-*myc* gene. This implies some crucial feature associated with these gene segments, presumably a transcription control element(s) brought into play on the translocated c-*myc* allele. (Indeed, a corollary is that the 5' and 3' c-*myc* breakpoints are a natural consequence of the organization of V, D, and J elements of the three Ig loci on their respective chromosomes). Recent studies on Ig-associated enhancers in mouse have implicated possible elements (Meyer and Neuberger, 1989; Eccles *et al.*, 1990; Pettersson *et al.*, 1990), suggesting a general model for c-*myc* activation via BL chromosomal translocation. The location of known Ig-associated enhancer elements is shown in Fig. 2, together with the probable place for new heavy (H) and λ light (L) chain enhancers in man (by analogy with those found in mouse). The common feature is the presence of an Ig enhancer located on the same chromosome as the c-*myc* allele. Furthermore, these Ig enhancers tend

A t(8;14)(q24;32) endemic BL

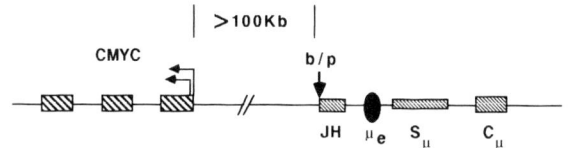

B t(8;14)(q24;32) sporadic BL

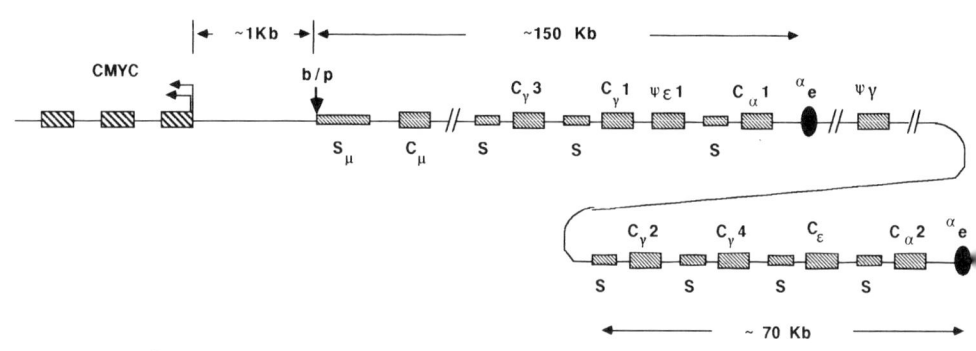

C t(2;8)(p12;q24)

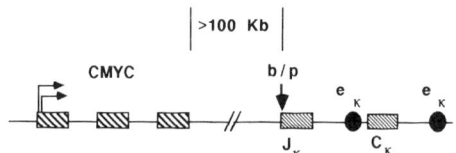

D t(8;22)(q24;q11)

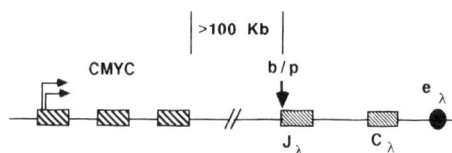

to occur at a considerable distance from the translocated c-*myc* gene; for example, there is a roughly comparable distance between the κ L chain enhancers and c-*myc* (in variant BL breakpoints) and between the Ig H μ enhancer and c-*myc* in the majority of endemic BLs (Fig. 2). Indeed, the comparison of endemic and sporadic BL breakpoints is even more suggestive. In the former, the breakpoints on 14q32 tend to be in the J$_H$ elements (Haluska et al., 1986), thus placing the Ig H μ enhancer greater than 100 kb from c-*myc*. In the latter, the breakpoints on 14q32 tend to be in the Ig H switch regions (Pelicci et al., 1986), thereby deleting the Ig H μ enhancer (Rabbitts et al., 1983b) from the translocated c-*myc* allele; however, the putative enhancer found downstream of the C$_α$ gene would be placed at a similar distance from the translocated c-*myc* gene as the μ enhancer in endemic BL. We propose, therefore, that the common feature of the BL translocations is the placement of an Ig enhancer element at some distance (usually greater than 100 kb) from the translocated c-*myc* gene to induce expression of this gene (but not necessarily at a high level). This distal enhancer placement appears to solve the c-*myc* paradox, because it explains the two seemingly bizarre features of BL translocations. First, the distal breakpoints with respect to c-*myc*, and second, the breakpoints both upstream and downstream of c-*myc*.

Interestingly, there is a T cell equivalent of the BL c-*myc* translocation. This t(8;14)(q24;q11) involves breakage 3' of c-*myc* and joining within the TCRα joining segments (Williams et al., 1984; Erikson et al., 1985; Finger et al., 1986; Shima et al., 1986; Bernard et al., 1988). This presumably operates like the proposed BL enhancer activation above, with the TCRα enhancer instead of the Ig enhancers. The common feature is again transcriptional disregulation via enhancer

---

FIG. 2. Ig enhancer location and c-*myc* breakpoints. The region within and immediately adjacent to the c-*myc* gene, located on chromosome 8q24, suffers chromosome translocations in Burkitt's lymphoma involving the three immunoglobulin loci. The figure depicts the variation seen in different translocations and illustrates the proposed common acquisition of an Ig enhancer element at some distance from the translocated c-*myc* gene. (A) The endemic (malaria belt) BL is mainly EBV$^+$ and translocations normally occur far upstream of c-*myc* and within the Ig H J$_H$ segments, bringing the Ig μ enhancer upstream of c-*myc*. (B) Sporadic BL (usually EBV$^-$) has translocation points near or within the first exon of c-*myc* and within the switch region of the Ig C$_H$ genes. The putative IgA enhancer(s) are thereby brought into the region upstream of c-*myc*. (C and D) Variant translocations involving the Ig κ (C) or λ (D) L chain genes break downstream from the c-*myc* gene at variable but large distance from the 3' end. The known and putative Ig L chain enhancers are indicated. The dual c-*myc* promoters are depicted by right-angled arrows; b/p represents translocation breakpoints.

replacement, resulting in inappropriate expression of the c-*myc*-encoded protein.

B. GENE FUSION RESULTING FROM CHROMOSOME TRANSLOCATION

Fusion of two transcription units was first described in cancer-associated chromosomal abnormalities in chronic myelogenous leukemia, wherein it was discovered that the c-*abl* protooncogene was consistently joined to another gene (the so-called *bcr* gene on chromosome 22) in the Philadelphia chromosome translocation t(9;22)(q34;q11) (de Klein *et al.*, 1982). Molecular studies of another specific translocation, t(6;9)(p23;q34), associated with acute nonlymphocytic leukemia (von Lindern *et al.*, 1990), have implicated a fusion gene product in the translocation. More recently, the acute promyelocytic leukemia carrying t(15;17)(q21;q11–22) has been shown to result from the fusion of the retinoic acid receptor α gene (17q21) and another previously unknown transcription unit from chromosome 15 (de Thé *et al.*, 1990; Borrow *et al.*, 1990). The latter locus, designated *myl* (de Thé *et al.*, 1990), has not yet been characterized but presumably its presence in the fusion mRNA is contributary to the oncogenic effect of the translocation.

In the lymphoid series of tumors, two quite different fusion genes have been studied. In the first case, a fusion was found between an Ig V_H segment and a TCRα J segment after inversion of chromosome 14, giving the *Ig T* fusion gene (Baer *et al.*, 1985; Denny *et al.*, 1986). The mechanism of this inversion was fully characterized by establishment of the structure of the junctions of the inversion chromosome (Baer *et al.*, 1987a), showing that the abnormal chromosome was created by the VDJ recombinase. Although the inversion had involved an entire chromosome arm (i.e., tens of millions of base pairs) and was associated with a T cell lymphoma, this chromosomal abnormality is probably not involved in the pathogenesis of that particular tumor. Attempts to find similar *Ig T* fusion genes in other T cell tumors have as yet failed, and indeed a second 14q32 locus (defined as 14q32.1 in tumor samples by *in situ* hybridization of probes from the 14q32 breakpoint region derived from both inversion and translocation 14) (Baer *et al.*, 1987b; Mengle-Gaw *et al.*, 1987, 1988; Davey *et al.*, 1988; Bertness *et al.*, 1990) has been described. Furthermore, two other abnormalities have been described in the tumor with the *Ig T* gene: first, a deletion within the *pvt-1*-like region adjacent to c-*myc* (Rabbitts *et al.*, 1986), and second, a different translocation t(7;9) (Reynolds *et al.*, 1987) (see Table I). The clustered nature of breakpoints at 14q32.1 in other T cell tumors has certainly strengthened the view that this region encodes the affected "protooncogene," but, as yet, no gene has been ascribed to the region.

The second fusion gene abnormality in lymphoid leukemia is the recently described cloning of the breakpoint of t(1;19)(q23;p13.3) (Mellentin et al., 1989a; Kamps et al., 1990; Nourse et al., 1990). This interesting translocation fuses the E2A gene for the Ig enhancer-binding proteins E12 and E47 (from chromosome 19p) with a transcribed gene designated *prl-1* (from chromosome 1). The *prl-1* gene has been shown to possess a homeodomain and the normal protein thus presumably has DNA-binding capacity. The translocation breakpoints of several t(1;19) examples fall into a single intron of *E2A*, and the 3' region of this gene, which carries helix–loop–helix (HLH) and basic DNA binding motifs, i.e , the b-HLH domain, is removed, to be replaced by the *prl-1* gene putative DNA-binding homeodomain. The fusion gene created by the t(1;19) therefore is most likely a chimeric transcription factor in which the activation domain of *E2A* has been linked to a new DNA-binding domain. The possible consequences to the affected cell are discussed below.

C. Helix–Loop–Helix Oncogenes in Chromosome Translocations

Apart from the c-*myc* oncogene, which contains both b-HLH and leucine zipper motifs (Murre et al., 1989a,b; Lassar et al., 1989; see also review by Collum and Alt, 1990) and the t(1;19) fusion gene, new protooncogenes at the junctions of chromosome translocations have been described that possess DNA-binding and protein dimerization motifs (Table I). Principally involved are the *lyl-1* gene (Mellentin et al., 1989b) and the *tal/SCL/TCL5* gene (Begley et al., 1989a,b; Bernard et al., 1989; Finger et al., 1989; Chen et al., 1990a,b), which appear to belong to a family of b-HLH genes because they share 84% homology in the relevant domain (Chen et al., 1990a,b; Begley et al., 1989b). The *lyl-1* gene (chromosome 19p13) was translocated with the TCRβ gene in a translocation t(7;19)(q35;p13) associated with a T-ALL tumor line (Mellentin et al., 1989b). The transcription unit for the protooncogene *lyl-1* was broken in the translocation such that the first, apparently noncoding, exon 1 was lost from the gene and presumably the normal functional promoter as well. The size of the *lyl-1* mRNA in the translocation-bearing cell line is smaller than the wild type, so the truncation of the *lyl-1* promoter may well be compensated, in the tumor, by the acquisition of a TCRβ transcriptional apparatus on the derivative chromosome. [It may be noteworthy that there appears to be little or no normal-size transcript in the tumor line (Cleary et al., 1988), implying the possiblity of a feedback of this protein onto its own promoter. If so, this feedback presumably operates via the putative DNA-binding b-HLH motif.]

The *lyl-1*-related gene, *tal/SCL/TCL5* [a gene located on chromosome 1p34 and translocated with the TCRδ from 14q11 in the t(1;14)(p32;q11)], is particularly interesting because of the seemingly high frequency of abnormalities in this gene in T-ALL (Brown *et al.*, 1990). Although the disruption of the gene via chromosome translocation in T-ALL is rare (3% of patients in one study) (Chen *et al.*, 1990b), it has been found that about a quarter of all T-ALL cases have a common 90-kb deletion with fairly precise common end points, one of which frequently occurs in the region of 1p32, at which the majority of translocations break. The *tal/SCL/TCL5* transcription unit seems to be disrupted in most of the 1;14 translocations, but it is not clear to what extent the coding region of the gene is damaged or whether, like c-*myc* in some BLs, for instance, a noncoding exon(s) is removed. Like lyl-1, tal/SCL/TCL5 possesses an amphipathic helix–loop–helix motif adjoined to a basic domain, which is presumed to be functioning in DNA binding related to transcription. However, the coding region with the b-HLH domain is not altered by the translocation. The precise effect of the lesion in the translocation tumors, or more importantly in those tumors with the common deletion, is thus obscure at present.

D. TRANSCRIPTIONAL DISRUPTION OF THE LIM DOMAIN ONCOGENES BY TRANSLOCATION

A new and different family of presumptive protooncogenes has been discovered with the detailed analysis of the *rhombotin* (or *Ttg-1*) gene. This gene was first found at the site of a rare translocation in T-ALL, t(11;14)(p15;q11), in association with the TCRδ locus from 14q11 (Boehm *et al.*, 1988a), and was found to encode a cysteine-rich protein (McGuire *et al.*, 1989; Boehm *et al.*, 1990b) that contains two apparently duplicated cysteine-rich domains (CRR domains) (Boehm *et al.*, 1990a). This abnormality has been seen cytogenetically several times, but only two breakpoints have been studied as yet. The first of these is found in the cell line RPMI 8402 (Boehm *et al.*, 1988a), and the second is from a patient with T-ALL, which disrupts the *rhombotin* transcription unit in a manner analogous to that in the RPMI 8402 cell line (Boehm *et al.*, 1991c). This translocation is, therefore, a rare but consistent lesion in some forms of T-ALL tumors. The 14q11 breakpoint region in both tumors is within the TCRδ locus and apparently involves a recombinase sequence-specific error (Boehm *et al.*, 1988a). The detailed analysis of this protein and its very high conservation among species as diverse as man, mouse, and fly (Boehm et al., 1990a) finally led to the determination of a homology of the CRR domain with the so-called LIM domains of some transcription regulators (see below).

The effect of the translocation on expression of the *rhombotin* gene is far from clear, but most likely it is not enhancer control from the TCR locus. Truncation of the normal promoter region is more likely, which results in the loss of control elements that modulate the developmental switch of expression, either a switch between the two promoters or switch in levels of mRNA (Boehm et al., 1991b). Normally, the major site of expression of rhombotin is the brain, as determined by transgenic mouse (Greenberg et al., 1990) and *in situ* hybridization experiments (Boehm et al., 1991a,b). Some other somatic tissues, including T cells, have low-level expression (Boehm et al., 1991b). These observations have contributed to the idea that genes important for differentiation can be important in the oncogenic process when subverted by translocation, a particularly relevant feature being the observation that the *rhombotin* gene is expressed mainly in postmitotic neurons in the developing and adult brain (Greenberg et al., 1990). High levels of *rhombotin* mRNA in the translocation-bearing T cell tumors thus presumably reflect loss of a rigid transcription control.

At least two other genes homologous to *rhombotin* have been found (Boehm et al., 1991c) and these LIM domain genes has now been characterized (Boehm et al., 1991c). One of these genes is also on chromosome 11, but at 11p13, and is involved in the frequent translocation t(11;14)(p13;q11). The region of the 11p13 band where most of the translocation breakpoints occur is very small (designated T-ALL^{bcr}) (Boehm et al., 1988b; Foroni et al., 1990; Sanchez-Garcia et al., 1991) and all breakpoints so far examined occur within 26 kb (Sanchez-Garcia et al., 1991) and all occur upstream of the *rhom-2* promoter. The LIM domain gene, designated *rhom-2* by virtue of its homology to *rhombotin*, apparently involved in this translocation is found about 26 kb on the telomeric side of the T-ALL^{bcr} and, like the *rhombotin* LIM domain oncogene, it occurs on the derivative chromosome with the V$_{\alpha/\delta}$ segments (Boehm et al., 1988b). The gene starts at the previously described *HTF^1* (Foroni et al., 1990), which is about 8 kb from the nearest translocation breakpoint.

E. Oncogene Location after Translocation: The Right or the Wrong Chromosome?

The abnormalities that involve the rearranging genes involve an oncogene "activated" after the formation of the abnormal chromosome. It is generally assumed that the Ig or TCR loci exert an effect on the oncogene at the junction of the derivative chromosome via enhancer elements. However, there is no *a priori* reason for this assumption. Oncogene activation via translocation or inversion does occur via Ig- or TCR-mediated transcriptional modulation (i.e., most usually enhancement); (see Table I), but increasing numbers of affected genes are located on the derivative chromosome lacking known Ig or TCR enhancer elements. For example, it is significant that, like the two LIM domain oncogenes on chromosome 11 (see above) affected by the distinct t(11;14), the presumptive protooncogene *tal/SCL/TCL5*, affected by the t(1;14), in some cases occurs on the derivative chromosome with the V_δ/V_α segments and not on that with the C_δ enhancer. This may still reflect the presence of an as-yet undetected enhancer-type element in the TCRδ locus, but alternatives such as promoter truncation in the case for the *rhombotin* translocation (see above) or exon deletion as in *tal/SCL/TCL5* are equally likely. This is schematically summarized in Fig. 3. The enhancer control model of the translocated c-*myc* gene in BL exemplifies the upstream/downstream enhancer placement (Fig. 3A and B) after translocation. The translocation of *tal/SCL/TCL5* can also be upstream (Fig. 3C) or downstream (Fig. 3B) of the gene, but only the latter brings into operation the TCR known enhancer element. In the described *rhombotin* and *rhom-2* translocations, the affected genes lie adjacent to the V elements after translocation (Fig. 3C).

In addition, a new candidate oncogene on 10q24 appears to acquire this type of orientation after chromosome translocation. Molecular analyses of translocations to 10q24 involving TCRβ [t(7;10)(q35;q24) (Boehm *et al.*, 1989)] and TCRδ [t(10;14)(q24;q11) (Kagan *et al.*, 1987, 1989; Lu *et al.*, 1990; Zutter *et al.*, 1990)] have shown that two disparate 10q24 breakpoint regions are involved. Recent analyses of the transcription units in this region indicates that a gene is located on the derivative 14q⁻ chromosome (Zutter *et al.*, 1990), which again corresponds to the segment of TCRδ locus with the V segments. Analyses of the 10q24 gene shows that a promoter truncation mechanism is the most likely and reveals a gene encoding a homeobox protein designated *hox-t* (Kennedy *et al.*, 1991).

V. Effects of Chromosomal Abnormalities on Lymphoid Cells

The initial descriptions of dimerization (homodimerization or heterodimerization) of transcription factors have given a new understanding of the intricate process of control of gene expression, without the need for a vast array of factors controlling the expression of yet more

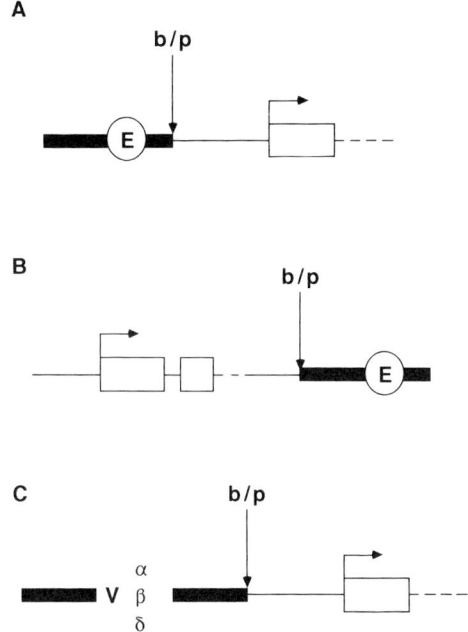

FIG. 3. Structural features of reciprocal translocation chromosomes. The two main organizational consequences of chromosome translocations. The "activated" oncogene is depicted in each case as the open box, with promoters indicated by right-angled arrows, and the incoming chromosomal segment (black box) with or without an enhancer (E). The breakpoint (b/p) position for the chromosome abnormality is shown by the vertical arrow. (A) Enhancer control of translocated genes via breakpoints upstream of the gene. Examples include c-*myc*, *bcl-2*, *bcl-3*, and *lyl-1*. (B) Enhancer control of translocated genes via breakpoints downstream of the gene. Examples include c-*myc*, *bcl-2*, and *tal/SCL/TCL5*. (C) Promoter modulation or truncation by translocation junctions at the 5' end of a gene. Examples include rhombotin, *rhom-2*, *tal/SCL/TCL5*, and *hox-t*.

factors (Abel and Maniatis, 1989). In addition, the results begin to explain the recurring finding of genes with dimerization/DNA-binding capacity at the junction of chromosomal translocations in lymphoid cells. A general picture emerges in which protein dimerization (specifically heterodimerization) and DNA binding play a crucial role in the tumorigenic activity of these newly discovered oncogenes.

A. FUNCTIONAL CHIMERISM INVOLVING DNA BINDING AND PROTEIN DIMERIZATION

The involvement of c-*myc* in translocations, particularly those found in BL, is a paradigm for protooncogene activation via chromosomal abnormality. The c-*myc* gene had been previously identified via its equivalent retroviral gene v-*myc* and was found to be next to translocations in B and, in the latter case, in T cell tumors (see Table I). The c-*myc* translocation apparently provides in each case a transcription enhancer somewhat distal to the translocated c-*myc* gene (see above). The effect of this will be to provide the cell with c-*myc*-encoded protein at unusual times in the cell cycle. The c-*myc*-encoded protein is known to bind DNA, both random sequences and a specific motif CACGTG (Blackwell et al., 1990), although the location of the latter motif in the genome is not yet known. The protein region with this binding activity includes an HLH motif plus the basic region, which contacts DNA on the NH_2-terminal side of the HLH motif (the so-called b-HLH domain). In addition, the c-*myc*-encoded protein has a leucine zipper (ZIP) motif (see above), which, together with the HLH domain, is presumably involved in protein–protein interaction. Therefore, the c-*myc*-encoded protein can presumably form heterodimers with other proteins and perform a specific DNA-binding role (Blackwell et al., 1990). Indeed, recent data show that a protein, designated Max (Blackwood and Eisenman, 1991), binds to the c-*myc*-encoded protein, presumably forming a complex involved in DNA binding and thus in transcription regulation. The disruption of transcription, afforded by the chromosomal translocation, will inappropriately provide substrate (i.e., c-Myc protein) for dimerization, causing disequilibrium of specific DNA-binding complexes. Figure 4 diagrammatically illustrates a possible scheme in which different protein dimerizations occur in normal cell cycle equilibria. Any disbalance in quantity of one component (e.g., excess or wrongly timed presence of c-Myc protein) will have an obvious effect on the overall equilibrium. The production of mRNA from genes controlled by c-Myc dimers/heterodimers, and presumably other members of the equilibrium, could thereby be effected with potential effects on growth characteristics (i.e., transformation).

Clearly this complex interaction network could be applicable to other systems. For example, the role of the b-HLH motif in both protein dimerization and in specific DNA binding has been established (Murre et al., 1989a,b; Lassar et al., 1989) and HLH proteins are thus apparently involved in specific transcriptional cascades, sometimes in differentiation pathways (Lassar et al., 1989). Further, as described above, the inappropriate synthesis of proteins bearing this motif is likely to disrupt the equilibrium of transcriptional regulators. Thus proteins such as those encoded by *lyl-1* and *tal/SCL/TCL5* probably form heterodimers with other proteins in the conductance of their

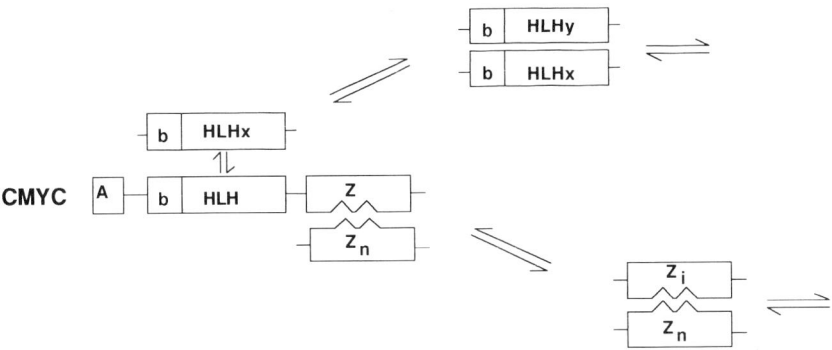

FIG. 4. Equilibrium of protein–protein interaction in transcription networks illustrated by c-Myc. A complex pattern of protein interactions can be envisaged through both HLH and leucine zipper (ZIP) domains. The c-Myc protein is an example that may apply to other protein–protein interaction equilibria. c-Myc has activation (A), b-HLH, and ZIP domains and can bind both DNA and proteins. Conceivably different HLH (HLH_x) and ZIP (Z_n) proteins bind the respective domains of c-Myc, modifying the putative transcriptional activity of the c-Myc protein. In turn, these factors may interact with other HLH (HLH_y) or ZIP (Z_i) proteins, and so on. This would represent a transcriptional equilibrium that would be able to elicit normal fine control of gene expression. Any imbalance of protein concentrations (e.g., c-Myc after translocation) could upset these equilibria.

normal role in transcriptional regulation. Chromosome translocation can disequilibrate the system by providing too much protein, or perhaps too little in some cases. If there is a concentration-dependent equilibrium of interactions in the cell, then a disbalance of one component will be deleterious.

A different type of transcription disruption is potentially exemplified by the fusion gene formed in the t(1;19) (Mellentin et al., 1989b; Kamps et al., 1990; Nourse et al., 1990). This involves the breakage of two genes, the products of which presumably normally act in transcription in separate ways, and potentially the creation of two new hybrid transcription factors. One of these has a new DNA recognition domain (i.e., the homeodomain of the *prl-1* gene) fused to a potential protein dimerization domain (i.e., in *E2A*). The fusion gene created by the t(1;19) therefore is most likely a chimeric transcription factor in which the activation and protein dimerization domains of *E2A* are linked to a new DNA-binding domain (Fig. 5). In addition, because both HLH sequences and the ZIP region (see below) are involved in protein interactions, the chimeric protein may bind to proteins not normally associated with the Prl-1 DNA-binding homeodomain. A likely out-

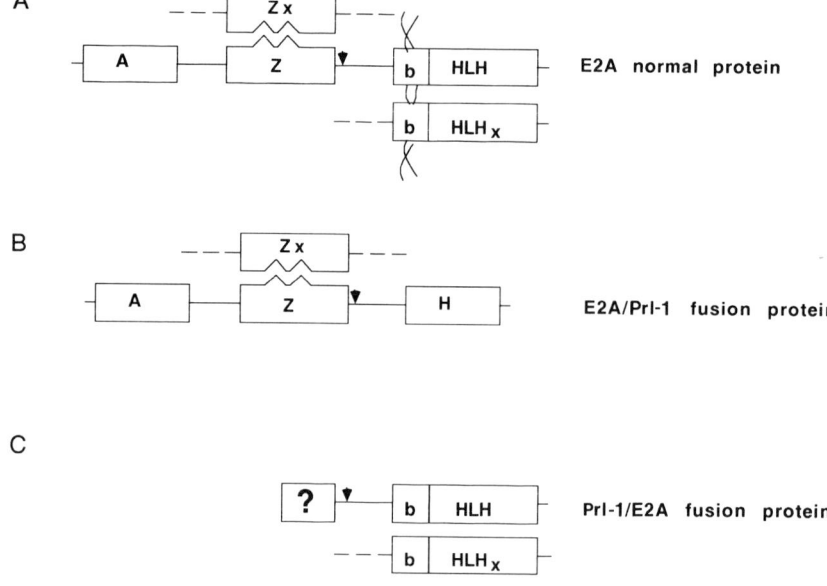

FIG. 5. Protein chimerism resulting from t(1;19). The translocation t(1;19) involves breakage within the *E2A* gene (chromosome 19) and within the *prl-1* gene (chromosome 1), resulting in at least one chimeric protein with potentially altered transcriptional activity. (A) A hypothetical picture of normal E2A protein, which consists of activation (A), ZIP (Z), and b-HLH domains, binding to other zipper (Z_x) and HLH (HLH_x) proteins (this interaction may be a single protein with Z and HLH domains) and interacting with the DNA helix. The position within the E2A protein affected by the translocation is indicated by the arrowhead. (B) The E2A/Prl-1 fusion product after translocation has the activation and ZIP domains of E2A linked to the homeodomain of the Prl-1 protein, which presumably confers altered DNA specificity and maintains potential binding to the Z_x protein. (C) A second possible consequence of the t(1;19) is a Prl-1/E2A fusion protein in which the b-HLH domain of E2A would be linked to an as-yet undefined part of Prl-1 (indicated by question mark), which again could have the normal binding with the putative HLH_x protein. A new transcriptional activation protein might therefore result.

come is that the genes to which the homeodomain usually binds are aberrantly controlled after chromosomal translocation, upsetting the balance of normal protein production in the cells with the translocation.

Figure 5 illustrates the complex potential effects on protein interactions after formation of chimeric transcription factors. For instance, E2A is a ubiquitously expressed b-HLH protein that is believed to form specific heterodimers with other b-HLH protein(s) to achieve high-affinity sequence-specific DNA binding (Murre et al., 1989b). It also carries a leucine zipper motif in the N-terminal part of the molecule; the leucine zipper is believed to function as an independent protein dimerization domain (Abel and Maniatis, 1989), sometimes associated with a basic region believed to bind to DNA (such as in c-*fos* and c-*jun*). A hypothetical complex is thus normally formed between the E2A protein and other b-HLH and leucine zipper proteins (Fig. 5A). In contrast, Prl-1 is expressed in a tissue-specific manner; a chimeric protein thus introduces a different DNA-binding specificity (normally spatially restricted) into a universally expressed transcription unit (Fig. 5B) as a consequence of the chromosome translocation t(1;19). Although not yet analyzed, another interesting consequence of the E2A/Prl-1 fusion could be the formation of the reciprocal protein (Fig. 5C). Here, the b-HLH domain is detached from the cognate ZIP domain, which may alter its DNA-binding specificity in the b-HLH heterodimer complex. It is thus formally possible that this alteration ("loosened" specificity of the DNA-binding domain of E2A) is a contributing factor to tumorigenesis. Clearly, the analysis of the *prl-1* gene structure is required before this possibility can be tested. Thus one protein may bind to homeobox sequences (probably adjacent to a specific set of developmentally regulated genes) and cause a disruptive effect to the tumor cell gene expression patterns and the other may bind to DNA sites via the b-HLH domain, perhaps as a normal heterodimer, but participate in the transcription process aberrantly because of the presence of the Prl-1 fragment.

A distinct, yet related process has been suggested to explain the involvement of the LIM domain oncogenes (Boehm et al., 1990a; Rabbitts and Boehm, 1990), possibly by causing disequilibrium in the availability (i.e., the concentrations) of DNA-binding factors. The first identified LIM domain oncogene was the *rhombotin* gene (Boehm et al., 1988a). *Rhombotin* has two LIM domains that account for most of the protein but has no identifiable DNA-binding motif (although it remains formally possible that DNA binding is mediated by the LIM domains themselves). However, other so-called LIM proteins have both LIM domains and homeodomains (presumed to be the DNA recognition part of the gene products) and transcriptional activation domains (Freyd et al., 1990; Karlsson et al., 1990; Way and Chalfie, 1988). The lack of any such recognizable DNA-binding motif in the rhombotin protein has led to the idea that the CRR or LIM domain may function in protein–protein interactions (Boehm et al., 1990a; Rabbitts and Boehm, 1990) rather than in direct DNA binding, thereby offering a means to modulate transcription. Thus LIM domains might form complexes with other LIM domains, either as homodimers or as heterodimers, and LIM-only proteins may function to form inactive (or even enhanced) transcription complexes (Fig. 6a). Precedents for such negative control of transcription have been described. In the process of myogenesis, the Id protein binds competitively to certain b-HLH proteins (such as MyoD), thereby inactivating their transcriptional activity (Benezra *et al.*, 1990). A similar role has been suggested for the *extramacrochaetae* gene of *Drosophila* (Ellis *et al.*, 1990; Garrell and Modolell, 1990); both *Id* and *extramacrochaetae* genes encode HLH proteins but lack the presumed DNA-binding basic domains (Fig. 6B). Such competitive activation or inactivation might therefore explain so-called dominant negative mutations (Hershkowitz, 1987).

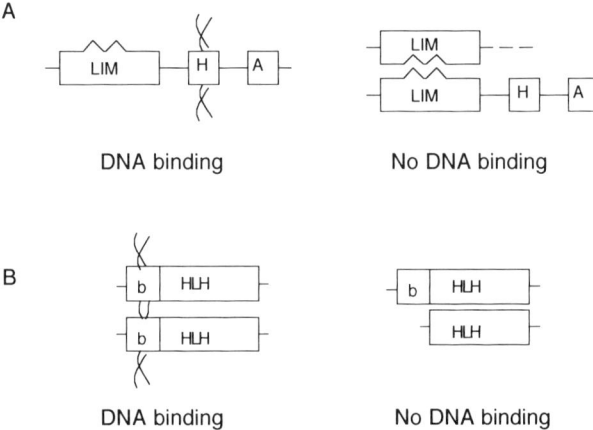

FIG. 6. Competitive protein dimerization in transcription regulation. Dimerization of proteins is a mechanism for negative regulation of transcription by functionally inactivating molecules involved in transcription. Models are presented for the operation of LIM (A) and HLH (B) domains in this negative regulation process. (A) The LIM proteins have been described with either LIM domains only, or LIM domains plus homeodomains (H) and activation (A) domains. The latter proteins presumably bind DNA via the homeodomain. Because proteins such as rhombotin only have LIM domains, it has been suggested that the LIM domain interacts with other LIM domains to produce complexes that are inactive (or conceivably more active) in transcription. Relative concentrations of the various components will thus clearly modulate the system. Thus variation in rhombotin levels, by the known chromosome translocation, for example, can disrupt the equilibrium. (B) Competitive protein interactions by HLH domains has been described, for example, for the Id protein, which has been shown to interact with proteins such as myoD, which controls myogenesis. Thus the interaction of the HLH of Id with myoD HLH produces an inactive complex that is incapable of transcriptional activation. (see text for specific references).

B. Developmentally Regulated Translocation Oncogenes

The majority of chromosomal translocations described here do not involve previously known oncogenes. One reason seems to be that these genes are involved in cellular differentiation processes, rather than cell division per se, which make their potential for acting in the transformation process limited (these might be termed "low-grade oncogenes" compared to the "high-grade" ones encoded by retroviruses). For example, it seems likely that the family of LIM domain genes [including both *rhombotin* and the 11p13 *rhom-2* LIM domain oncogenes (see above)] will have members with both LIM-only genes and LIM plus homeodomain-containing genes. A complex of interactions between such factors may explain some differentiating events both in thymus and other organ development, and explain the specific involvement of *rhombotin* and the *rhom-2* gene in T-ALL disease. Similar arguments apply to the t(1;19) E2A/Prl-1 fusion protein and to the other b-HLH proteins, such as those encoded by *tal/SCL/TCL5*. Particularly interesting will be studies on the target genes activated by the E2A/Prl-1 fusion protein because presumably at least some of these will be genes normally developmentally regulated via Prl-1 factor binding with its homeodomain. How inappropriate expression of those genes contributes to the tumor phenotype is an interesting question. Similarly interesting is the precise involvement of the *rhombotin* gene in tumorigenesis, particularly when viewed from the standpoint that activating translocations are found in T cell tumors, yet the gene is normally expressed in such disparate cells as postmitotic neurons of the central nervous system (Greenberg *et al.*, 1990; Boehm *et al.*, 1991b).

C. Other Unusual Proteins Affected by Chromosome Translocations

Not all the analyzed translocation genes are apparently transcription factors (Table I). Two gene products have recently been studied whose function has not been deduced as yet, but whose role in transcription (at least directly) is unlikely, viz. the Bcl-2 and Bcl-3 proteins affected by t(14;18)(q32;q21) and t(14;19)(q23;p13.3), respectively. The *bcl-2* gene has been widely studied in the B cell tumor, follicular lymphoma (Tsujimoto *et al.*, 1987; Cleary *et al.*, 1986; Bakhshi *et al.*, 1987). Translocation breakpoints are found both upstream and downstream of the gene (located on chromosome 18q21) in association with the Ig H chain J$_H$ region [implicating the VDJ recombinase in the mechanism of translocation (Tsujimoto *et al.*, 1987), but some without signal-sequence specificity (Bakhshi et al., 1987)]. The consequences of translocation on the gene are not clear. No dramatic mutations can be seen in the gene (Seto et al., 1988), but the consistent connection of the Ig H μ enhancer upstream or downstream of bcl-2 suggests enhancer control of bcl-2 transcription is the key feature after translocation (Table I). It now seems that the Bcl-2 protein is associated with the inner mitochondrial membrane (Hockenberry et al., 1990), which therefore is the first example of this category of cellular protein being involved in tumorigenesis. The biological characteristics of the bcl-2 gene and the consequences of "activation" of this gene in follicular lymphoma are of extreme interest, as this seems to belong to a class of gene that functions in the G_0 stage of cell cycle, somehow conferring the ability of cells to survive for long periods in culture (Vaux et al., 1988). Deregulated bcl-2 in transgenic mice induces a polyclonal expansion of small resting B cells (of IgM–IgD phenotype) that appear to have the propensity for prolonged cell survival (McDonnell et al., 1989). In addition, bcl-2 can cooperate with c-myc in doubly transgenic mice to produce hyperproliferation of pre-B and B cells, with resulting tumor development (Strasser et al., 1990). The mechanism of action of bcl-2 is still obscure, although recent experiments have indicated that bcl-2 blocks the apoptotic death of a pre-B lymphocyte cell line (Hockenberry et al., 1990).

The "activation" of the bcl-3 gene via the t(14;19) is probably also via Ig enhancer control (McKeithan et al., 1987, 1990). Although it is not easy to assess the significance of mRNA levels in tumor cells because the exact normal cell counterpart is not available, comparative Northern hybridization experiments suggest significantly higher levels of bcl-3 mRNA in translocation-bearing cell lines than in other lines (Ohno et al., 1990). This is consistent with enhancer control of the translocated gene. The two studied cases of this translocation involve the S_α region upstream of $C_{\alpha 1}$ (McKeithan et al., 1990), so the putative IgA enhancer discussed above for c-myc is implicated. The function of Bcl-3 protein is unknown. The protein sequence, however, has some interesting features, presumably related to its function. The central portion (sandwiched between proline-rich N-terminal and C-terminal domains) is composed of seven tandem copies of an ~35 amino acid motif, the so-called CDC-10 motif (Ohno et al., 1990), also found in cell cycle genes of yeast, and cell–cell interaction genes in Drosophila and Caenorhabditis elegans. It therefore appears that the CDC-10 motif is a versatile protein–protein interaction domain that is used by proteins of varying function.

VI. The Development of T Cell Leukemia: A Paradigm for Clonal Evolution and Tumor Development

The study of chromosomal abnormalities in human T cell leukemias has illuminated some of the features of tumor progression. The creation of the abnormal T cell clone with a chromosomal translocation occurs before or after the cell has entered the thymus usually as an early event, in many cases due to aberrant activity of the VDJ recombinase.

The contribution of the chromosome abnormality to the tumor phenotype may only be slight as, for example, in CLL. Translocations involving genes such as the b-HLH genes or the LIM domain oncogenes may affect transcription equilibria, whereas translocations involving genes such as *bcl-2* or *bcl-3* may produce cells whose life expectancy is enhanced relative to normal neighbors. In this respect, some chromosomal abnormalities may be analogous to the effect of Epstein–Barr virus (EBV) on B cells in Burkitt's lymphoma or adult T cell leukemia virus (ATLV) on T cells in adult T cell leukemia (ATL), viz. to produce a cell with an immortalized phenotype, and providing a target cell in which further damaging genetic changes occur.

The study of translocations and inversions in T cell tumors from patients with ataxia telangiectasia (AT) has provided some important insights into the conversion of preleukemia to overt leukemia. AT is a multisymptomatic autosomal recessive disorder in which patients have a propensity to the development of T cell leukemia carrying specific translocations or inversions. These include inv14(q11;q32) or t(14;14)(q11;q32), and more rarely, t(X;14)(q28;q11). All studied examples of these abnormalities have TCRα at the 14q11 breakpoints (Baer *et al.*, 1987b; Mengle-Gaw *et al.*, 1987, 1988; Davey *et al.*, 1988; Bertness *et al.*, 1990). [In addition, a T cell tumor in an AT patient was found to carry a translocation, t(7;14)(q35;q32), involving TCR J$_\beta$ (Russo *et al.*, 1988).] The 14q32 breakpoints have been found to be in the proximal part of the chromosome band [designated 14q32.1 (see above)] and are similar in non-AT T cell neoplasias and in AT-associated neoplasias (Kennaugh *et al.*, 1986; Baer *et al.*, 1987b). A study of two AT-afflicted sisters showed one with a nonleukemic clonal expansion defined by a t(14;14) in about 50–70% of peripheral T lymphocytes (Sherrington *et al.*, 1991). The breakpoint of the translocation t(14;14) occurs in the same region of 14q32.1 as those of leukemic patients. In addition, the size of the t(14;14) T cell clone within the nonleukemic patient's blood has been stable over 5 years. Because the clone of T cells has the same translocation as overt tumors, it is consistent with the idea of the preleukemic nature of the clone and supports the view that often the translocations will occur in preleukemia.

Because the chromosomal abnormalities are usually acquired early in thymic maturation of T cells, further normal differentiation is possible in which T cell receptor gene rearrangement (on the allelic chromosome) proceeds to produce a functional receptor on the surface. This allows the preleukemic T cells to escape the possiblity of negative selection in the thymus. Conversely, if any T cell with an acquired abnormality subsequently fails in thymus selection processes, this cell will be eliminated in the usual way and will not survive as a potential precursor of the overt tumor. Afflicted T cells that are positively selected and can leave the thymus form a pool of cells responsive to different antigens in the periphery and also a pool that is a target for subsequent mutations to occur.

Thus the early acquisition of T cell chromosome abnormalities, particularly in CLL, seems to be followed by a long period of accumulation of clonal derivatives. Presumably, the longevity of the initial translocation clone provides a greater target for secondary mutations, which may result in the emergence of the clinically significant malignant clone. If such genetic changes accumulate, an overt tumor may develop. It remains uncertain at the single-cell level whether this eventuality will occur, thus any given chromosome abnormality may never manifest itself during the life span of the afflicted individual. Translocations in ALL may, on the other hand, be more immediately deleterious.

References

Abel T., and Maniatis, T. (1989). *Nature (London)* 341, 24.
Adams, J. M., Harris, A. W., Pinkert, C. A., Corcoran, L. M., Alexander, W. S., Cory, S., Palmiter, R. D., and Brinster, R. L. (1985). *Nature (London)* 318, 533.
Alt, F. W., and Baltimore, D. (1982). *Proc. Natl. Acad. Sci. U.S.A.* 79, 4118.
Baer, R., Chen, K.-C., Smith, S. D., and Rabbitts, T. H. (1985). *Cell* 43, 705.
Baer, R., Forster, A., and Rabbitts, T. H. (1987a). *Cell* 50, 97.
Baer, R., Heppell, A., Taylor, A. M. R., Rabbitts, P. H., Boullier, B., and Rabbitts, T. H. (1987b). *Proc. Natl. Acad. Sci. U.S.A.* 84, 9069.
Bakhshi, A., Wright, J. J., Graninger, W., Seto, M., Owens, J., Cossman, J., Jensen, J. P., Goldman, P., and Korsmeyer, S. J. (1987). *Proc. Natl. Acad. Sci. U.S.A.* 84, 2396.
Begley, C. G., Aplan, P. D., Davey, M. P., Nakahara, K., Tchorz, K., Kurtzberg, J. Hershfield, M. S., Haynes, B. F., Cohen, D. I., Waldmann, T. A., and Kirsch, I. R. (1989a). *Proc. Natl. Acad. Sci. U.S.A.* 86, 2031.
Begley, C. G., Aplan, P. D., Denning, S. M., Haynes, B. F., Waldmann, T. A., and Kirsch, I. R. (1989b). *Proc. Natl. Acad. Sci. U.S.A.* 86, 10128.

Benezra, R., Davis, R. L., Lockshon, D., Turner, D. L., and Weintraub, H. (1990). *Cell* **61**, 49.
Bernard, O., Cory, S., Gerondakis, S., Webb, E., and Adams, J. M. (1983). *EMBO J.* **2**, 2375.
Bernard, O., Larsen, C.-J., Hampe, A., Mauchauffé, M., Berger, R., and Mathieu-Mahul, D. (1988). *Oncogene* **2**, 195.
Bernard, O., Guglielmi, P., Jonveaux, P., Cherif, D., Gisselbrecht, S., Mauchauffe, M., Berger, R., Larsen, C.-J., and Mathieu-Mahul, D. (1989). *Genes, Chromosomes, Cancer* **1**, 1.
Bertness, V. L., Felix, C. A., McBride, O. W., Morgan, R., Smith, S. D., Sandberg, A. A., and Kirsch, I. R. (1990). *Cancer Genet. Cytogenet.* **44**, 47.
Blackwell, T. K., Kretzner, L., Blackwood, E. M., Eisenman, R. N., and Weintraub, H. (1990). *Science* **250**, 1149.
Boehm, T., and Rabbitts, T. H. (1989a). *Eur. J. Biochem.* **185**, 1.
Boehm, T., and Rabbitts, T. H. (1989b). *FASEB J.* **3**, 2344.
Boehm, T., Baer, R., Lavenir, I., Forster, A., Waters, J. J., Nacheva, E., and Rabbitts, T. H. (1988a). *EMBO J.* **7**, 385.
Boehm, T., Buluwela, L., Williams, D., White, L., and Rabbitts, T. H. (1988b). *EMBO J.* **7**, 2011.
Boehm, T., Mengle-Gaw, L., Kees, U.R., Spurr, N., Lavenir, I., Forster, A., and Rabbitts, T. H. (1989). *EMBO J.* **8**, 2621.
Boehm, T., Foroni, L., Kennedy, M., and Rabbitts, T. H. (1990a). *Oncogene* **5**, 1103.
Boehm, T., Greenberg, J. M., Buluwela, L., Lavenir, I., Forster, A., and Rabbitts, T. H. (1990b). *EMBO J.* **9**, 857.
Boehm, T., Gonzalez-Sarmiento, R., Kennedy, M., and Rabbitts, T. H. (1991a). *Proc. Natl. Acad. Sci. U.S.A.*, **88**, 3927.
Boehm, T., Spillantini, M.-G., Sofroniew, M. V., Surani, M. A., and Rabbitts, T. H. (1991b). *Oncogene*, **6**, 695.
Boehm, T., Foroni, L., Kaneko, Y., Perutz, M. F., and Rabbitts, T. H. (1991c). *Proc. Natl. Acad. Sci. U.S.A.* **88**, 4367.
Borrow, J., Goddard, A. D., Sheer, D., and Soloman, E. (1990). *Science* **249**, 1577.
Brown, L., Cheng, J.-T., Chen, Q., Siciliano, M. J., Crist, W., Buchanan, G., and Baer, R. (1990). *EMBO J.* **9**, 3343.
Cesarman, E., Dalla-Favera, R., Bentley, D., and Groudine, M. (1987). *Science* **238**, 1272.
Champagne, E., Takihara, Y., Sagman, U., de Sousa, J., Burrow, S. Lewis, W. H., Mak, T. W., and Minden, M. D. (1989). *Blood* **73**, 1672.
Chen, Q., Cheng, J. T., Tsai, L.-H., Schneider, N., Buchanan, G., Carroll, A., Crist, W., Ozanne, B., Siciliano, M. J., and Baer, R. (1990a). *EMBO J.* **9**, 415.
Chen, Q., Yang, C. Y.-C., Tsan, J. T., Xia, Y., Ragab, A. H., Peiper, S. C., Carroll, A., and Baer, R. (1990b). *J. Exp. Med.* **172**, 1403.
Cheng, J.-T., Yang, C. Y.-C., Hernandez, J., Embrey, J., and Baer, R. (1990). *J. Exp. Med.* **171**, 489.
Cleary, M. L., Smith, S. D., and Sklar, J. (1986). *Cell* **47**, 19.
Cleary, M. L., Mellentin, J. D., Spies, J., and Smith, S. D. (1988). *J. Exp. Med.* **167**, 682.
Collum, R. G., and Alt, F. W. (1990). *Cancer Cells* **2**, 69.
Croce, C. M. (1987). *Cell* **49**, 155.
Croce, C. M., Thierfelder, W., Erikson, J., Nishikura, K., Finan, J., Lenoir, G. M., and Nowell, P. C. (1983). *Proc. Natl. Acad. Sci. U.S.A.* **80**, 6922.
Davey, M. P., Bertness, V., Nakahara, K., Johnson, J. P., McBride, O. W., Waldmann, T. A., and Kirsch, I. R. (1988). *Proc. Natl. Acad. Sci. U.S.A.* **85**, 9287.

Davis, M. M., and Bjorkman, P. J. (1988). *Nature (London)* **334**, 395.
Davis, M., Malcolm, S., and Rabbitts, T. H. (1984). *Nature (London)* **308**, 286.
deKlein, A., van Kessel, A. G., Grosveld, G., Bartram, C. R., Hagemijer, A., Bootsma, D., Spurr, N. K., Heisterkamp, N., Groffen, J., and Stephenson, J. R. (1982). *Nature (London)* **300**, 765.
de Thé, H., Chomienne, C., Lanotte, M., Degos, L., and Dejean, A. (1990). *Nature (London)* **347**, 558.
Denny, C. T., Hollis, G. F., Hecht, F., Morgan, R., Link, M. P., Smith, S. D., and Kirsch, L. R. (1986). *Science* **234**, 197.
Eccles, S., Sarner, N., Vidal, M., Cox, A., and Grosveld, F. (1990). *New Biologist* **2**, 801.
Ellis, H. M., Spann, D. R., and Posakony, J. W. (1990). *Cell* **61**, 27.
Erikson, J., Finger, L., Sun, L., Ar-Rushdi, A., Nishikura, K., Minowada, J., Finan, J., Emanuel, B. S., Nowell, P. C., and Croce, C. M. (1985). *Science* **232**, 884.
Finger, L. R., Harvey, R. C., Moore, R. C. A., Showe, L. C., and Croce, C. M. (1986). *Science* **234**, 982.
Finger, L. R., Kagan, J., Christopher, G., Kurtzberg, J., Hershfield, M. S., Nowell, P. C., and Croce, C. M. (1989). *Proc. Natl. Acad. Sci. U.S.A.* **86**, 5039.
Foroni, L., Boehm, T., Lampert, F., Kaneko, Y., Raimondi, S., and Rabbitts, T. H. (1990). *Genes, Chromosomes, and Cancer* **1**, 301.
Freyd, G., Kim, S. K., and Horvitz, H. R. (1990). *Nature (London)* **344**, 876.
Garrell, J., and Modolell, J. (1990). *Cell* **61**, 39.
Graham, M., and Adams, J. (1986). *EMBO J.* **5**, 2845.
Greenberg, J. M., Boehm, T., Sofroniew, M. V., Keynes, R. J., Barton, S. C., Norris, M. L., Surani, M. A., Spillantini, M.-G., and Rabbitts, T. H. (1990). *Nature (London)* **344**, 158.
Haluska, F. G., Finver, S., Tsujimoto, Y., and Croce, C. M. (1986). *Nature (London)* **324**, 158.
Hamlyn, P. H., and Rabbitts, T. H. (1983). *Nature (London)* **304**, 135.
Harvey, R. C., Marteneire, C., Sun, L. H. K., Williams, D., and Showe, L. C. (1989). *Oncogene* **4**, 341.
Henglein, B., Synovzik, H., Groitl, P., Bornkamm, G. W., Hartl, P., and Lipp, M. (1989). *Mol. Cell. Biol.* **9**, 2105.
Hershkowitz, I. (1987). *Nature (London)* **329**, 219.
Hockenberry, D., Nunez, G., Milliman, C., Schreiber, R. D., and Korsmeyer, S. J. (1990). *Nature (London)* **348**, 334.
Kagan, J., Finan, J., Letofsky, J., Besa, E. C., Nowell, P. C., and Croce, C. M. (1987). *Proc. Natl. Acad. Sci. U.S.A.* **84**, 4543.
Kagan, J., Finger, L. R., Letofsky, J., Finan, J., Nowell, P. C., and Croce, C. M. (1989). *Proc. Natl. Acad. Sci. U.S.A.* **86**, 4161.
Kamps, M. P., Murre, C., Sun, X.-h, and Baltimore, D. (1990). *Cell* **60**, 547.
Kaneko, Y., Frizzera, G., Shikano, T., Kobayashi, H., Maseki, N., and Sakurai, M. (1989). *Leukemia* **3**, 886.
Karlsson, O., Thor, S., Norberg, T., Ohlsson, H., and Edlund, T. (1990). *Nature (London)* **344**, 879.
Kennaugh, A. A., Butterworth, S. V., Hollis, R., Baer, R., Rabbitts, T. H., and Taylor, A. M. R. (1986). *Hum. Genet.* **73**, 254.
Kennedy, M., Boehm, T., Kees, U., Gonzalez-Sarmiento, R., and Rabbitts, T. H. (1991). Submitted.
Lasser, A. B., Buskin, J. N., Lockshon, D., Davis, R. L., Apone, S., Hauschka, S. D., and Weintraub, H. (1989). *Cell* **58**, 823.

Leder, P., Battey, J., Lenoir, G., Moulding, C., Murphy, W., Potter, H., Steward, T., and Taub, R. (1983). *Science* **222**, 765.
Leder, A., Pattengale, P. K., Kuo, A., Stewart, T. A., and Leder, P. (1986). *Cell* **45**, 485.
Lefranc, M.-P., Forster, A., Baer, R., Stinson, M. A., and Rabbitts, T. H. (1986). *Cell* **45**, 237.
Lu, M., Dubé, I., Raimondi, S., Carroll, A., Zhao, Y., Minden, M., and Sutherland, P. (1990). *Genes, Chromosomes, Cancer* **2**, 217.
McDonnell, T. J., Deane, N., Platt, F. M., Nunez, G., Jaeger, U., McKearn, J. P., and Korsmeyer, S. J. (1989). *Cell* **57**, 79.
McGuire, E. A., Hockett, R. D., Pollock, K. M., Bartholdi, M. F., O'Brien, S. O., and Korsmeyer, S. D. (1989). *Mol. Cell. Biol.* **9**, 2124.
McKeithan, T. W., Rowley, J. D., Shows, T. B., and Diaz, M. O. (1987). *Proc. Natl. Acad. Sci. U.S.A.* **84**, 9257.
McKeithan, T. W., Ohno, H., and Diaz, M. O. (1990). *Genes, Chromosomes & Cancer* **1**, 247.
Mellentin, J. D., Murre, C., Donlon, T. A., McCaw, P. S., Smith, S. D., Carroll, A. J., McDonald, M. E., Baltimore, D., and Cleary, M. L. (1989a). *Science* **246**, 379.
Mellentin, J. D., Smith, S. D., and Cleary, M. L. (1989b). *Cell* **58**, 77.
Mengle-Gaw, L., and Rabbitts, T. H. (1987). *EMBO J.* **6**, 1959.
Mengle-Gaw, L., Willard, H. F., Smith, C. I. E., Hammarström, L., Fischer, P., Sherrington, P., Lucas, G., Thompson, P. W., Baer, R., and Rabbitts, T. H. (1987). *EMBO J.* **6**, 2273.
Mengle-Gaw, L., Albertson, D. G., Sherrington, P. D., and Rabbitts, T. H. (1988). *Proc. Natl. Acad. Sci. U.S.A.* **85**, 9171.
Meyer, K. B., and Neuberger, M. S. (1989). *EMBO J.* **8**, 1959.
Morse, B., South, V. J., Rothberg, P. G., and Astrin, S. M. (1989). *Mol. Cell. Biol.* **9**, 74.
Murphy, W., Sarid, J., Taub, R., Vasicek, T., Battey, J., Lenoir, G., and Leder, P. (1986). *Proc. Natl. Acad. Sci. U.S.A.* **83**, 2939.
Murre, C., McCaw, P. S., and Baltimore, D. (1989a). *Cell* **56**, 777.
Murre, C., McCaw, P. S., Vaessin, H., Caudy, M., Jan, L. Y., Jan, Y. N., Cabrera, C. V., Buskin, J. N., Hauschika, S. D., Lassar, A. B., Weintraub, H., and Baltimore, D. (1989b). *Cell* **58**, 537.
Nourse, J., Mellentin, J. D., Galili, N., Wilkinson, J., Stanbridge, E., Smith, S. D., and Cleary, M. L. (1990). *Cell* **60**, 535.
Oettinger, M. A., Schatz, D. G., Gorka, C., and Baltimore, D. (1990). *Science* **245**, 1517.
Ohno, H., Takimoto, G., and McKeithan, T. W. (1990). *Cell* **60**, 991.
Pelicci, P.-G., Knowles, D. M., Magrath, I., and Dalla-Favera, R. (1986). *Proc. Natl. Acad. Sci. U.S.A.* **83**, 2984.
Pettersson, S., Cook, G. P., Brüggemann, M., Williams, G. T., and Neuberger, M. S. (1990). *Nature (London)* **344**, 165.
Rabbitts, T. H., and Boehm, T. (1990). *Nature (London)* **346**, 418.
Rabbitts, T. H., Hamlyn, P. H., and Baer, R. (1983a). *Nature (London)* **306**, 760.
Rabbitts, T. H., Forster, A., Baer, R., and Hamlyn, P. H. (1983b). *Nature (London)* **306**, 806.
Rabbitts, T. H., Baer, R., Davis, M., Forster, A., Hamlyn, P. H., and Malcolm, S. (1984a). *Curr. Top. Microbiol. Immunol.* **113**, 166.
Rabbitts, T. H., Forster, A., Hamlyn, P., and Baer, R. (1984b). *Nature (London)* **309**, 592.
Rabbitts, T. H., Baer, R., Buluwela, L., Mengle-Gaw, L., Taylor, A. M., and Rabbitts, P. H. (1986). *Cold Spring Harbor Symp. Quant. Biol.* **LI**, 923.

Reynolds, T. C., Smith, S. D., and Sklar, J. (1987). *Cell* **50**, 107.
Royer-Pokora, B., Fleischer, B., Ragg, S., Loos, U., and Williams, D. (1989). *Hum. Genet.* **82**, 264.
Russo, G., Isobe, M., Pegoraro, L., Finan, J., Nowell, P. C., and Croce, C. M. (1988). *Cell* **53**, 137.
Sanchez-Garcia, I., Kaneko, Y., Gonzalez-Sarmiento, R., Campbell, K., White, L., Boehm, T., and Rabbitts, T. H. (1991). *Oncogene* **6**, 577.
Schatz, D. G., Oettinger, M. A., and Baltimore, D. (1989). *Cell* **59**, 1035.
Seto, M., Jaeger, U., Hockett, R. D., Graninger, W., Bennett, S., Goldman, P., and Korsmeyer, S. J. (1988). *EMBO J.* **7**, 123.
Sherrington, P., Thick, J., Taylor, M., and Rabbitts, T. H. (1991). Submitted.
Shima, E. A., LeBeau, M. M., McKeithan, T. W., Minowada, J., Showe, L. C., Mak, T. W., Minden, M. D., and Rowley, J. D. (1986). *Proc. Natl. Acad. Sci. U.S.A.* **83**, 3439.
Showe, L. C., Ballantine, M., Nishikura, K., Erikson, J., Kaji, H., and Croce, C. M. (1985). *Mol. Cell. Biol.* **5**, 501.
Stern, M.-H., Lipkowitz, S., Aurias, A., Griscelli, C., Thomas, G., and Kirsch, I. R. (1989). *Blood* **74**, 2076.
Strasser, A., Harris, A. W., Bath, M. L., and Cory, S. (1990). *Nature (London)* **348**, 331.
Strominger, J. L. (1989). *Science* **244**, 943.
Taub, R., Kirsch, I., Morton, C., Lenoir, G., Swan, D., Tronick, S., Aaronson, S., and Leder, P. (1982). *Proc. Natl. Acad. Sci. U.S.A.* **79**, 7837.
Taub, R., Kelly, K., Battey, J., Latt, S., Lenoir, G. M., Tantravahi, U., Tu, Z., and Leder, P. (1984). *Cell* **37**, 511.
Tighe, L., Forster, A., Clark, D. M., Boylston, A. W., Lavenir, I., and Rabbitts, T. H. (1987). *Eur. J. Immunol.* **17**, 1729.
Tonegawa, S. (1983). *Nature (London)* **302**, 575.
Tsujimoto, Y., Gorham, J., Cossman, J., Jaffe, E., and Croce, C. M. (1985a). *Science* **229**, 1390.
Tsujimoto, Y., Jaffe, E., Cossman, J., Gorham, J., Nowell, P. C., and Croce, C. M. (1985b). *Nature (London)* **315**, 340.
Tsujimoto, Y., Bashir, M. M., Givol, I., Cossman, J., Jaffe, E., and Croce, C. M. (1987). *Proc. Natl. Acad. Sci. U.S.A.* **84**, 1329.
Vaux, D. L., Cory, S., and Adams, J. M. (1988). *Nature (London)* **335**, 440.
von Lindern, M., Poustka, A., Lerach, H., and Grosveld, G. (1990). *Mol. Cell. Biol.* **10**, 4016.
Way, J. C., and Chalfie, M. (1988). *Cell* **54**, 5.
Williams, D. L., Look, A. T., Melvin, S. L., Roberson, P. K., Dahl, G., Flake, T., and Stass, S. (1984). *Cell* **36**, 101.
Yoffe, G., Schneider, N., Van Dyk, L., Yang, C. Y.-C., Siciliano, M., Buchanan, G., Capra, J. D., and Baer, R. (1989). *Blood* **74**, 374.
Zutter, M., Hockett, R. D., Roberts, C. W. M., McGuire, E. A., Bloomstone, J., Morton, C. C., Deaven, L. L., Crist, W. M., Carroll, A. J., and Korsmeyer, S. J. (1990). *Proc. Natl. Acad. Sci. U.S.A.* **87**, 3161.

This article was accepted for publication on 4 March 1991.

Interleukin-2, Autotolerance, and Autoimmunity

GUIDO KROEMER, JOSÉ LUIS ANDREU, JOSÉ ANGEL GONZALO, JOSÉ C. GUTIERREZ-RAMOS*, AND CARLOS MARTÍNEZ-A.

Centro de Biología Molecular (CSIC), Universidad Autónoma de Madrid, Campus de Cantoblanco, 20049 Madrid, Spain

Insomma se il modello non riesce a trasformare la realtá, la realtá dovrebbe riuscire a trasformare il modello. . . . Se le cose stanno cosí, il modello dei modelli dovrá servire a ottenere dei modelli trasparenti, diafani, sottili come ragnatele; magari addirittura a dissolvere i modelli, anzi a dissolversi.

(Italo Calvino, 1983)

In summary, if the model does not succeed in transforming reality, reality must succeed in transforming the model. . . . If this is how things stand, the model of models must serve to achieve transparent models, diaphanous, subtle as spider webs, or perhaps even to dissolve models, or indeed to dissolve itself.

I. Introduction

During the past decade, an ever increasing number of cytokines has been molecularly cloned and investigated with respect to their involvement in physiological and pathological processes. Cytokines are polypeptide mediators released by mobile or sessile cells into the microenvironment; they orchestrate the complex processes of inflammation, immune reaction, and hematopoiesis (Arai *et al.*, 1990). It may be anticipated that the continuous progress in cytokine and cytokine receptor biochemistry will allow intervention in a variety of hematological and immunological disorders via therapeutic administration of recombinant or synthetic cytokine agonists and antagonists, thus revolutionizing medicine in the near future as did the advent of antibiotics during the 1940s and 1950s. The interleukin-2 (IL-2) and interleukin-2 receptor (IL-2R) system is probably the best characterized of these lymphocytotrophic hormones, both in terms of molecular biochemistry and function (Möller, 1986; Smith, 1988). Within the peripheral immune system, production of endogenous IL-2 is confined to a restricted population of $CD4^+$ T lymphocytes, mainly unprimed (T_{H0}) and cells of the inflammatory (T_{H1}) phenotype. Secretion of IL-2 probably reflects a commitment event in T cell activation and is the final outcome of various converging second-messenger pathways that integrate signals from multiple cell surface receptors linking the T cell with the microenvironment. Antigen-specific, clonal proliferation of

---
* Present address: Basel Institute for Immunology, CH-4005 Basel, Switzerland.

peripheral T lymphocytes is initiated through a process of signal transduction, wherein the specific interaction of antigen/major histocompatibility complex (MHC) molecules and the CD3/T cell receptor (TCR) complex plus CD4 triggers the expression of IL-2 and its homologous receptor (IL-2R). Subsequent autocrine IL-2/IL-2R interaction allows the T cell to undergo proliferation. In addition, IL-2 exerts multiple, pleiotropic effects via specific cell surface receptors expressed on a wide array of immunologically relevant cells, including intrathymic pro-T, T, B, and natural killer (NK) cells and null lymphocytes and cells of the monocyte/macrophage series. IL-2 thus occupies a central position in the physiology of the immune system and plays a pivotal role in the function of T, B, and NK effector cells, as well as in thymocyte maturation. By virtue of its immunomodulatory effects, it has attracted the attention of clinical immunologists and immunopathologists. Lymphocytes treated *ex vivo* with human recombinant IL-2 (rIL-2), as well as IL-2 per se, are used clinically in the treatment of solid and disseminated cancer, and the application of rIL-2 in the therapy of immunodeficiencies and viral infections is currently under investigation. Similarly, IL-2 derivatives of monoclonal antibodies (mAbs) targeted to the IL-2R are tested with respect to their capacity to eradicate IL-2R-bearing tumor cells and to intervene in the IL-2/IL-2R system to obtain immunosuppressive effects.

Abnormalities in the endogenous production of IL-2, *in vitro* inducibility of IL-2 secretion, expression of IL-2 receptors, or IL-2 responsiveness have been reported for a variety of disorders, including acquired and inherited immunodeficiencies, infections, cancer, transplantation crisis, graft-versus-host disease, and age-associated immune dysfunctions. This survey will focus on alterations concerning the IL-2/IL-2R system in autoimmune disease, its role in the maintenance and abrogation of autotolerance, as well as the possibility to induce or mitigate autoimmune reactions by specific interventions in the IL-2/IL-2R pathway.

The notion that manifest autoimmunity has a multifactorial and polygenetic etiology in which several genetic defects, eventually combined with environmental insults, intervene as predisposing or etiological factors is the cornerstone of our present concepts of diverse diseases such as insulin-dependent diabetes mellitus (IDDM), multiple sclerosis, rheumatoid arthritis (RA), or scleroderma (Shoenfeld and Schwarts, 1984; Kroemer *et al.*, 1990). Although it is beyond the scope of this review to enumerate exhaustively all defects implicated in the etiopathogenesis of autoimmune diseases, it should be remembered that the catalog of putative autoimmunity-inducing factors is extremely heterogeneous, ranging from certain MHC haplotypes to TCR locus alleles, genetically determined defects in T cell maturation, endogenous viruses, endocrine defects, target organ defects, and viral and bacterial infections (Möller 1990). In support of this view, experimental autoimmunity can only be induced in certain, susceptible strains and virtually all animal models developing spontaneous autoimmune disease have been shown to rely on a polygenetic basis. Moreover, the genetics of human autoimmune diseases appears to be extremely complex, and "susceptibility" loci linked to the development of certain autoimmune diseases, e.g., certain MHC alleles, are only identified by statistical operations showing that they augment the relative risk to develop disease. The notion of a plurietiological basis of manifest autoimmunity has theoretical as well as practical implications. First, this concept may be reconciled with the idea that immunological self-tolerance relies on several complementary mechanisms arranged in a fail-safe hierarchy. Only in the case of failure of several of the "safety valves" that normally maintain the *horror autotoxicus* (e.g., clonal deletion, peripheral tolerance, suppressive circuits, and idiotypic network) does autoaggression erupt. Second, it becomes conceivable that a given autoimmune manifestation does not have exactly the same etiology in different individuals, but constitutes the final outcome of distinct, partially overlapping combinations of predisposing factors and/or external stimuli. Accordingly, for example, development of systemic lupus erythematosus (SLE) or diabetes mellitus in various animal strains has rather different genetic bases, and human autoimmune diseases could be aggregates of clinical symptoms rather than single nosological entities.

In view of these conceptual considerations, it is important to evaluate carefully the effects of IL-2 on each mechanism that normally ensures tolerance and precludes autoaggression. Moreover, it is of great relevance whether abnormalities in the IL-2/IL-2R system reflect a level of immune dysfunction common to individuals suffering from similar or related autoimmune symptoms, and whether such imbalances in the IL-2/IL-2R system contribute as primary coetiological factors to the initiation of autoimmunity, or, on the contrary, are secondary events possibly involved in the self-perpetuation of autoaggressive inflammatory processes.

II. Physiology of Interleukin-2 and Its Receptor

A. Mechanics of the IL-2/IL-2R System

Interleukin-2 (formerly T cell growth factor, TCGF) is a polypeptidic mediator described in every mammalian species (Gillis *et al.*, 1978) investigated in this respect, as well as in *Gallus domesticus*

(Schauenstein and Kroemer, 1987), *Xenopus laevis*, and *Ciprinius carpi* (Watkins and Cohen, 1985). The components of the IL-2/IL-2R system of the mouse and humans are sufficiently homologous to allow cross-species receptor–ligand associations and bioactivity (Doi *et al.*, 1989; Yamaguchi *et al.*, 1989). IL-2 has been cloned both on the cDNA and genomic level and both murine and human IL-2 are available as recombinant products (Kashima *et al.*, 1985; Taniguchi *et al.*, 1983). Human IL-2 is a 15-kDa glycoprotein that exhibits a certain biochemical heterogeneity due to variable glycosylation and sialylation of the peptide backbone. IL-2 has a carbohydrate-binding (lectin) domain with specificity for high-mannose glycopeptides, which may play a critical role in the clearance and intracellular routing of this molecule (Sherblom *et al.*, 1989). IL-2 contains one disulfide bridge whose disruption destroys biological activity (Yamada *et al.*, 1987). According to crystallographic data, the core structure of IL-2 is made up of six antiparallel α helices (designated A to F) and contains no segments of β secondary structure (Brandhuber *et al.*, 1987). The topographical relation among individual helices and functional domains responsible for receptor binding has been partially elucidated. Attempts are under way to modify the IL-2 molecule to improve its biological effectiveness or to change its pharmacokinetics. Thus, substitution of certain amino acids by site-directed mutagenesis gives rise to superagonists or antagonists of IL-2 (Zurawski *et al.*, 1990), and attachment of poly(ethylene glycol) enhances the serum half-life of the molecule (Katre, 1990).

One of the particular features of the IL-2 receptor is the existence of at least two distinct receptor components designated IL-2Rβ (heavy chain, p70–75), a member of the cytokine receptor superfamily (Bazan, 1990), and IL-2Rα (light chain, p55). According to the cluster of differentiation (CD) nomenclature, the IL-2Rα chain is designated CD25. In primates it is also termed Tac. The minimal IL-2R capable of signal transduction contains the IL-2Rβ glycoprotein, which binds its polypeptidic ligand with intermediate affinity ($K_d \sim 10^{-9}$ M). The high-affinity IL-2R ($K_d \sim 10^{-11}$ M) is a heterodimer composed of the p75 β chain and the inducible low-affinity ($K_d \sim 10^{-8}$ M) p55 IL-2Rα chain (reviewed by Smith, 1989). IL-2 internalization and signal transduction are mediated by both high- and intermediate-affinity forms of IL-2R, but not by low-affinity IL-2R. The membrane half-life of unoccupied high-affinity receptors is 150 minutes, whereas subsequent to IL-2 binding these receptors disappear 10 times more rapidly due to IL-2Rβ-mediated internalization of the receptor–ligand complex. IL-2Rα chains manifest a much lower turnover rate of ≥6 hours, whether or not they are occupied by IL-2 (Smith, 1988). This probably contributes to the accumulation of IL-2Rα on activated T cells. When present in excess, the α chain rapidly associates with IL-2 (Smith, 1988) and may serve to pass IL-2 to preexisting α/β heterodimers (Saragovi and Malek, 1990) or isolated β chains by lateral diffusion, thus forming a ternary IL-2/IL-2Rα/β complex (Kamio et al., 1990). Although the IL-2Rα has no signaling function, its presence on the cell surface augments the IL-2 sensitivity, i.e., it shifts the dose–response curve for IL-2 to lower concentrations by virtue of its ability to augment the receptor affinity. Replacement of the short cytoplasmic (13 amino acids) and transmembrane regions of IL-2Rα with the corresponding regions of the insulin receptor does not affect its function, indicating that the high-affinity IL-2R is generated by noncovalent association of both chains solely at the extracellular regions (Hatakeyama et al., 1987). Substitution of Asp 20 of human IL-2 or deletion of Phe 124 results in a molecule that is unable to interact with IL-2Rβ, but retains the ability to bind to IL-Rα (Collins et al., 1988). Thus, different regions of the IL-2 molecule are recognized by the two receptor components. Probably, the amino-terminal helix A interacts with the IL-2Rβ, and helices B and E react with IL-2Rα (Zuraski et al., 1990).

In addition to the two IL-2R constituents described above, several groups have postulated the existence of further IL-2R components (reviewed by Waldmann, 1989). Zurawski et al. (1990) reported that mouse IL-2 molecules mutated in Gln 141 (located in the carboxy-terminal α helix F) bind with reduced affinity to T cells expressing the high-affinity IL-2 receptor, yet bind normally to transfected fibroblasts expressing only the α and β chains of the receptor, thus suggesting the existence of a third receptor component specifically expressed in T cells. Transfection experiments with a human β chain cDNA in both fibroblasts and T cells resulted in functional β chain expression only in the T lymphocytes (Hatakeyama et al., 1989a), indicating that additional T cell-specific IL-2R subunits may be critical to IL-2-mediated signal transduction. In mice, p22, p40 (Saragovi and Malek, 1990), and p100 subunits (Sharon et al., 1990) have been described. In humans, a putative p64 IL-R component has been tentatively named IL-2Rγ chain (Takeshita et al., 1990). A p70 molecule probably is a precursor of IL-2Rβ (p75) that lacks posttranslational processing such as N-linked glycosylation and sialylation (Asao et al., 1990a). Another p70–75 molecule that is recognized by the mitogenic antibody YTA-1 appears to be physically associated with IL-2Rβ (Sugie et al., 1990). Intercellular adhesion molecule-1 (ICAM-1; CD54, p95) has been shown to interact physically with IL-2Rα within the cell membrane (Burton et al., 1990) and may speculatively focus IL-2 receptors to the site of homotopic T–T interactions mediated by ICAM-1 and its ligand LFA-1 (CD11a/CD18).

To date, only limited information is available concerning the mechanism of IL-2R-mediated signal transduction, although mutational studies revealed that a restricted cytoplasmic region of the IL-2Rβ chain proximal to the inner surface of the cell membrane is essential for triggering cell proliferation. Signal transduction may be abrogated in truncated IL-2Rβ chain mutants without affecting the ligand internalization function of IL-2Rβ, indicating that the internalization process may not couple directly with the intracellular signaling pathway (Hatakeyama et al., 1989b). The large (286 amino acids) cytoplasmic domain of IL-2Rβ lacks an obvious tyrosine kinase domain. IL-2 induces tyrosine phosphorylation of IL-2Rβ within minutes (Asao et al., 1990b) and stimulates phosphorylation of unidentified substrates on serine/threonine or tyrosine residues (Ferris et al., 1989). Protein kinase C, phosphatidylinositol turnover, calcium mobilization, and cyclic nucleotidyl monophosphates do not participate in IL-2 signaling (Mills et al., 1986, 1988; Tigges et al., 1989). Additional signals transduced by unknown second-messenger systems appear to be involved (Tigges et al., 1989). In cloned T helper clones, IL-2 induces within 20 hours transcription of genes encoding proteins involved in ribosomal function (e.g., elongation factor-2 and ribosomal phosphoprotein P1), various glycolytic enzymes, cytoskeletal proteins, unidentified proteins, and a product homologous to DNA-binding proteins that bind to the Y box of the class II MHC promoter and the epidermal growth factor receptor enhancer (Sabath et al., 1990). Moreover, it induces rapid (within 1 hour) transcription of the genes encoding the transferrin receptor and cyclin, a co-factor of DNA polymerase-δ (Smith, 1989), as well as of the cellular oncogenes c-*fos*, c-*myc*, c-*pim*, and c-*myb* (Dautry et al., 1988). These latter products may be involved in mediating the effect of IL-2 on the transition from G_1 to the S phase of the cell cycle.

B. Detection and Distribution of Cells Producing IL-2 or Expressing IL-2R

IL-2 has been initially quantitated by measuring its biological effects. Most IL-2 bioassays measure the growth-promoting effect of logarithmic dilutions of IL-2-containing culture supernatants on IL-2-dependent cell lines. After a short-term culture period (2–3 days), either DNA synthesis is measured by assessing the incorporation of radiolabeled thymidine and thymidine analogs (Gillis *et al.*, 1978) or mitochondrial activity is quantitated by determining the metabolization of tetrazolium derivatives giving rise to colored cleavage products (Mosmann, 1983). The dose–response curves are compared to a standard curve to allow the expression of IL-2 activity in multiples of the dose that induces half-maximal stimulation of cell growth (ED_{50}) or arbitrarily defined units. Nonetheless, the disadvantages of such bioassays are considerable: limited sensitivity, long duration, and low specificity due to responses to other stimulatory lymphokines or inhibitory factors contained in culture supernatants. Recently, enzyme-linked immunoassays based on the use of specific monoclonal antibodies have become commercially available. In addition, IL-2 gene transcription may be measured by Northern blots. More sensitive assays that allow for the quantitation of IL-2 mRNA based on the polymerase chain reaction have been described (Wang *et al.*, 1989; Murray *et al.*, 1990). This latter method is suitable for determining IL-2 production by *ex vivo* explanted cells that have not been stimulated by *in vitro* culture. For the estimation of frequencies of IL-2-producing cells, either limiting dilution assays or *in situ* hybridization techniques are available (McGuire and Rothenberg, 1987). Furthermore, reverse hemolytic assays (Lewis *et al.*, 1990) and spot enzyme-linked immunosorbent assays (ELISAs) (Skidmore *et al.*, 1989) have been developed to determine production of lymphokines by individual cells.

IL-2 is produced by T cell precursors at a circumscript developmental stage (after intrathymic immigration, before acquisition of cell surface expression of CD4 and/or CD8). Among peripheral T cells, significant IL-2 secretion appears to be confined to activated $CD4^+$ T helper cells of the unprimed (T_{H0}) or inflammatory type (T_{H1}). In the murine system, "naive" T cells that only produce IL-2 differentiate *in vitro* to either T_{H1} or T_{H2} (helper type) cells that markedly differ in their lymphokine secretion pattern. T_{H1} clones produce IL-2, interferon-γ (IFN-γ), granulocyte/macrophage colony-stimulating factor (GM-CSF), IL-3, tumor necrosis factor-β (TNF-β, formerly lymphotoxin), and TNF-α (Mosmann and Coffman, 1989). T_{H1}-type cells are implicated in delayed-type hypersensitivity, cytotoxicity, and polyclonal B cell responses. In contrast, T_{H2} clones secrete GM-CSF, IL-3, IL-4, IL-5, IL-6, and IL-10, but not IL-2 or IFN-γ. T_{H2}-type cells are required for the induction of IgE responses and provide help to polyclonal and specific antibody responses. T_{H1} and T_{H2} cells also differ in their ability to act as accessory cells, in their accessory cell requirements, and in the presence of IL-1 receptors. Abortive IL-2 gene transcription occurs in T_{H2} cells only upon stimulation in the presence of cycloheximide (Muñoz et al., 1989). Some but not all CD8$^+$ cell clones produce low levels of IL-2 (McGuire et al., 1988). Based on Northern blot analyses, low IL-2 gene expression has been equally reported for B cells that have been stimulated with a combination of phorbol ester and calcium ionophore (Taira et al., 1987).

IL-2R components are evaluated by incubating cells with radiolabeled recombinant IL-2, measuring the bound radioactivity, and Scatchard analysis of the data (reviewed by Smith, 1988). Moreover, increasing amounts of mAbs directed against the two receptor components (Tanaka et al., 1985; Takeshita et al., 1989; Suzuki et al., 1989) permit the determination of IL-2R components on individual cells using cytofluorometric or cytohistochemical methods. Soluble IL-2Rα chains are quantitated by means of commercially available ELISAs.

Whereas IL-2 production is restricted to few lymphocyte subsets, components of the IL-2R are found on a wide array of immunologically relevant cells. IL-2Rβ is constitutively expressed on quiescent monocytes/macrophages (Espinoza-Delgado et al., 1990) and on cells of the lymphoid lineage, including lymphoid cell precursors present in the thymus (J. C. Gutierrez, unpublished observation), mature resting CD4$^+$ or CD8$^+$ T cells (Nishi et al., 1988; Yagita et al., 1989), and natural killer cells (Siegel et al., 1987). Heterodimeric IL-2R$\alpha\beta$ is constitutively expressed on the CD16$^-$ NK subset (Caligiuri et al., 1990). Triggering via the CD3/T cell receptor complex induces IL-Rα expression by peripheral blood T cells, and stimulation via a variety of different receptors induces functional IL-2R in intrathymic T cell precursors, lymph-borne dendritic cells, oligodendrocytes, B lymphocytes, CD16$^+$ NK cells, mast cells, and cells of the macrophage/monocyte series, including hepatic Kupffer's cells and Langerhans' cells from the skin. In addition, IL-2Rα chains have been detected on mouse pituitary cells by immunochemical methods (Smith, L. R., et al., 1989).

C. IL-2 AND IL-2R GENE REGULATION

In peripheral T cells, IL-2 induction depends on a series of requirements, including specific, i.e., TCR-mediated, and nonspecific signals. The antigen, recognized via the TCR, has to be presented in the form of peptides bound to the antigen-binding groove of the class II major histocompatibility complex on the surface of an antigen-presenting cell (APC). These peptides and the MHC class II molecule are corecognized by the TCR and CD4. In addition, the APC has to deliver nonspecific, so-called costimulatory signals to the T cell, the nature of which has thus far not been elucidated (Fig. 1). If one of these conditions is not fulfilled, either IL-2 induction does not take place or, on the contrary, the IL-2 gene is silenced and may be reactivated only after a certain lag period (reviewed by Jenkins et al., 1987; Mueller et al., 1989; Schwartz, 1990). Triggering of the TCR/CD3–CD4 complex via an appropriate class II–antigenic peptide combination is coupled to phosphoinositol hydrolysis and Ca^{2+} mobilization (reviewed by Möller, 1987; Ashwell and Klausner, 1990; Altman et al., 1990). Nonetheless, increases in intracellular calcium concentration $[Ca_i^{2+}]$ and protein kinase C (PKC) activation by themselves are suboptimal second messengers in the induction of IL-2 transcription, and further, unknown signaling pathways appear to be involved (June et al., 1989; Mueller et al., 1989). Inversely related to the strength of the primary signal delivered via the TCR, secondary signals are required to allow maximal IL-2 production. Such agonistic signals could be mediated by a growing list of activating molecules on the T cell surface, such as certain lymphokine receptors (e.g., for IL-1 and IL-6) (Klarnet et al., 1989; Vink et al., 1990), the transmembrane proteins CD1, CD2 (via LFA-3 interaction), CD5, CD28 (via interaction with the B cell activation antigen B7/BB-1), CD69, Tp45, Tp90, Tp103, Tp135–145, and the transferrin receptor, and, in the mouse, the glycosylphosphatidylinositol-anchored activation antigens Thy-1 and TAP (Ly6, CD59) (reviewed by Geppert et al., 1990; Altman et al., 1990). Two biochemical pathways of IL-2 regulation may be distinguished, one that acts posttranscriptionally by increasing the half-life of IL-2 mRNA in a cyclosporin A-resistant fashion and may be triggered by CD28 stimulation, another in a cyclosporin A-sensitive fashion that operates on the transcriptional level and involves increases in $[Ca_i^{2+}]$ and PKC activation (June et al., 1989).

A highly cooperative interaction between various nuclear factors, each of which follows different activation schedules, has to occur to allow activity of the enhancer of the IL-2 and IL-2Rα genes (reviewed by Muegge and Durum, 1989; Crabtree, 1989; Ullman et al., 1990; Fig. 1). More than five sequence motifs to which functionally relevant nuclear proteins bind have been identified in the 5' flanking region of the IL-2 gene, namely, an NFAT-1-binding site, an NF-κB-like motif, two AP-1-like binding sites, and a site for NF-IL-2A. Two enhancers serve as TCR-responsive elements, one which is bound by a factor specifically induced in T-activated T cells (NFAT-1), the other by the constitutively produced factor NF-IL-2A, an Oct-1-like factor that probably suppresses IL-2 transcription in resting T cells and must be modified or cooperate with other binding factors to allow activity of the IL-2 promoter. Protein kinase C activation allows NF-κB to be re-

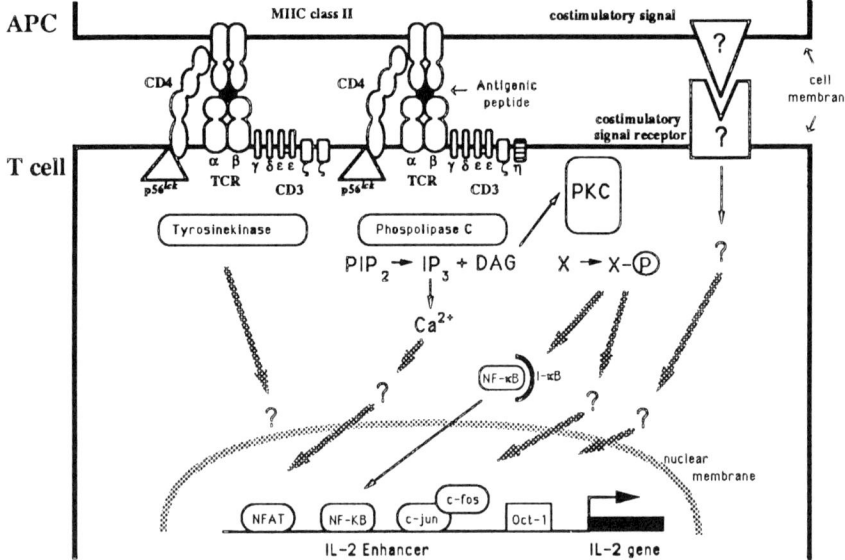

FIG. 1. Activation of the IL-2 gene in T_{H1} cells. The antigenic peptide is bound to a groove of the class II molecule created by the two amino-terminal domains of the heterodimeric MHC structure on the antigen-presenting cell (APC). The MHC in conjunction with antigen is recognized by the two polymorphic variable regions of the α and β chains of the T cell receptor (TCR)/CD3 complex plus the coreceptor CD4, giving rise to a ternary CD4–TCR/CD3–class II complex. Cooperative binding stabilizes the TCR/CD3 association with its ligand antigen/MHC (trans-interaction) and furthers the TCR/CD3 cis-interaction with CD4. On the inner leaflet of the membrane, CD4 associated with p56^{lck}, now brought into the vicinity of the TCR/CD3 complex, allows phosphorylation of the CD3 ζ (and possibly γ and ε chains) on tyrosine residues. Among different types of CD3 complexes, the $\zeta\eta$ heterodimer appears to be responsible for coupling TCR occupancy to activation of phospholipase C that catalyzes the hydrolysis of phosphatidylinositol-4,5-biphosphate (PIP$_2$) into 1,2-sn-diacyglycerol (DAG) and inositol-1,4,5-triphosphate (IP$_3$) (Berridge and Irvine, 1989). The ζ_2 homodimer may couple to activation of tyrosine kinases, including p56^{lck} associated with CD4 and other members of the src encoded family of tyrosine kinases, e.g., Fyn and/or other signaling pathways (Ashwell and Klausner, 1990). At this step, phosphatases (e.g., CD45) may intervene in an interactive network of protein tyrosine phosphorylation and dephosphorylation. IP$_3$ raises [Ca$_i^{2+}$] by opening calcium channels in the plasma membrane and releasing Ca^{2+} ions from sequestered intracellular stores (Gardner, 1989). DAG activates a serine/threonine kinase protein kinase C (PKC) isozyme (Nishizuka, 1986) that depends on the presence of [Ca$_i^{2+}$] and phospholipid and phosphorylates the CD3 γ chain among other yet unidentified intracellular substrates (X). Costimulatory signals are only provided by APCs, depend on direct cell contact, and theoretically may operate by direct functional and/or physical interaction with the TCR/CD3 complex, via increasing the interaction between T lymphocyte and antigen-presenting cells, or by its own second-messenger cascade (Mueller et al., 1989). In the absence of a costimulatory signal, long-term inactivation of IL-2 gene transcription follows TCR occupancy (Schwartz, 1990). Candi-

leased from an inhibitory binding molecule (I-κB) present in the cytoplasm and to translocate to the nucleus (Lenardo et al., 1989) and induces c-Fos, which interacts with c-Jun (AP-1) via leucine zippers, leading to the formation of a complex that binds to the phorbol-responsive element with an affinity 300-fold higher than does c-Jun alone (Muegge et al., 1989). c-Jun is specifically induced by IL-1. Analysis of deletion mutants suggests that all the protein-binding sites must be occupied to allow activity of the IL-2 enhancer. Signals that activate only one of the enhancer elements will not give rise to IL-2 transcription.

The promoter region of the IL-2Rα gene consists of a minimum of five positive regulatory elements and at least one negative element (reviewed by Muegge and Durum, 1989). As does the IL-2 gene, the IL-2Rα promoter contains an NF-κB-like element, but other regulatory sequence motifs (UE-1, SP-1, serum response element) are not found in the IL-2 gene. The fact that IL-2 and IL-2Rα share at least one regulatory element (Hoyos et al., 1989) may explain why these two gene products are often coexpressed. Unlike the IL-2 gene, which is strictly dependent on triggers from the antigen receptor plus accessory signals, IL-2Rα is expressed after activation with IL-1, IL-5, phorbol myristate acetate (PMA), or binding of ligands to the TCR alone, without a requirement of second signals. Phorbol 12-myristate 13-acetate, the transactivator protein (Tax) from the type I human T cell leukemia virus (HTLV-I), and tumor necrosis factor induce nuclear proteins that interact with a κB-like sequence (Lowenthal et al., 1989). IL-1 and IL-2 induction of murine IL-2Rα transcription depends on a segment in the 5' flanking region upstream of such cis-acting regulatory elements (Plaetinck et al., 1990). In contrast to IL-2, for IL-2Rα transcription a signal from the antigen receptor is sufficient to activate expression in T cells.

---

date costimulatory signal receptors include CD2, LFA-1, CD28, and cytokine receptors (see text). Via undefined signaling events, these pathways modify and activate the ubiquitous Oct-1 (NF-IL-2A) factor Durand et al., 1988) and induce the transcription of early-activating genes such as c-fos (Muegge et al., 1989) or NFAT-1, whose product is specific for T cells (Shaw et al., 1988). Protein kinase C modifies a cytoplasmic inhibitor of NF-κB (I-κB), which releases NF-κB, allowing it to move to the nucleus (Hoyos et al., 1989; Lenardo and Baltimore, 1988). These and other factors bind to their respective transcription-regulating elements (e.g., c-Fos binds to the phorbol ester-responsive element in conjunction with c-Jun), and cooperatively enhance IL-2 gene transcription (Ullman et al., 1990). In addition, IL-2 mRNA stability is subjected to regulation (not shown).

D. IL-2 AND THYMOCYTE DIFFERENTIATION

Early during ontogeny, the IL-2/IL-2R system operates in an antigen-nonspecific fashion in T cell development, i.e., during the process in which precursor cells recruited from fetal liver or bone marrow acquire T cell-specific surface structures in an ordered process in the thymus. In the course of the quantitatively most important pathway of intrathymic differentiation, a pro-T cell (i.e., the earliest intrathymic precursor that still lacks surface expression of the CD3/TCR complex, CD1, CD2, CD4, and CD8) passes to the $CD2^+$ pre-T cell stage ($CD2^+$ CD^{3-} CD^{4-} CD^{8-}), then acquires surface expression of the α/βTCR/CD3 complex. Subsequently, it coexpresses the coreceptor molecules CD4 and CD8 ("double-positive stage"), and finally loses either CD4 or CD8, thereby presenting a mature "single-positive" phenotype (reviewed by Toribio, 1988; Strominger, 1989). IL-2 appears to be involved in thymocyte maturation in an obligatory fashion. IL-2Rα blockade by mAbs during fetal life precludes development of mature ($CD3^+$ $CD4^+$ $CD8^-$ or $CD3^+$ $CD4^-$ $CD8^+$) T cells in mice (Tentori et al., 1988) and, in the postnatal phase, inhibits thymic regeneration after sublethal irradiation (Zuñiga-Pflücker and Kruisbeek, 1990). Furthermore, the intrathymic expression of the human IL-2Rα in transgenic mice results in the accumulation of T cell precursors (Gutierrez-Ramos et al., 1989b). Incubation of fetal thymus organ cultures with anti-IL-2Rα reduces cell recovery, including the number of immature ($CD3^-$ $CD4^-$ $CD8^-$) T cells (Jenkinson et al., 1987; Zuñiga-Pflücker et al., 1990). In contrast, suspension cultures of murine and human fetal thymocytes exhibit IL-2-promoted proliferation in vitro (Hardt et al., 1985; Toribio et al., 1988, 1989). Upon in vitro culture with IL-2, pro-T cells cocultured with stromal cells acquire the surface markers of immature thymocytes (predominant phenotype: α/βTCR$^+$ $CD3^+$ $CD4^-$ $CD8^-$) in a stepwise process (Toribio et al., 1988).

In the fetal thymus, when thymocyte ontogeny is synchronized due to the invasion of the thymic anlage by T cell precursors in waves, clear-cut peaks in IL-2 and IL-2Rα expression are observed (Shimonkevitz et al., 1987; Jenkinson et al., 1987; Carding et al., 1989). Intrathymic IL-2Rα expression follows an analogous kinetics and distribution in the mammalian and chicken embryo, indicating the conservation of a principal thymocyte differentiation program over large evolutionary intervals (Fedecka-Brunner et al., 1991). IL-Rα is expressed at a circumscript precursor stage, preceding the loss of the capacity to reconstitute a thymus and marking the initiation of a burst of cell proliferation and of TCR rearrangement and expression (Plearse *et al.*, 1989; Egerton *et al.*, 1990). Constitutive IL-2Rβ expression is already found on human thymocyte precursors committed to either the NK or T cell lineage, but these cells lack IL-2 and IL-2Rα gene transcripts (Toribio *et al.*, 1989). After migration into the thymus, it is probably the interaction with thymic stromal cells that induces IL-2 and IL-2Rα expression on pro-T cells, thus allowing formation of high-affinity L-2Rαβ and autocrine proliferation via the IL-2/IL-2R pathway (de la Hera *et al.*, 1989a; Toribio *et al.*, 1989). On differentiating thymocytes, a panel of monomorphic surface molecules (e.g., CD2, CD7, CD28, and CD45R, the receptors for IL-1, IL-3, IL-9, TNF, granulocyte/macrophage colony-stimulating factor, and soluble CD23) are candidates for mediating nonspecific triggering of IL-2 production upon interaction with stromal cells by direct cell contact and/or soluble products (reviewed by Martínez-A., 1990). IL-2 itself has the capacity of stimulating IL-2Rα expression via the IL-2Rβ (Toribio *et al.*, 1989). At the pro-T cell stage, the IL-2/IL-2R pathway operates throughout ontogeny—although such cells are outnumbered during the neonatal period by more mature populations—promoting polyclonal antigen-nonspecific expansion and surface expression of differentiation markers such as CD2, CD3, CD4, and CD8 (Jenkinson *et al.*, 1987; Toribio *et al.*, 1988, 1989; Zuñiga-Pflücker *et al.*, 1990).

After TCR gene rearrangement, when commitment to the T cell lineage is ensured, cellular activation and the IL-2/IL-2R pathway are subjected to TCR-controlled regulation. The inducibility of both IL-2 and IL-2Rα transcription critically depends on the maturation stage of the thymocyte. Thus, the bulk of TCR/CD3$^+$ cells, especially of the CD4$^+$ CD8$^+$ stage, which is subjected to clonal selection processes, may not be stimulated to express IL-2 or IL-2Rα (Rothenberg *et al.*, 1988; Boyer *et al.*, 1989). The IL-2/IL-2R pathway may be stimulated again in mature (CD3$^+$ CD4$^+$ CD8$^-$ or CD3$^+$ CD4$^-$ CD8$^+$) thymocytes, although the induction requirements differ from extrathymic T lymphocytes (Boyer *et al.*, 1989) and the signaling requirements for IL-2 induction appear to shift during cell differentiation. In contrast to peripheral T cells, thymic T cell precursors may not be activated to produce IL-2 with a combination of Ca^{2+} ionophore plus phorbol ester, but require additional IL-1 stimulation (Rothenberg *et al.*, 1988).

IL-2 and IL-4 skew the quantitative outcome of human pro-T cell cultures differentiating *in vitro* toward a predominance of lymphocytes bearing α/β and γ/δ TCRs, respectively (Bárcena *et al.*, 1990, 1991a). The IL-4-triggered proliferation and differentiation of γ/δ T cell precursors is not blocked by the anti-IL-2Rα antibodies and does not require thymocyte–stroma cell interactions. However, it is inhibited by IL-2, suggesting that the relative concentrations of different lymphokines present in the thymic microenvironment may ultimately favor the preponderance of α/β or γ/δ T cells (Bárcena et al., 1991b).

E. Effects of IL-2 on the Peripheral Immune System and IL-2 Toxicity

In peripheral immune cells, the IL-2/IL-2R system is embedded into an interactive network in which multiple factors regulate IL-2 and IL-2R expression, are influenced in their secretion by IL-2, or exert postreceptive antagonistic and agonistic effects on IL-2-responsive cells (Table I). IL-2 receptors are distributed over a wide spectrum of immunologically relevant cells (see Section IB). By consequence, IL-2 may exert multiple, pleiotropic effects. In activated B cells, IL-2 directly favors switching the μ heavy-chain mRNA transcription from the membranous to the secretory form, which, together with the induction of J chain transcription, ultimately results in the secretion and assembly of pentameric immunoglobulin M (Nakanishi et al., 1984; Tigges et al., 1989). Monocytes are stimulated by IL-2 in their cytotoxic function (Malkovsky et al., 1987), and NK cells, both in their growth and in their effector function (Trinchieri et al., 1984). Practically all T cell subsets that display the high-affinity IL-2R$\alpha\beta$ after activation respond to IL-2 by undergoing cell division (Smith, 1988). As an exception to this rule,

TABLE I
Position of IL-2 in the Cytokine Network[a]

Property	Cytokine or cytokine receptor
Induced by IL-2 in vivo	IL-6, IFN-γ, TNF-α
Induced by IL-2 in vitro	IL-1α, IL-1β, IL-5, IL-6, IFN-γ, TNF-α, TNF-β, GM-CSF
Induces IL-2 production	IL-1, IL-6, IL-7
Inhibits IL-2 production	IL-10
Induces IL-2Rα	IL-1, IL-5, IL-7, IFN-γ, TNF-α
Down-regulates IL-2Rα	IL-4
Inhibits IL-2 action	IL-4, TGF-β
Synergizes with IL-2	IL-6, IL-7, TNF-α (thymocyte proliferation)

[a] The table summarizes interactions discussed throughout the test and is not intended to encompass all interactions between the IL-2/IL-2R system in the cytokine network. Moreover, neglected here are interactions between the IL-2/IL-2R system with peptide mediators classified as hormones (endorphins, ACTH, CRF, etc.) and only one predominant effect per lymphokine is listed, which may differ depending on the cell type and the activation schedule studied.

some HTLV-I-transformed T cell lines that express IL-2Rβ or into which β chain cDNA has been transfected (Sugamura *et al.*, 1985; Tsudo *et al.*, 1989) respond to IL-2 with a negative signal, i.e., inhibition of DNA synthesis. Because of the heterogeneity of T lymphocytes, high doses of exogenous IL-2 may stimulate a panel of distinct, in part even antagonizing, immune functions. Thus, IL-2 potentiates the function of cytotoxic T effector cells (Kern *et al.*, 1981; Fearon *et al.*, 1990), stimulates or down-regulates T cell help provided to humoral effector cells (Stötter *et al.*, 1980; Kennedy *et al.*, 1987), enhances suppressor cell function (Holda *et al.*, 1986; Hirohata *et al.*, 1989), and elicits the production of other lymphokines (IL-6, IFN-γ, and TNF-α) (Heslop *et al.*, 1989; Kasid *et al.*, 1989), which, in turn, may exert pleiotropic effects *in vivo*. It is the quantity and activation state of individual T cell subsets in a mixed population that determine the net outcome of IL-2 stimulation. Accordingly, depending on the experimental model, IL-2 may exert parodoxical, stimulatory as well as inhibitory effects on *in vivo* immune responses, such as graft-versus-host reaction (Sykes *et al.*, 1990) and delayed-type hypersensitivity (Hancock *et al.*, 1987; Kradin *et al.*, 1989). It will be discussed later that exogenous IL-2 may either induce autoimmune disease or have beneficial effects on spontaneously developing systemic autoimmunity.

Nonetheless, IL-2 appears to have predominantly immunostimulatory properties *in vivo*. This notion is supported by two lines of evidence. On one hand, antibodies directed against the inducible IL-2Rα chain have immunosuppressive, but never stimulatory, effects. Anti-IL-2Rα antibodies inhibit delayed-type hypersensitivity reactions (Diamantstein *et al.*, 1988) and suppress allograft reactions both in rodents (Kirkman *et al.*, 1985; Granstein *et al.*, 1986) and in humans (Soulillou *et al.*, 1987). Clinical trials have shown that the human steroid-resistant acute graft-versus-host reaction responds to xenogeneic anti-IL-2Rα mAbs (Hervé *et al.*, 1990). On the other hand, a variety of different states of immunodepression are associated with an infraphysiological IL-2 production. Inherited states of immunodeficiency such as severe combined immunodeficiency (Weinberg and Parkman, 1990; Disanto *et al.*, 1990), Nezelof syndrome, Bruton's agammaglobulinemia, and Wiskott–Aldrich syndrome are characterized by a defective IL-2 secretion from peripheral blood lymphocytes (López-Botet *et al.*, 1982; Flomenberg *et al.*, 1983). Similarly, acquired immunodeficiencies provoked by total lymphoid irradiation (Bass and Strober, 1990), thermal injury (Teodorczyk-Injeyan *et al.*, 1986), graft-versus-host reactions (Welte *et al.*, 1984), therapy with cyclosporin A, glucocorticoids, or cyclophosphamide (Merluzzi *et al.*, 1983), old age (Thomen and Weigle, 1981; Alex-Martínez *et al.*, 1988), parasitosis (Harel-Bellan *et al.*, 1983), solid or disseminated tumors at advanced stages (Koch *et al.*, 1984; Kay and Kaplan, 1986), monoclonal gammopathies (Massaia *et al.*, 1990), or infection with human immunodeficiency virus (HIV) (Wainberg *et al.*, 1983) are accompanied by defects in IL-2 secretion. In many of these states of immunodeficiency, IL-2 ameliorates *in vitro* parameters of cellular immune function (e.g., mitogen responsiveness and cytotoxic function). In humans, IL-2 has thus been proposed for the therapy of severe combined immunodeficiency (SCID) (Pahwa *et al.*, 1989) and AIDS (McElrath *et al.*, 1990). In AIDS patients, local administration of recombinant IL-2 acts systematically, stimulating immunity to recall antigens, and thus may be of potential benefit in the defense against opportunistic pathogens encountered in HIV infection. Local IL-2 has also been reported to reduce the local parasite load in patients with lepromatous leprosy (Kaplan *et al.*, 1989) or disseminated cutaneous leishmaniasis (Akuffo *et al.*, 1990). In experimental models it improves resistance to a wide range of pathogens, e.g., herpes simplex virus (Kohl *et al.*, 1989), *Mycobacterium avium* (Bermudez *et al.*, 1989), *Klebsiella pneumoniae* (Iizawa *et al.*, 1988), and septicemia-causing *Escherichia coli* (Goronzy *et al.*, 1989). In immunodeficient nonresponders to hepatitis B vaccination, low-dose IL-2 induces systemic immune responses against hepatitis B surface (HBs) antigens (Meuer *et al.*, 1989). After autologous bone marrow transplantation, IL-2 infusions activate peripheral T cell reconstitution in patients (Blaise *et al.*, 1990).

One of the most spectacular and clinically exploitable effects of IL-2 is its capacity to render the function of $CD8^+$ cytotoxic T cells (CTLs) independent of $CD4^+$ T cell help (Fearon *et al.*, 1990) and to enhance CTL-mediated tumor regression (Rosenberg *et al.*, 1987). This effect is observable in very different protocols. In animal models, a wide-ranging panel of disseminated or solid cancers regresses when peripheral lymphocytes and tumor-infiltrating lymphocytes are treated with IL-2 *ex vivo* prior to reinjection, when IL-2 is adminstered *in vivo* in form of the recombinant protein, or when the IL-2 gene is transfected into tumor cells. Recombinant IL-2 is now widely used in the experimental treatment ("adoptive immunotherapy") of human advanced melanoma and renal cell carcinoma, as well as a variety of additional neoplasias that do not respond to conventional surgical, radiological, and cytostatic therapy. In patients, the optimal effective dose of IL-2 is very close to the maximum tolerable dose, and IL-2 has a number of toxic side effects, including fever, diarrhea, diffuse erythroderma, thrombocytopenia, anemia, eosinophilia, prerenal azotemia, hyperbilirubinemia, neuropsychiatric symptoms, and a capillary leak syndrome characterized by peripheral and pulmonary edema, hypotension, oliguria, and left ventricular dysfunction (Lotze et al., 1986; Ognibene et al., 1988; Huang et al., 1990). Of these toxicities, the cardiovascular side effects are dose limiting. The vascular leakage syndrome involves a systemic activation of the complement system (Thijs et al., 1990) and could be the consequence of TNF secretion elicited by IL-2, because passive immunization against TNF partially abrogates this syndrome in mice (Fraker et al., 1989). Interestingly, an idiopathic systemic capillary leak syndrome spontaneously developing in man also is associated with hyperactivation of the IL-2/IL-2R system (Cicardi et al., 1990). The IL-2 treatment-associated esosinophilia is probably mediated by IL-5, whose production is elicted by IL-2 (Yamaguchi et al., 1990; Macdonald et al., 1990). IL-2 initiates metabolic responses associated with critical illness, including increases in the serum concentrations of adenocorticotropic hormone (ACTH), cortisol, and catecholamines (Michie et al., 1988). Bacteremia with nonopportunistic pathogens, another unusually frequent complication of IL-2 treatment, may be attributed to an acute, profound, and reversible defect in neutrophil chemotaxis (Klempner et al., 1990). Another frequent side-effect of IL-2 is the induction of autoimmune phenomena (*vide infra*).

F. COMPARTMENTALIZATION REDUCES PLEIOTROPY OF IL-2

In view of its critical role in the regulation of peripheral tolerance (see Section III,C) and its multiple, pleiotropic effects (Paul, 1989), the function of IL-2 must be tightly controlled to avoid systemic effects that would endanger the specificity of an immune response. Accordingly, various mechanisms guarantee compartmentalization of IL-2, i.e., chronological and topographical restriction of IL-2 production, bioavailability, and responsiveness, thus avoiding systemic ("endocrine") IL-2 effects, at least under physiological conditions (reviewed by Kroemer et al., 1991).

1. IL-2 Production

As discussed above, secretion of IL-2 and expression of the high-affinity IL-2 receptor are developmentally controlled during intrathymic T cell differentiation. In the periphery, IL-2 is produced only by a defined population of mature T lymphocytes in which the IL-2 gene is transcribed or silenced depending on the combination of antigenic and nonspecific activation signals to which the cell is exposed. IL-2 production is inhibited by a series of mediators produced during inflammation. Glucocorticoids elicited in the course of systemic immune responses selectively inhibit IL-2 production by T_{H1} cells (Daynes and Araneo, 1989). Prostaglandins locally released from macrophages inhibit IL-2 secretion by elevating intracellular cAMP levels (Tilden and Balch, 1982; Minakuchi et al., 1990). T_{H1} and T_{H2} cells exhibit reciprocal negative regulation—T_{H2} cells by secreting IL-10 ("cytokine synthesis inhibitory factor" (Moore et al., 1990), which down-regulates IL-2 production, and T_{H1} cells via IFN-γ, which inhibits T_{H2} growth.

2. IL-2 Bioavailability

Under normal circumstances, IL-2 acts either as an autocrine growth factor (Meuer et al., 1984) (e.g., T_{H1} cells and pro-T cells) or in a paracrine fashion, where specific T_H cells release IL-2-containing vesicles in apposition to the contact site with the interacting cell (Sitkovsky and Paul, 1988). The signals that induce IL-2 gene transcription and its mRNA are short-lived. Similarly, the half-life of IL-2 is short (Mier and Gallo, 1982). IL-2 is rapidly and specifically cleared by the kidney (Donohue and Rosenberg, 1983). Moreover, it is antagonized by soluble inhibitors (Kucharz and Goodwin, 1988) and is subjected to proteolytic processes. Finally, IL-2 is "consumed" by IL-2Rβ-bearing cells, which may serve as a "lymphokine sink" (Lo et al., 1989).

3. IL-2 Responsiveness

High-affinity IL-2R expression is activation dependent in most cell types. Accelerated ligand-mediated internalization of the intermediate or high-affinity (β or $\alpha\beta$) IL-2R (Robb and Greene, 1987; Hatakeyama et al., 1987) provides an intrinsic control over antigen-initiated, but IL-2-dependent, T cell clonal expansion and favors the return to the state of resting, unresponsive cells (Smith, 1988). Chronic stimulation of murine T_{H1} (but not T_{H2}) clones with high IL-2 doses induces a refractory period, during which TCR-triggered phosphatidylinositol cleavage and Ca^{2+} influx are inhibited (Schell and Fitch, 1989). The T_{H2} product IL-4 exerts antagonizing effects on IL-2-driven human B lymphocyte differentiation, as well as on B (Tigges et al., 1989), T (Martinez et al., 1990), and NK cell proliferation (Nagler et al., 1988). IL-4 triggers down-regulation of both the IL-2R chains on human B lymphocytes and on αCD3-stimulated peripheral T cells (Martinez et al., 1990). Occupancy of the TCR at the time that IL-2 interacts with the IL-2R blocks T_{H1} cells and cytotoxic T lymphocytes from entering the S phase (Nau *et al.*, 1987; Williams *et al.*, 1990). In addition, the IL-2-dependent growth of T cells is specifically regulated by the extent of aggregation of CD4/CD8 with the TCR (Jönsson *et al.*, 1989). By consequence, any T lymphocyte whose TCR recognizes structures other than MHC (or the right class of MHC molecules—class I for CD8, class II for CD4) will primarily engage in recognition events leading to its own inactivation. A special case of this type of interaction may be the putative recognition of idiotypes among T cells or T and B cells, perhaps accounting for the suppressed state of normal heterogeneous lymphocyte populations (Bandeira *et al.*, 1987).

III. IL-2 and Autotolerance

A. ANTITOLERANCE EFFECT OF IL-2 *in Vivo*

In many experimental systems, IL-2 reverts immunological tolerance (nonresponsiveness). IL-2 abrogates immune tolerance in privileged sites such as the eye and the uterus. Alloantigens placed into the anterior chamber of the eye elicit an antigen-specific suppression of delayed-type hypersensitivity responses. Systemic administration of IL-2 during the first 72 hours of alloantigen implantation abrogates this tolerization process (Niederkorn, 1987). IL-2 inoculated after mating, but before fetal implantation, reduces pregnancy viability in mice, a phenomenon probably involving immunological mechanisms (Tezabwala *et al.*, 1989). In mice that exhibit an MHC-controlled immunological low responsiveness to sperm whale myoglobulin, IL-2 raises the antibody response to levels similar to those observed in high-responder mice (Kawamura *et al.*, 1985). Similarly, the long-lasting state of tolerance induced by the intravenous injection into neonatal CBA mice of lymphoid cells from (CBA × C57BL)F_1 hybrids is brought to an end by exogenous IL-2 (Malkovsky *et al.*, 1985). From the above-mentioned findings it appears clear that IL-2 may abrogate immunological tolerance, at least in certain situations. Malkovsky and Medawar (1984) formulated a theory according to which tolerance is established in conditions of IL-2 paucity. However, in 1984, the mechanisms responsible for the maintenance of autotolerance were more hazy than at present, and no attempts were made to implicate IL-2 in a particular tolerization mechanism.

According to current understanding, immunological self-tolerance relies on several complementary mechanisms arranged in a fail-safe hierarchy. There are basically three mechanisms for T cells to avoid autoimmunity associated with self-recognition. First, the bulk of T cells expressing T cell receptors with high affinity for self-structures succumb to programmed cell death in the thymus. The physical elimination of such developing T cells is called "clonal deletion." Second, autoaggressive T cells that have not been eliminated by clonal deletion are thought to be functionally inactivated upon contact with specific antigen in the thymus (central anergy) or in the periphery (peripheral tolerance, or anergy) (Blackman et al., 1990; Ramsdell and Fowlkes, 1990); i.e., they are activated in a way that does not permit a subsequent response to antigen exposure. Third, self-reactive T cells that enter the periphery could be functionally inactivated by other T cells or by antibodies (immunosuppression). The idiotypic–antiidiotypic network and suppressor cell circuits are thought to intervene at further levels of autotolerance preservation. In the following paragraphs it will be discussed whether IL-2 interferes with clonal deletion or anergization processes.

B. PUTATIVE INTERFERENCE OF IL-2 WITH CLONAL DELETION

As discussed in Section II,C, the inducibility and induction requirements of IL-2 and IL-2Rα expression change during thymocyte development. Speculatively, an interruption of the IL-2/IL-2R pathway may be related to programmed cell death occurring during intrathymic T cell selection. At the double-positive stage, when clonal deletion is generally thought to occur, IL-2 and IL-2Rα are not expressed on the great majority of cells, nor is their expression inducible. Only the fraction of double-positive cells that lacks CD1 expression (about 5% of $CD4^+$ $CD8^+$ mouse thymocytes) responds to IL-2 in vitro (de la Hera et al., 1989b). Recently, it has been reported that a stromal cell clone with nursing activity supported the growth and differentiation of murine $CD4^+$ $CD8^+$ thymocytes in response to IL-2 (Nishimura et al., 1990). It remains elusive whether those $CD3^+$ $CD4^+$ $CD8^+$ cells that survive positive and negative selection (approximately 1% of all thymocytes) express IL-2 and/or the IL-2R. Interestingly, CD3 cross-linking, which is thought to mimic the signal leading to clonal deletion, inhibits CD2-initiated IL-2 (not IL-2Rα) expression in human thymocytes, but not in peripheral blood lymphocytes (Ramarli et al., 1987).

It has been shown that IL-2 may interfere with apoptotic T cell death in various experimental systems. IL-2 protects murine CTLL-2 cells from glucocorticoid-induced oligonucleosome-length DNA fragmentation (apoptosis) and cell death (Nieto and López-Rivas, 1989). Similarly, IL-2 protects human medullary $CD1^-$ single-positive ($CD4^+$

CD8⁻ or CD4⁻ CD8⁺) thymocytes from anti-CD3-induced apoptotic death (Nieto *et al.*, 1990). In contrast, incubation of total murine thymocytes with IL-2 fails to reduce anti-CD3-triggered DNA fragmentation (Tadakuma *et al.*, 1990). Culture of immature CD3⁺ CD4⁻ CD8⁻ thymocytes from different mouse strains in the presence of IL-1 and IL-2 does not abrogate strain differences in the selection of the V_β repertoire, but modifies this repertoire toward a more mature and more selected pattern (Papiernik and Pontoux, 1990), thus providing indirect evidence against an interference with clonal deletion. The observation that the TCR-triggered suicide process of T cell hybridomas is paralleled by IL-2 production just prior to apoptosis further argues against the possibility that IL-2 might exert a general rescue effect from apoptotic T cell death (Odaka *et al.*, 1990).

C. THE IL-2/IL-2R SYSTEM—A SERVOMODULATOR REGULATING FUNCTIONAL TOLERANCE

Anergization (i.e., acquisition of antigen nonresponsiveness) of T lymphocytes may become manifest in different ways, and various phenotypes of anergic, tolerized T cells have been described: (1) T cells that have lost the capacity of transcribing the IL-2 gene but respond to exogenous IL-2 (Schwartz, 1990); (2) T lymphocytes that have lost IL-2 responsiveness (Gajewski *et al.*, 1989; Williams *et al.*, 1990); (3) T cells that down-regulate the expression of the coreceptor molecule CD8 in the mouse (von Boehmer, 1990), e.g., cells expressing a potentially autoreactive transgenic TCR recognizing male-specific antigen and that have not been clonally eliminated; and (4) T cells that express subnormal densities of the TCR after *in vitro* tolerization with high antigen doses in the human system (Lamb *et al.*, 1987).

As reviewed by Schwartz (1990), silencing of the IL-2 gene T_{H0} or T_{H1} cells results from the stimulation of the α/βTCR/CD3 complex plus CD4 by class II-bound antigen in the absence of one or several yet-unknown costimulatory signals normally delivered by antigen-presenting cells. *In vitro* stimulation of T cells in a sterile setting (e.g., with MHC molecules incorporated into liposomes, monoclonal antibodies to the CD3ε chain coated to plastic surfaces, macrophages that have lost their costimulatory function due to chemical fixation, antigen exposed by "nonprofessional" antigen-presenting cells such as keratinocytes, or IFN-γ-stimulated MHC class II⁺ fibroblasts) induces a long-lasting antigen-nonresponsive state characterized by a selective lack of IL-2 production, but nearly normal IL-3 or IFN-γ secretion and intact IL-2Rα expression and IL-2 responsiveness. The induction of anergy is mimicked by a short rise in $[Ca_i^{2+}]$, provided that protein kinase C activity is low, and is blocked by chelation of calcium ions, cyclophosphamide, and inhibition of protein synthesis by cycloheximide (Jenkins et al., 1987; Mueller et al., 1989). The costimulatory signal that prevents clonal anergization is provided by "professional" APCs (B cells, macrophages, and monocytes), cannot be replaced by APC culture supernatants, and requires direct intercellular interactions (see Fig. 1). At present, it is not known whether structures facilitating APC–T cell interactions or delivering signals after binding to a receptor on the T_{H1} cell operate as costimulants. The presence of IL-2 during tolerization does not prevent the development of anergy (Jenkins et al., 1987). Nevertheless, once established, the anergic state may be reversed by stimulation of cells with high IL-2 concentrations, suggesting that it is maintained by a stable regulatory factor that is diluted out with multiple rounds of cell division (Schwartz, 1990).

In vivo evidence in favor of selective long-lasting silencing of the IL-2 gene as a mechanism of clonal anergy is provided by transgenic mice that express MHC-encoded molecules under the control of the rat insulin promoter. This promoter allows expression of a given transgene exclusively in pancreatic β cells. Hyperexpression of syngeneic or allogeneic class I or abnormal expression of class II antigens in pancreatic islets does not induce autoimmune insulitis, but entails a nonautoimmune form of diabetes, probably as a consequence of metabolic perturbations that provoke the degeneration of Langerhans' cells (reviewed by Burkly et al., 1990). In mice lacking endogenous class II (I-E) expression, ectopic expression of class II (I-E) molecules induces I-E nonreactivity of both thymic and peripheral T cells (Lo et al., 1989). Nevertheless, T cells that express products of V_β gene families specific for as-yet unidentified antigens presented by I-E molecules ($V_\beta 5$ and $V_\beta 17a$) are not deleted. Such T cells fail to produce IL-2 after cross-linking with $V_\beta 5$- and $V_\beta 17a$-specific antibodies, whereas control cells from nontransgenic animals secrete IL-2 (Burkly et al., 1989). When an insulin-promoter H-$2K^b$ (class I) transgene is expressed on a MHC disparate (H-$2K^d$) background, a mere peripheral T cell tolerance develops. Splenocytes, not thymocytes, fail to kill H-$2K^b$-bearing targets. Exogenous IL-2 reverses the unresponsiveness of these spleen cells *in vitro* (Morahan et al., 1989).

In addition to the transgenic models, selective silencing of the IL-2 gene occurs in nondeletional tolerance induced *in vivo* by different protocols. If mice carrying the *b* allele of the minor lymphocyte stimulating locus (Mls-1^b) are injected intravenously with Mls-1^a splenocytes, the host lymphocytes lose their capacity to respond to Mls-1^a stimulator cells *in vitro*. Nevertheless, the percentage of potentially

Mls-1^a-reactive cells (e.g., those cells that utilize genes from the $V_\beta 6$ TCR gene family) does not decrease. The anergic $V_\beta 6^+$ CD4$^+$cell population fails to produce IL-2 (but expresses IL-2R) *in vitro* upon stimulation, in marked contrast to $V_\beta 6^+$ cells from naive mice (Rammensee *et al.*, 1989). Injection of the superantigen *Staphylococcus* enterotoxin B (SEB), which preferentially stimulates T cells carrying the $V_\beta 8$ gene product, results in selective anergy of $V_\beta 8^+$ cells that fail to secrete IL-2 upon *in vitro* exposure to SEB, whereas splenocytes from unprimed controls do so (Kawabe and Ochi, 1990). Jones *et al.* (1990) have recently shown that T cells from neonatally thymectomized (neoTx) mice that bear a potentially autoreactive repertoire (e.g., I-E-reactive TCR $V_\beta 11^+$ cells in I-E-bearing mice) fail to produce IL-2 in response to appropriate stimulators.

Two examples illustrate that intrathymic induction of clonal anergy also results in selective down-regulation of IL-2 production. First, a transgenic γ/δTCR that specifically recognizes the *b* allele of a MHC class-I-related (TL) product (TL^b) has been expressed in mice that carry either a TL^b or TL^d background. Transgenic TCR$^+$ thymocytes are not deleted in TL^b mice in spite of their self-reactivity. In contrast to transgenic TCR$^+$ cells from TL^d mice, however, they have lost their capacity to produce IL-2 in response to TL^b stimulator cells, and proliferative responses to TL^b are defective unless exogenous IL-2 is added (Bonneville *et al.*, 1990). Second, in radiation bone marrow chimeras, donor T cells develop functional tolerance to radioresistant thymic epithelial antigens. If cells from a donor that carries the Mls-1^b allele and lacks expression of I-E MHC class II products are transferred into a Mls-$1^{a \times b}$ I-E$^+$ irradiated host, autoreactive $V_\beta 6a^+$ (specific for MLs-1^a) and $V_\beta 17^+$ (specific for I-E) donor cells are not deleted in the thymus. When incubated with appropriate Mls-1^a and I-E+ stimulator cells, thymocytes from such chimeras do not respond by proliferation, although they express IL-2Rα. Addition of exogenous IL-2 partially restores the defective proliferative response to stimulation with anti-$V_\beta 17a$ and anti-$V_\beta 6$ antibodies (Ramsdell *et al.*, 1989).

Finally, IL-2 abrogates states of tolerance characterized by a downregulation of TCR molecules or CD4/CD8 receptors. In the human system, *in vitro* high-dose tolerance, which results in TCR downregulation, is abolished by IL-2 (Essery *et al.*, 1988). Teh *et al.* (1989) reported that a certain percentage of cells expressing a transgenic male-specific (H-Y) α/βTCR are not deleted in male mice, although they must be autoreactive. The peripheral transgene-expressing T cell population in male (autoreactive) but not in female (nonautoreactive) mice is composed almost entirely of cells that have down-regulated the

CD8 accessory molecule required for efficient H-Y recognition. Interestingly, injection of a vaccinia virus construct that releases human IL-2 into the peritoneal cavity of male mice induces expression of CD8 in peritoneal T cells, splenocytes, and lymph node cells bearing the transgenic TCR, suggesting that they might reacquire their function (Table II) (de Cid et al., 1991). Analogous results have been obtained *in vitro* culture in the presence of recombinant IL-2 (unpublished results).

It has to be stated that the antitolerance effect of IL-2 is limited to T_{H1} and possibly CTL cells (Ksander and Streilin, 1990) and that peripheral tolerance mechanisms affecting T_{H2} and B cells are probably not influenced by IL-2. Thus, for instance, the cytokines IL-1α, IL-β, and TNF-α, but not IL-2, abrogate the ability of deaggregated human γ globulin to induce a state of antigen-specific tolerance in mice (Gahring and Weigle, 1990).

D. Nonspecific Killing Induced by IL-2

Several reports suggest that IL-2 induces nonspecific cytotoxic reactions. A $CD4^+$ class II MHC–ovalbumin-specific murine T cell clone once activated with specific antigen kills innocent bystander cells, probably in response to autocrine IL-2. Addition of cyclosporin A,

TABLE II
EFFECT OF IL-2 ON MICE BEARING A MALE-SPECIFIC
TRANSGENIC T CELL RECEPTOR[a]

	Percentage among peritoneal cells in mice treated with	
Phenotype/Sex	WT.VV	IL-2.VV
$F23.1^+$ $CD4^+$ $CD8^-$		
Male	1.8 ± 0.5	4.0 ± 0.6
Female	7.3 ± 0.8	5.6 ± 0.6
$F23.1^+$ $CD4^-$ $CD8^+$		
Male	6.8 ± 0.7	23.3 ± 3.0
Female	2.4 ± 0.5	4.1 ± 0.5

[a] Mice bearing a male-specific T cell receptor (TCR) (von Boehmer, 1990) were injected with 1×10^7 plaque-forming units of recombinant IL-2/vaccinia virus (IL-2.VV) or wild-type vaccinia virus (WT.VV) starting from the sixth week of age. Treatment was repeated three times every 2 weeks and peritoneal cells were analyzed by two-color immunocytofluorometry at 12 weeks of age. Note that the percentage of T cells bearing the $V_{\beta}8$ epitope recognized by F23.1 (present on the transgenic TCR) and the accessory molecule CD8 augments in response to IL-2 in male mice.

which blocks IL-2 secretion, inhibits specific target-induced bystander killing. A low dose of exogenous IL-2 mimics the effect of specific target recognition and induces the ability to lyse bystander target cells (Gromkowski *et al.*, 1988). Similarly, Moriyama *et al.* (1988) showed that high IL-2 doses induced four out of six long-term murine $CD8^+$ clones specific for trinitrophenyl-modified self to develop an anomalous cytotoxicity against syngeneic and allogeneic tumor cells. In addition, IL-2 has been reported to activate NK cells that may lyse autologous targets, e.g., islet cells (Pukel *et al.*, 1987). When pancreatic islets infiltrated with lymphocytes are recovered from spontaneously diabetic NOD mice or from donors treated with streptozotocin and are cultured in the presence of IL-2, $CD8^+$ $\alpha/\beta TCR^+$ cells bearing a large granular lymphocyte morphology emerge. These lymphocytes destroy islet cells and other cells such as fibroblasts in a nonspecific way (Kay *et al.*, 1989). Similarly, human blood mononuclear cells activated with IL-2 (lymphokine-activated killer, or LAK, cells) lyse thyroid epithelial cells (Migita *et al.*, 1989) and autologous macrophages (Streck *et al.*, 1990).

Whether these nonspecific killing phenomena are also observed *in vivo*, e.g., during intravenous administration of high doses of IL-2, thus explaining some of the toxic effects of IL-2, remains to be confirmed.

IV. IL-2 in Autoimmunity—Phenomenology

A. IL-2 AND IL-2R AT THE SITE OF THE AUTOIMMUNE ATTACK

IL-2 production and IL-2R-expressing cells are detected at the site of autoimmune reactions, for example, in the salivary gland of patients with Sjögren's syndrome (Fox *et al.*, 1985), in the thyroid of OS chickens (Kroemer *et al.*, 1985a), in chronic active plaques in brains of patients with multiple sclerosis (Hoffmann *et al.*, 1986), in autoimmune diseases of the eye (Hooks *et al.*, 1988), and in the synovial fluid of RA patients (Lemm and Warnatz, 1986). Ontogenetic studies revealed that the IL-2R$^+$ cells predominate during the initial phase of autoimmune infiltration but decrease during advanced disease stages (Kroemer *et al.*, 1985a). *Ex vivo* explanted cells from RA joint fluids transcribe more efficiently the genes coding for IL-2 and IL-2Rα than do blood lymphocytes. IL-2 mRNA persists after stimulation with phytohemagglutinin (PHA) and phorbol ester for several days, in contrast to the transient (24 hours) presence of IL-2 mRNA in stimulated blood lymphocytes (Buchan *et al.*, 1988). The presence of IL-2 bioactivity in RA synovial fluids is a matter of debate, although monospecific immunoassays detect local IL-2 (Yamagata et al., 1988), possibly because local inhibitors may mask IL-2 bioactivity (Miossec et al., 1987; Symons et al., 1988). On the contrary, Yamagata et al. (1988) reported a decrease in IL-2 inhibitor in rheumatoid joint fluids.

In agreement with the local presence of cells expressing IL-2R and/or producing IL-2, many autoreactive T cell clones secrete IL-2 upon contact with autoantigen (Haskins et al., 1989). IL-2-dependent autoreactive T cell lines or clones have been isolated in experimentally induced as well as in spontaneously developing autoimmune diseases, such as experimental autoimmune encephalomyelitis (Ben-Nun et al., 1981), thyroiditis (Maron et al., 1983), uveoretinitis (Caspi et al., 1986), adjuvant and collagen-induced arthritis (Holoshitz et al., 1983: Holmdahl et al., 1985), experimental and spontaneous diabetes (Koevari et al., 1983; Kolb et al., 1985), rheumatoid arthritis (Wilkins et al., 1984), and myasthenia gravis (Hohlfeld et al., 1984). Many of these IL-2-dependent cell lines are capable of mediating autoimmunity when injected into normal histocompatible recipients. Pretreatment of the cells or the recipients with mAbs directed against IL-2Rα prevents the adoptive transfer of the disease (vide infra). Collectively, these data indicate that IL-2 producer lymphocytes and IL-2-receptive cells are involved in the initiation of autoimmune lesions.

B. *In Vitro* IL-2 PRODUCTION IN AUTOIMMUNE DISEASE

In numerous studies, *in vitro* IL-2 production has been measured upon activation of peripheral lymphocytes with nonspecific polyclonal stimuli (Con A, PHA, and anti-CD3 mAbs), class II-expressing autologous cells (so-called autologous mixed-lymphocyte reaction), alloantigens, or trinitrophenyl (TNP)-modified self. In many autoimmune diseases the IL-2 activity present in the conditioned media of such short-term cultures (24–72 hours) was significantly decreased (Table III), a finding that correlates with low proliferative responses and deficient cytotoxic T and NK cell response. Because exogenous IL-2 in part reverses the deficient cell-mediated *in vitro* responses, the "IL-2 defect" was held responsible for the decreased *in vitro* blastogenesis (Linker-Israeli et al., 1983; Emery et al., 1984; Volk and Diamantstein, 1986). In several instances, *in vitro* IL-2 production negatively correlates with disease activity (Kitas et al., 1988;Selmaj et al., 1988; Eisenstein 1988), although in some diseases the *in vitro* IL-2 defect is also detectable during remissions, e.g., in lupus erythematosus (Murakawa et al., 1985). A negative correlation between spontaneous *in vitro* IgG secretion and IL-2 production by peripheral blood lymphocytes (PBLs) in response to PHA plus phorbol ester has been described in

TABLE III
In Vitro IL-2 Production in Response to the T Cell Mitogens Con A or PHA

Observation	Disease	Reference
IL-2 deficiency	Lupus-prone mice	Altman et al., 1981; Wofsy et al., 1981; Dauphinée et al., 1981
	Human SLE	Alcocer-Varela and Alarcón-Segovia, 1982
	Human IDDM	Zier et al., 1983
	BB rat with IDDM	Elder and MacLaren, 1983
	Chronic active hepatitis, primary biliary cirrhosis	Saxena et al., 1986
	Multiple sclerosis	Selmaj et al., 1988
	Graves' disease	Eisenstein et al., 1988
	Sjögren's syndrome	Miyasaka et al., 1984
	Human RA	Cathely et al., 1986; Kitas et al., 1988
	Rat adjuvant arthritis	Stünkel et al., 1988
	Juvenile diabetes	Roncarlo et al., 1988
	NOD mouse with IDDM	Yokono et al., 1989
	Rheumatic fever	Alarcón-Riquelme et al., 1990
	UCD200 chickens with PSS	Gruschwitz et al., 1991
IL-2 hypersecretion	Obese strain of chickens with autoimmune thyroiditis	Kroemer et al., 1985a,b
	Human PSS	Umehara et al., 1988
	Primary IgA nephropathy	Schena et al., 1989

patients with SLE (Huang et al., 1988). These authors devised a lymphocyte activity score (logarithm of the quotient of IgG secretion and IL-2 production) that correlated with clinical parameters. It is an unresolved issue whether the *in vitro* IL-2 defect affects all patients suffering from a given autoimmune disease, because IL-2 levels are highly variable both in control and patient populations (Kitas et al., 1988). In lupus-prone mice, the *in vitro* IL-2 defect is age dependent and accompanies the development of manifest autoimmunity or lymphoproliferation (Dauphinée et al., 1981; Via and Shearer, 1988a). Thus in (NZB × NZW)F_1 mice the apparent *in vitro* IL-2 defect is corrected by lethal irradiation and reconstitution with bone marrow of BALB/c *nu/nu* donors, a procedure that also postpones the development of manifest autoimmunity (Ikehara et al., 1989).

Twin studies revealed that the IL-2 deficiency of diabetic patients is not genetically determined (Kaye et al., 1986). On the other hand, lymphocytes from normal individuals with the histocompatibility antigens HLA-B8 and DR3, i.e., a MHC haplotype associated with several autoimmune diseases including Sjögrens's syndrome, myasthenia gravis, Graves' disease, and coeliac disease, display a decreased responsiveness to T cell mitogens (McCombs et al., 1986). The impaired response to suboptimal concentrations of phytohemagglutinin and concanavalin A is associated with deficient L-2 production (Hashimoto et al., 1989) that may be mediated at least in part by decreased IL-1 synthesis (Hashimoto et al., 1990). Because this abnormality occurs in normal individuals with an autoimmunity-predisposing MHC, but without evidence of autoimmunity, it presumably precedes the autoimmune manifestations and may reflect an alteration of immune regulation that predisposes to immune-mediated disease. Along the same line, familial occurrence of impaired IL-2 activity has been reported for patients with systemic lupus erythematous (Sakane et al., 1989) and rheumatic fever (Alarcón-Riquelme et al., 1990). Consanguineous relatives of patients displayed lower IL-2 production than unrelated age- and sex-matched controls. Although these studies suggest a genetic basis of the *in vitro* IL-2 defect, it is not excluded that environmental effects (e.g., streptococcal infections in the case of rheumatic fever) contribute to the familial occurrence of this abnormality. Among 25 first-order relatives of SLE patients, 14 were found to exhibit low IL-2 production (Sakane et al., 1989). Thus, the incidence of the *in vitro* IL-2 defect is much higher than that expected for lupus (2–5%), indicating that abnormal behavior of T lymphocytes does not automatically entail autoimmune disease, but, at most, confers a predisposition that, combined with other genetically determined or environmental factors, induces SLE.

Several examples illustrate that decreased *in vitro* IL-2 production in autoimmunity is not a general rule. Lymph node cells from less than 6-week-old MRL/Mp-*lpr/lpr* mice—that is, prior to the manifestation of autoimmunity and lymphadenopathy—produce IL-2 in the absence of stimuli when cultured *in vitro* for 5–7 days, whereas control cells do not (Weston et al., 1987). This phenomenon depends on the interaction between the *lpr/lpr* gene and MRL/Mp background genes and is not observable in short-term cultures. The elevated IL-2 production in such conditions may be due to a primary abnormality of antigen-presenting cells from MRL/Mp-*lpr/lpr* mice (Weston et al., 1988). Using the polymerase chain reaction technique, it has been demonstrated that fresh $CD4^+$ lymphocytes from MRL/Mp-*lpr/lpr* mice contain as much IL-2 mRNA as MRL/Mp-+/+ controls, indicating that these cells are not defective in IL-2 production *in vivo* (Murray et al., 1990). Once removed from their microenvironment, $CD4^+$ cell lines from MRL/Mp-*lpr/lpr* mice appear to be hyperreactive and can grow as long-term lines producing IL-2 and other lymphokines (IL-3, IL-4, and IL-5) in the absence of exogenous antigen or mitogen (Rosenberg *et al.*, 1984). Spontaneous IL-2 hypersecretion is observed in rats or mice with experimental type II collagen autoimmune arthritis, but not adjuvant arthritis (Phadke *et al.*, 1986). Unfractionated blood mononuclear cells or purified $CD4^+$ cells from patients with progressive systemic sclerosis (PSS) hyperproduce IL-2 upon PHA stimulation in the presence of indomethacin. This hyperproduction is more accentuated in patients with visceral manifestations such as pulmonary sclerosis (Umehara *et al.*, 1988) and is accompanied by increased IL-4 and B cell growth factor (BCGF) production (Famularo *et al.*, 1990). In the Obese strain (OS) of chickens with spontaneous Hashimoto-like disease, a genetically determined, autosomal-dominant IL-2 hypersecretion in response to Con A or PHA becomes manifest before autoimmunity develops (Kroemer *et al.*, 1985b, 1987; Kroemer and Wick, 1989; Wick *et al.*, 1989). Spontaneous and PHA-induced IL-2 hypersecretion has also been found in patients with primary IgA nephropathy (Schena *et al.*, 1989).

The causes of low *in vitro* IL-2 production observable in many autoimmune diseases are apparently complex and a number of possible cellular and molecular mechanisms have been blamed to account for the "IL-2 defect." *In vitro* IL-2 production is known to be the final outcome of a cascade of events involving monocyte/macrophage activation, release of IL-1 from monocytes, interactions between T lymphocytes and accessory cells, and direct mitogen-dependent activation of T cells. Moreover, different T cell subsets may exert mutual regulation. Soluble factors suppress IL-2 production, and IL-2 is "consumed" by $IL-2R^+$ cells. By consequence, the net IL-2 production assessed by determining the IL-2 bio- or immunoactivity of culture supernatants may be subjected to a series of modulations intervening at different levels of this cascade. Accordingly, primary or autoantibody-mediated abnormalities concerning IL-1 production, suppressor T cells and macrophages, and numeric or functional T cell defects have been thought to underlie IL-2 hypoproduction.

1. *Low IL-1*

In SLE patients, low IL-1 production by monocytes was found to correlate with low IL-2 levels, suggesting that a deficient IL-1 production could account for the IL-2 defect (Alcocer-Varela *et al.*, 1983; Linker-Israeli *et al.*, 1983). However, in MRL/Mp-*lpr/lpr* mice in which IL-1 also is deficient (Donnelly *et al.*, 1990), addition of exogenous IL-1 did not normalize IL-2 secretion (Wofsy et al., 1981). Studies addressing the correlation of *in vitro* production of lymphokines in humans suggest that IL-1 and IL-2 production are regulated by independent genetic systems (Endres, et al., 1989).

2. Suppressor Macrophages

Activated macrophages from NOD mouse splenocyte cultures have been reported to release prostaglandin E, which in turn suppresses IL-2 production in response to Con A (Yokono et al., 1989). Similarly, in the BB rat, activated macrophages act as suppressor elements (Prud'homme et al., 1984). In both animal models, as well as in Sjögren's syndrome, addition of indomethacin to cultures partially repairs the IL-2 defect (Miyasaka et al., 1986).

3. Autoantibodies

Autoantibodies have also been speculated to interfere with IL-2 production. Miyagi et al. (1989) reported that purified IgG fractions of some SLE sera inhibit IL-2 production at two distinct phases of IL-1-dependent IL-2 production—first by binding to macrophages and inhibiting IL-1 production, and second by inhibiting IL-1 binding to T cells. Similarly, SLE patients have reportedly developed antibodies to Tac-antigen-positive T cell lines (Sano et al., 1986). Lipopolysaccharide (LPS)-stimulated B cells from different strains of lupus-prone mice [(NZB × NZW)F_1, BXSB] produce antibodies directed against IL-2 (Ishizaka and Tsuji, 1989).

4. Decrease of IL-2 Producing Cells

Preferential homing of IL-2-producing T_{H1} cells into the site where the autoimmune attack is launched might entail a decrease in circulating IL-2 producers (Roncarlo et al., 1988). In MRL/Mp-*lpr/lpr* mice, it has been postulated that an increase of T cells bearing the abnormal lpr phenotype may dilute out IL-2-secreting cells (Simon et al., 1984).

5. Suppressor T Cells

Linker-Israeli et al. (1985) demonstrated that $CD8^+$ cells or a cell-free supernatant from $CD8^+$ lymphocytes from SLE patients actively inhibits IL-2 production by normal lymphocytes, but Murakawa et al. (1985) failed to demonstrate the involvement of suppressor cells or inhibitory molecules in SLE-associated defective IL-2 production. Naides (1986) reported that MRL/Mp-*lpr/lpr* lymph node cells could suppress IL-2 production by trinitrophenyl-primed MRL/Mp−+/+ lymph node cells, and Via and Shearer (1988a) showed that CD4$^+$ "suppressor" cells inhibit the IL-2 production of congenic MRL/Mp-+/+ CD4$^+$ T helper cells to MHC self-restricted antigens. Because these CD4$^+$ suppressor cells (possibly T$_{H2}$ cells) are detected late in the course of autoimmunity, these authors interpreted their presence not as a primary initiating event in the development of autoimmunity, but rather as a compensatory mechanism representing a frustrated attempt to down-regulate excessive T$_H$ activity (Via and Shearer, 1988b). In this context, it may be worthwhile to mention that T$_{H2}$ cells secrete IL-10, which inhibits IL-2 production by T$_{H1}$ cells (Moore et al., 1990).

6. Signal Transduction

It has been suggested that the IL-2 producer abnormality might reside at the level of the mitogen receptor signal transduction of the T cell. Bypassing the early steps of signal transduction by using phorbol ester in conjunction with calcium ionophore partially restores the ability of MRL/Mp-*lpr/lpr* splenocytes to produce IL-2 (Koizumi et al., 1986). Both in (NZB × NZW)F$_1$ and in MRL/Mp-*lpr/lpr* mice, addition of PMA corrects the *in vitro* defect in IL-2 production in response to Con A (Santoro et al., 1983), but this is not the case in BB rats (Prud'homme et al., 1984). Similarly, addition of PMA repaired the deficiency in PHA-stimulated IL-2 production by cells from patients with inactive SLE (Murakawa and Sakane, 1988), and, in part, in individuals with Sjögren's syndrome (Miyasaka et al., 1986). SLE T cells display significantly lower peak increases in intracellular free calcium (Ca$_i^{2+}$) than do controls after stimulation with anti-CD3 mAb or PHA, whereas calcium ionophore produces increases in Ca$_i^{2+}$ in SLE patients similar to that seen in normals (Sierakowski et al., 1989). This suggests that a component of the defect responsible for the *in vitro* IL-2 defect is proximal to protein kinase C activation and Ca^{2+} mobilization and may involve impaired TCR-mediated signal transduction.

The common notion that autoimmune disease results from a hyperreactivity of the immune system is incompatible with an "IL-2 defect." This paradox could be resolved if a minimum of IL-2 is indispensable for the induction of suppressor pathways (Volk and Diamantstein, 1986; Serreze and Leiter, 1988), the down-regulation of helper T cell activation (Kennedy et al., 1987), and the physiological elimination of altered, potentially autoantigenic or autoaggressive host cells by cytotoxic T cells (Smith and Talal, 1982). However, the many hypotheses explaining why IL-2 is defective in autoimmune disease and how this defect may be implicated in the disease process probably have been constructed in vain, simply because the fundamental premise is wrong. As will be discussed in the following section, decreased *in vitro* IL-2 production by freshly isolated lymphocytes exposed to lectins is not representative for the *in vivo* situation.

C. *In Vivo* IL-2 Production in Autoimmune Disease

The advent of sensitive and specific IL-2 bio- and immunoassays has allowed measurement of IL-2 contained in serum, thus revealing that several autoimmune diseases associated with defective *in vitro* IL-2 production were characterized by elevated serum IL-2 levels. In SLE, PSS, and multiple sclerosis, determinations of serum IL-2 by means of bioassays or commercial enzyme-linked immunoassays revealed abnormally high IL-2 serum concentrations (Table IV). In SLE, the IL-2 bio- and immunoactivity was completely and specifically removed after passage through an anti-IL-2 immunoabsorbent column and, under nonreducing conditions, occurs in a higher molecular weight fraction (around 50K) than native IL-2, possibly because it is bound to a carrier protein. To explain the discrepancy between the *in vitro* IL-2 defect and the apparent *in vivo* IL-2 hyperproduction, Huang *et al.* (1988) suggested that precommitment and preactivation of peripheral T cells *in vivo* result in a transient exhaustion of IL-2 secretion. In support of this hypothesis, PHA responses of T cells from patients with active SLE are restored if cells are allowed to "rest" *in vitro*. The capacity of circulating SLE T cells to secrete IL-2, which was decreased by a factor of 10 at the beginning, become normal if either total T cells or purified $CD4^+$ cells are maintained in culture 2 to 3 days prior to stimulation with PHA and PMA (Huang *et al.*, 1986). Con A-activated normal T cells become thoroughly refractory to Con A or IL-2 in *in vitro* culture (Wilde *et al.*, 1984; Kehrl *et al.*, 1986) and an IL-2 secretion defect can be induced experimentally in rats by injecting large doses of T cell mitogen (Ridge *et al.*, 1986). The most conclusive piece of evidence in favor of the hypothetical exhaustion of T cells is provided by the intriguing negative correlation between high serum

TABLE IV
Autoimmune Diseases Associated with Elevated Serum IL-2 Levels

Autoimmune disease	Correlation with	Reference
Multiple sclerosis	Relapse	Adachi *et al.*, 1989
	Chronic progressive disease	Trotter *et al.*, 1988
SLE	Active disease	Huang *et al.*, 1988
PSS	Extent of skin involvement	Kahaleh and Leroy, 1989
UCD200 chickens with PSS	—	Gruschwitz *et al.*, 1991

IL-2 activities and low *in vitro* IL-2 production in lupus patients, in multiple sclerosis, and in an animal model for progressive systemic sclerosis (Table IV). Moreover, immunosuppressive therapy of SLE with glucocorticoids and cyclosporin A—drugs for which there is well-documented suppression of IL-2 production—corrects the apparent *in vitro* IL-2 secretion defect, possibly by preventing exhaustion of T cells *in vivo* (Huang *et al.*, 1988). In Brown–Norway (BN) rats, a strain susceptible to mercury-induced autoimmunity, *in vivo* $HgCl_2$ treatment entails an apparent IL-2 defect, but BN lymphocytes produce IL-2 in response to *in vitro* $HgCl_2$ treatment (Baran *et al.*, 1988). Thus, $HgCl_2$ could stimulate and exhaust IL-2-producing cells *in vivo*.

As an alternative to the exhaustion theory, it is conceivable that production of IL-2 by a subpopulation of *in vivo*-activated lymphocytes elicits feedback regulation pathways that attempt to restore the homeostasis of the IL-2/IL-2R system by nonspecifically suppressing IL-2 production. In favor of this concept argues the fact that patients treated with high IL-2 doses exhibit depressed mitogen responses *in vitro*. Regardless of why IL-2 production is reduced *in vitro*, the fact that IL-2 serum is elevated in various autoimmune conditions always thought to be associated with an IL-2 defect documents that extreme caution should be exercised in the extrapolation from *in vitro* functional data to any *in vivo* situation.

It remains to be clarified whether all states of autoimmunity are accompanied by elevated *in vivo* production of IL-2, to what extent increased IL-2 levels are primary or secondary to activation of T cells by autoantigens, and whether genetic factors determine the capacity to produce IL-2. Increased serum IL-2 is not specific for autoimmune diseases, because it is also increased in sera from pregnant women (Favier *et al.*, 1990) and from patients with chronic Epstein–Barr virus syndrome (chronic fatigue syndrome) (Cheney *et al.*, 1989). As IL-2 induces soluble and membrane-bound IL-2Rα both *in vitro* and *in vivo* (Reem *et al.*, 1984; Lotze *et al.*, 1987), it cannot be ruled out that the high sIL-2R serum levels observed in most autoimmune diseases are secondary to a high systemic IL-2 concentration (see below). Alternatively, serum IL-2 and sIL-2R together may reflect abnormal polyclonal or autoantigenic stimulation of cells that express both the lymphokine and its receptor. As to the question of whether IL-2 production is influenced by hereditary factors, data obtained in different animal models perhaps are the most relevant. On the one hand, immunization with collagen type II plus complete Freund's adjuvant induces arthritis in susceptible rodent strains concomitant to an enhanced spontaneous IL-2 release by spleen cells (Phadke *et al.*, 1986).

On the other hand, in chickens afflicted with Hashimoto-like autoimmune thyroiditis, IL-2 hypersecretion is an autosomal dominant trait, detectable in F_1 hybrids with an unrelated strain (CB strain) and in 50% of the parental backcross (CB × F_1) generation (Kroemer et al., 1988). Because such F_1 and backcross animals do not develop thyroiditis, IL-2 hyperproduction is not secondary to the autoimmune process. In F_2 animals, IL-2 hyperproduction, together with other factors, predisposes to the development of thyroiditis (Kroemer et al., 1989, 1990) (Fig. 2). In different breeding experiments, high IL-2 production and high IL-2Rα expression did not segregate, suggesting that both abnormalities have a common origin or are functionally linked (Kroemer et al., 1988).

A rather particular situation is represented by human T cell lymphotropic virus type I (HTLV-I)-associated myelopathy or tropical spastic

---

FIG. 2. Immunogenetic analysis of T cell hyperreactivity in Obese strain (OS) chickens with Hashimoto-like autoimmune thyroiditis. (A) Concanavalin A response of peripheral blood lymphocytes derived from (OS × CB)F_1 hybrids, OS, normal healthy CB, and (F_1 × CB) backcross animals. Peripheral blood mononuclear cells (10^6) from 4-week-old donors were stimulated with 2.5 μg/ml Con A for 48 hours, followed by removal of 100 μl culture supernatant and a 6-hour 5-[^{125}I]-2-desoxyuridine pulse label. Each point represents the mean triplicate value of one individual animal. According to their Con A responsiveness, the (F_1 × CB) animals were classified in two groups of "high" (solid circles) and "low" (open circles) responders. High and low responders are equally distributed among (F_1 × CB) backcross animals homozygous for the MHC of CB or MHC$^{OS/CB}$ heterozygotes, indicating MHC independence. Because F_1 animals do not develop thyroiditis yet exhibit an elevated OS-like T cell proliferation, Con A hyperreactivity of OS chickens is not secondary to the autoimmune disease. (B) IL-2 bioactivity in culture supernatants of the same cultures as measured in A. Note that the same animals that exhibit Con A hyperreactivity also produce higher IL-2 levels than do CB controls. (C) Frequency of IL-2Rα$^+$ cells. In parallel cultures, peripheral blood mononuclear cells (same donors as in A and B) were subjected to a short-term Con A stimulation (4 hours) followed by cytofluorometric assessment of IL-2Rα$^+$ cells using a specific monoclonal antibody (INN-Ch16). Elevated IL-2R expression occurs in the same (OS × CB)F_1 × CB backcross animals that produce high amounts of IL-2 and hyperproliferate with Con A. (D) Correlation of thyroid infiltration and T cell hyperreactivity in (OS × CB)F_2 animals. Development of thyroiditis is determined by one recessive gene, so that roughly three-quarters of the animals exhibit a degree of lymphoid infiltration of the thyroid physiological for the chicken (<10% infiltrated area), whereas one-quarter of the animals show >10% destruction of the thyroid gland. Within these groups, the degree of thyroid infiltration appears to correlate with the dominantly inherited Con A hyperreactivity (solid circles), present in 75% of F_2 animals. The asterisk marks a statistically significant increase in the incidence of T cell hyperreactivity (χ^2 test, $p < 0.0.5$) and the mean values of Con A responses (Whitney–Man nonparametric test, $p < 0.05$) among animals developing "borderline" thyroiditis (5–10% thyroid infiltration) as compared to undiseased (OS × CB)F_2 controls. [Data from Kroemer et al. (1989, 1990).]

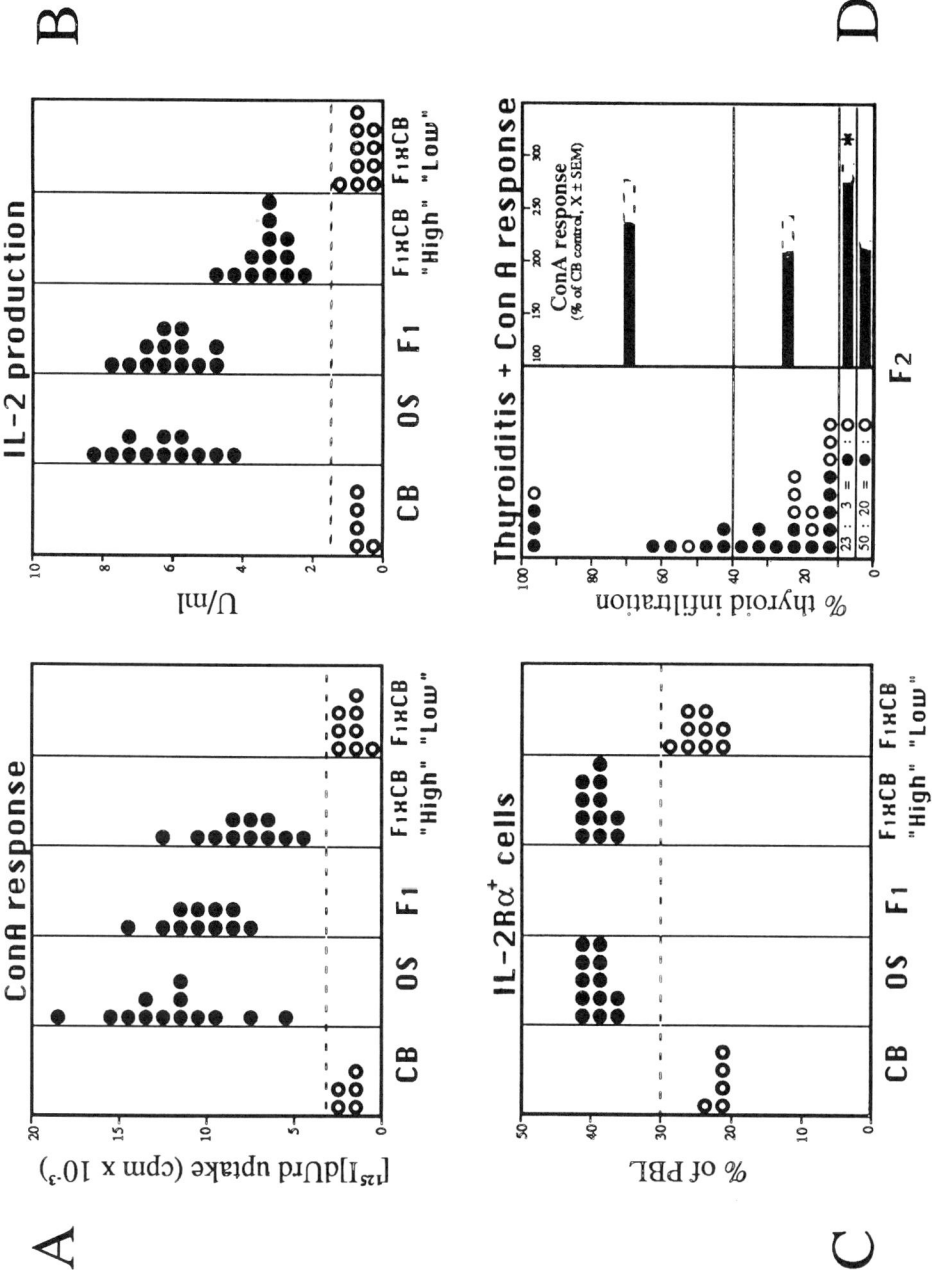

paraparesis, a demyelinating, possibly immune-mediated disease. By means of the polymerase chain reaction, Tendler et al. (1990) have analyzed the *ex vivo* mRNA expression of IL-2, IL-2Rα, and pX, a retroviral gene that encodes a transcriptional activator (p40^{tax}, or Tax) of certain host cellular genes including IL-2 and IL-2Rα (Hoyos et al., 1989). Among patients with HTLV-I-associated myelopathy and seropositive HTLV-I carriers, peripheral blood lymphocytes exhibit expression of pX correlating quantitatively with IL-2 and IL-2Rα transcription, as well as with the propensity to spontaneously proliferate *in vitro*. This proliferation is mediated by the IL-2/IL-2R pathway, as shown by its blockade by anti-IL-2Rα. Thus, HTLV-I-associated myelopathy represents an autocrine phase of HTLV-I infection in which the pX transactivator induces the coordinate expression of IL-2 and IL-2Rα by activating a NF-κB-like factor involved in the transcriptional regulation of both genes. Another pX product (p27^{rex}) has recently been shown to up-regulate IL-2Rα expression by protecting IL-2Rα mRNA from degradation (Kanamori et al., 1990). In addition, HTLV-I-infected T cells secrete a factor (T cell leukemia-derived factor, ADF) that acts as an IL-2Rα inducing agent and acts synergistically with IL-1 and IL-2 on cellular proliferation. This product is identical with a protein released from Epstein–Barr virus-infected B cells and exhibits a considerable homology to phospholipase C as well as to the disulfide-reducing enzyme thioredoxin found in prokaryotes and mammalian cells (Wakasugi et al., 1990).

These examples illustrate that hyperproduction of IL-2 may be either genetically determined or secondary to the exposure to autoantigen or viral infection. The fact that relapses and chronic progression of autoimmune diseases are accompanied by higher serum IL-2 levels suggests that IL-2 may be secreted by T cells during the autoimmune attack, although the exact cause/effect relationship remains to be elucidated. Thus, it remains elusive whether endogenous IL-2 stimulates T cells to launch an autodestructive attack or, on the contrary, if the inflammatory process initiates without significant participation of IL-2, which then is produced during the self-perpetuating disease phase. Future longitudinal and vertical studies will have to decide whether secretion of IL-2 may be genetically programmed and whether it *de facto* predisposes to the development of autoimmune diseases in humans. In this context it may be noteworthy that patients with high IL-2 plasma levels appear to be immunological high responders. Cornaby et al. (1989) published that, among 21 patients with preoperative levels in excess of 15 ng/ml IL-2, 19 (90%) rejected renal allografts, whereas only 23% (15 out of 70) of the patients with lower serum IL-2 concentrations required treatment for rejection. Thus, an acquired or inherited IL-2 hyperproduction may predispose to immune hyperreactivity.

D. ELEVATED SERUM LEVELS OF SOLUBLE IL-2Rα IN AUTOIMMUNE DISEASE

Upon *in vitro* activation, T cells do not only produce cell-associated IL-2Rα (p55), but also shed soluble IL-2Rα (sIL-2R) chains (Rubin *et al.*, 1985), which may be easily quantitated by means of immunoradiometric or immunoenzymatic techniques employing specific monoclonal antibodies. sIL-2R is released in large amounts by both $CD4^+$ and $CD4^-$ T cells after exposure to mitogenic or antigenic stimuli (Reske-Kunz *et al.*, 1987). sIL-2R is also present in the supernatants of Epstein–Barr virus-transformed lymphoblastoid cell lines, suggesting that any cell that expresses IL-2Rα is able to release sIL-2R (Nelson *et al.*, 1986). Two mechanisms may account for the shedding of α chain from the cell surface. Alternative splicing of the α chain mRNA may give rise to "anchor minus" protein, which is then released (Treiger *et al.*, 1986). However, the predominant mechanism of release is proteolytic cleavage at the cell surface (Robb and Kutny, 1987), as indicated by the high degree of heterogeneity in the amino terminus of IL-2Rα. Whereas the cell surface α chains have a relative molecular mass of 55 kDa, the soluble form is 10 kDa smaller due to lack of transmembrane and intracytoplasmic domains. In contrast to the transient expression of cell-associated IL-2R after *in vitro* lymphocyte activation, which peaks 48–72 hours poststimulation and declines thereafter, cumulative supernatant levels of sIL-2R continuously increase (Rubin *et al.*, 1985).

Apparently analogous to the *in vitro* findings, sIL-2R is detected at low levels in the serum of healthy donors, but immune activation occurring *in vivo* leads to a significant increase of IL-2Rα in body fluids. sIL-2R is enhanced in mice injected with lipopolysaccharide (Balderas *et al.*, 1987), in individuals rejecting transplants (Colvin *et al.*, 1987; Lawrence *et al.*, 1988), in leprosy patients (Tung *et al.*, 1987), in patients with HTLV-I-induced leukemias (Greene *et al.*, 1986), and in other Tac-expressing neoplasias (Waldmann, 1989). In HIV-seropositives cases, sIL-2R levels correlate inversely with $CD4^+$ cell counts (Prince *et al.*, 1988) and IL-2Rα is predominantly expressed on monocytes (Allen *et al.*, 1990). Nonspecific stress also appears to induce sIL-2R. Thus, patients on dialysis (Chatenoud *et al.*, 1986) and burn patients (Xiao *et al.*, 1988) display high sIL-2R in the serum. Moreover, treatment of cancer patients with IL-2 results in subsequent elevations in the number of IL-2Rα$^+$ blood lymphocytes and in the serum concentration of sIL-2R (Lotze et al., 1987), a phenomenon reminiscent of in vitro induction of IL-2Rα by IL-2 on mature T cells (Reem et al., 1984).

Elevated circulating sIL-2R has been found in many autoimmune diseases (Table V). This finding may reflect a state of ongoing lymphocyte activation, leading to commensurate induction and expression of IL-2R on circulating lymphocytes or, alternatively, on inflammatory cells locally implicated in the disease. The pathogenetic role for activated immune cells is substantiated by a more marked elevation of sIL-2R, relative to levels in paired sera, in body fluids derived from the site of the autoaggressive attack, i.e., synovial fluid in RA (Keystone et al., 1988), and in cerebrospinal fluid in multiple sclerosis (Trotter et al., 1988). This observation implies also that local production of sIL-2R may constitute an important source of serum sIL-2R in autoimmune disease.

TABLE V
Elevated Circulating Soluble IL-2Rα in Diseases with Known or Possible Autoimmune Genesis

Disease	Additional observations	Reference
Multiple sclerosis	Higher levels during relapse	Adachi et al., 1989
	Higher concentration in CSF	Trotter et al., 1988
SLE	Correlation with hypocomplementemia and proteinuria	Wolf and Brelsdorf, 1988
RA	No correlation with clinical disease, higher concentration in synovial fluid	Keystone et al., 1988
BXSB mice MRL/Mp-*lpr/lpr* (NZB × W)F$_1$	Correlates with lymphadenopathy and autoantibody titers	Balderas et al., 1987
Still's syndrome	—	Symons et al., 1988
Behçet's disease	—	Akoglu et al., 1990
PSS	Correlates with disease activity	Degiannis et al., 1988
Autoimmune chronic active hepatitis	Higher in active disease	Lobo-Yeo et al., 1990
Polymyositis	High in acute disease	Wolf and Bathge, 1990
Inflammatory bowel disease	Correlates with disease activity in colitis ulcerosa	Mahida et al., 1990
Primary IgA nephropathy	Elevated Tac on circulating T cells coexpressing DR	Schena et al., 1989
Sarcoidosis	Increased Tac also in bronchoalveolar lavage fluid	Lawrence et al., 1988
Kawasaki syndrome	Higher in patients with coronary aneurysms	Barron et al., 1990

The correlation of serum sIL-2R with disease activity is controversial. In RA, no significant correlation between sIL-2R and 15 different clinical variables could be established in a cohort study (Keystone et al., 1988). However, longitudinal studies revealed that peak levels coincide with active disease phases (Symons et al., 1988; Semenzato et al., 1988; Rubin et al., 1990). In human SLE, high correlations with parameters indicative of disease activity (low levels of the complement factors 3 and 4, and proteinuria) were found (Wolf and Brelsdorf, 1988). In the latter case it has been suggested that sIL-2R may be an earlier indicator of fluctuations in disease activity than either hypocomplementemia or proteinuria. Multivariate analysis confirmed a positive association of sIL-2R levels with skin scores and mortality, and a negative correlation with disease duration in PSS (Degiannis et al., 1988). Positive correlations with serologic manifestations of autoimmunity and development of lymphadenopathy have been reported for the lupus-prone mouse strains MRL/Mp-lpr/lpr, BXSB, and (NZB × NZW)F_1 (Balderas et al., 1987). In Sjögren's syndrome, disease progression to extraglandular involvement is accompanied by high sIL-2R (Manoussakis et al., 1989). In certain diseases for which an autoimmune component is discussed, sIL-2R is also increased. This is true for primary IgA nephropathy (Schena et al., 1989) and active pulmonary sarcoidosis (Lawrence et al., 1988). In contrast to the many autoimmune diseases associated with high sIL-2R levels, children with prediabetes or newly diagnosed diabetes contain lower sIL-2R levels than do healthy controls (Keller and Jackson, 1989).

The physiological significance of circulating sIL-2R is a matter of debate. sIL-2R may bind IL-2 with about the same activity as the corresponding surface molecule (10×10^{-8} M (Rubin et al., 1986; Robb and Kutny, 1987) and has been speculated to function as an IL-2 antagonist. However, there is no direct evidence that this is the case. On the contrary, sIL-2R, which binds IL-2 with an affinity three orders of magnitude lower than that of the $\alpha\beta$ heterodimeric IL-2R, might function as a transport protein and release IL-2 in the vicinity of a high-affinity IL-2R, thus protecting it from proteolytic clearance (Keystone et al., 1988). Moreover, the hypothesis that sIL-2R might antagonize IL-2 is invalidated by the fact that adult NZB mice have elevated serum sIL-2R levels (Balderas et al., 1987), but nonetheless lack detectable serum IL-2 inhibitor (Hardt et al., 1981). Similarly, the majority of patients with clinically active SLE and about half of the patients with RA exhibit decreased activity of an IL-2 inhibitor present in normal sera (Djeu et al., 1986). It remains to be established whether sIL-2R really subserves an immunomodulatory function or is merely a marker of cellular immune activation. Irrespective of these possibilities, measurement of sIL-2R could provide a valuable noninvasive approach to the analysis of disease-associated immune activation *in vivo* and give an objective indication or prediction of ongoing autoimmune disease activity. Moreover, it may provide a welcome addition to the clinician's armamentarium for the evaluation of therapeutic schedules, because successful treatment is accompanied by a decline in circulating sIL-2R levels. As the major cellular source of sIL-2R comprises T cells, B lymphocytes, as well as cells from the macrophage/monocyte lineage, quantification of sIL-2R may be expected to give a less precise image of T cell activation than do determinations of the T cell-specific product IL-2 in body fluids.

E. ABNORMALITIES IN IL-2R EXPRESSION AND IL-2 RESPONSIVENESS ON CIRCULATING LYMPHOCYTES

Various reports state that peripheral blood lymphocytes from patients with autoimmune disease exhibit elevated expression of IL-2R (Tac), e.g., in multiple sclerosis (Selmaj *et al.*, 1986), PSS (Freundlich and Jimenez, 1987), RA (Kitas *et al.*, 1988), primary IgA nephropathy (Schena *et al.*, 1989), and sarcoidosis, where they proliferate with IL-2 at a rate greater than cells from healthy donors (Konishi *et al.*, 1988). Evidence for elevated (Selmaj *et al.*, 1986) and prolonged (deFreitas *et al.*, 1986) expression of membrane-bound IL-2Rα in multiple sclerosis extends the notion that an activated cellular immune state parallels the evolution of autoimmune processes. Physiological concentrations of IL-2 sufficient only to saturate the high-affinity IL-2R induce abnormally high proliferative responses in SLE (Alcocer-Varela and Alarcón-Segovia, 1982). Elevated proliferative responses to IL-2 have been described for thymic lymphocytes from UCD200 chickens with PSS (van de Water *et al.*, 1990) and thymocytes of patients with myasthenia gravis. Hyperreactivity of peripheral blood cells from myasthenia gravis patients correlates with antiacetylcholine receptor autoantibody titers and decreases after thymectomy (Cohen-Kaminsky *et al.*, 1989).

In strict contrast to the above-mentioned findings, freshly prepared T cells from SLE patients show decreased responses to high doses of IL-2 and have subnormal levels of intermediate IL-2-binding molecules (Tanaka *et al.*, 1989). Similarly, PHA-stimulated blasts from patients with SLE express subnormal levels of high-affinity IL-2R (Ishida *et al.*, 1987). Murakawa *et al.* (1985) reported that $CD4^+$ cells from about one-half of the SLE patients showed a defective IL-2 responsiveness, whereas $CD8^+$ cells responded well to IL-2. The cytotoxic activity of IL-2-activated NK cells is defective in SLE (Froelich et al., 1989). In PBLs from SLE patients, a decrease in precursor frequencies for IL-2-responsive cells has been described. However, resting the cells for 3 days prior to activation and adding exogenous IL-2 during the initial activation step results in enhancement of precursor frequencies to normal levels (Warrington et al., 1989), indicating that many of the IL-2R defects found may be secondary to the decreased *in vitro* IL-2 production of cells derived from autoimmune donors (*vide supra*).

Enhanced IL-2 responsiveness has been reported for B cells. Whereas unfractionated B cells from immunologically normal control mice respond to IL-2 only after stimulation with lipopolysaccharide plus anti-Ig, MRL/Mp-*lpr/lpr* B cells respond to IL-2 following exposure to LPS alone, and B cells from young (NZB × NZW)F_1 mice proliferate with IL-2 in the absence of additional antigens (Lehmann et al., 1986). These B cells belong to the nonadherent fraction, show no spontaneous immunoglobulin secretion, bear a $CD5^+$ $B220^{dull}$ phenotype, and do not stain with antibodies directed against the IL-2Rα chain (Sekigawa et al., 1989; Hasegawa et al., 1989); however, they express the IL-2Rβ chain. Probably, these B lymphocytes are preactivated *in vivo* in order to facilitate the reception of the second signal of IL-2 for their proliferation. Similarly, B cells from SLE patients stimulated with IL-2 secrete elevated amounts of IgG as compared to normal controls (Huang et al., 1988).

In summary, abnormalities in the *ex vivo* expression of IL-2R components and in IL-2 responsiveness are encountered in peripheral lymphocytes of autoimmune donors. Most of these aberrations point to a hyperactivation of the cellular and humoral immune system.

V. IL-2 in Autoimmunity—*In Vivo* Interventions

A. IL-2R-Targeted Immune Intervention in Autoimmune Disease

During the last few years it has become abundantly clear that cells of the T helper phenotype are implicated in the pathogenesis of autoimmune diseases in an obligatory fashion. Accordingly, CD4-targeted mAbs have been used to intervene successfully in the development of various autoimmune diseases (reviewed by Auffray et al., 1991). Similarly, reagents have been designed to interfere specifically with the action of one of the principal T helper products, IL-2. Monoclonal antibodies directed against the inducible IL-2Rα have been employed for selective immunosuppression. Because IL-2Rα is only expressed by a minor population of NK cells and a small subset of T cells subsequent to antigen-specific activation, such mAbs may be expected to exert a more "specific" immunosuppressive effect than antibodies that recognize constitutive membrane determinants and whose suppressive effect could be detrimental for the host's defense against external pathogens. Attempts are being made to produce IL-2 derivatives that either function as IL-2 antagonists capable of displacing endogenous IL-2 from its receptor without IL-2R triggering, or, alternatively, target toxins to cells bearing the high-affinity IL-2R. In addition to physical receptor blockade, IL-2Rα mAbs may cause deletion of the specific T cells activated during the immune response by (1) binding to the IL-2R with subsequent complement fixation, (2) antibody-dependent cell-mediated cytotoxicity, and/or (3) opsonization and phagocytosis by the reticuloendothelial system (Mouzaki et al., 1987).

A large number of autoimmune diseases have been treated with anti-IL-2Rα mAbs. Anti-IL-2Rα inhibits the passive transfer of experimental autoimmune encephalitis and adjuvant arthritis. Thus, the mAb ART-18 protects Lewis rats from lethal doses of a syngeneic T cell line recognizing the encephalitogenic peptide sequence of myelin basic protein, but does not suppress subsequent T cell responses to ovalbumin (Wekerle and Diamantstein, 1986). Con A-activated splenocytes from adjuvant-sensitized Lewis rats fail to transfer arthritis to naive recipients treated with ART-18 (Diamantstein and Osawa, 1986). However, ART-18 does not influence active adjuvant arthritis (Stünkel et al., 1988). ART-18 also suppresses the adoptive transfer of experimental autoimmune neuritis, but treatment after appearance of clinical signs has no effect (Hartung et al., 1989). In collagen type II-induced arthritis of DBA/J mice (Banerjee et al., 1988), such a mAb reduces antitype II collagen antibody levels and has a beneficial effect on the incidence of arthritis. In diabetes-prone BB rats, temporary treatment with an anti-IL-2Rα mAb (ART-18) in combination with a subtherapeutic dose of cyclosporin A normalizes plasma glucose, enhances pancreatic insulin content, and reduces insulitis (Hahn et al., 1987, 1988). In newly diagnosed diabetic BB rats, this treatment results in the maintenance of Langerhans cell volume density and in normoglycemia in 70% of the animals. Treatment with an anti-IL-2Rα mAb reduces the autoimmune insulitis characteristic of diabetes-prone NOD mice and protects lupus-prone (NZB × NZW)F_1 mice from the development of glomerulonephritis (Kelley et al., 1988). Recently, positive results were reported for patients with therapy-resistant active erosive rheumatoid arthritis that had been treated for a limited period (3 months) with a rat IgG_{2b} anti-Il-2Rα mAb (Kyle et al., 1989). Collectively, these results underscore the importance of a small population of IL-2R$^+$ cells in the causation of autoimmune-induced tissue damage. The mode of action of anti-Il-2Rα may comprise direct effects on the "afferent limb" of the pathogenic immune response to the autoantigen by inhibiting helper/inducer T cells or the "efferent limb" by suppressing effector T cells. Moreover, such antibodies may indirectly potentiate endogenous inhibitory circuits because they spare suppressor cells (Schneider et al., 1986). The fact that anti-IL-2Rα antibodies are more effective in preventing autoimmunity than in ameliorating established diseases may reflect the primordial importance of IL-2 in the initiation, rather than in the perpetuation, of the pathogenetic process.

Anti-IL-2Rα mAbs function by depleting IL-2Rα-bearing cells and their cytotoxic potential as IL-2R-targeted drugs may be improved by conjugation with α- and β-emitting isotopes or toxins (reviewed by Waldmann, 1989). A chimeric 45-kDa recombinant protein composed of IL-2 fused to a modified *Pseudomonas* exotoxin lacking its cell recognition domain (IL-2-PE40) has been used to target the membrane penetration and ADP-ribosylation domains of the toxin to cells bearing the IL-2R. IL-2-PE40 inhibits the development of experimental autoimmune uveoretinitis in Lewis rats immunized with retinal S antigen (Roberge et al., 1989). IL-2-PE40 administered 9 days after active immunization, i.e., during the efferent phase of the immune response, results in a significant reduction of the incidence and severity of the organ-specific self-aggression over controls. IL-2-PE40 has no effect on the production of antibodies against the immunizing antigens, but it precludes the development of effector T cells as demonstrated by transfer experiments. Intraperitoneal application of IL-2-PE40 also effectively and specifically suppresses the development of mycobacterial adjuvant arthritis in Lewis rats when given before the establishment of clinical disease, whereas the toxin devoid of its IL-2R-binding moiety and an IL-2PE40 mutant lacking an enzymatically active ADP-ribosylation domain show no ability to suppress adjuvant arthritis (Case et al., 1989).

These studies point to the possibility of generating effective IL-2R-targeted immunosuppressants using the techniques of genetic engineering, thus hopefully opening new avenues toward a therapy of autoaggression that intervenes at a level of immune dysfunction common to different diseases.

B. Autoimmune Manifestations after *in Vivo* Application of Recombinant IL-2

A frequent event after IL-2 therapy is autoimmune thyroiditis (Table VI). Atkins et al. (1988) reported that 7 out of 34 patients (34%) treated for melanoma, renal cell carcinoma, or colon carcinoma with IL-2 and

TABLE VI
INDUCTION OF INFLAMMATORY PHENOMENA OF PROBABLE AUTOIMMUNE GENESIS IN
PATIENTS TREATED WITH RECOMBINANT IL-2

Autoimmune manifestation	Treatment	Reference
Hashimoto-like thyroiditis	IL-2 + LAK	Atkins et al., 1988
	IL-2	Hartmann et al., 1989; van Liessum et al., 1989; Beuzeboc et al., 1989
	IL-2 + IFNα-2a	Pirchert et al., 1990
Vitiligo	LAK + IL-2	Atkins et al., 1988
Dermal vasculitis	LAK + IL-2	
Pemphigus vulgaris	IL-2 + IFNβ	Ramseur et al., 1989
Encephalomyelitis	IL-2	Vecht et al., 1990
Acute multifocal hepatitis	IL-2 + LAK	Hoffman et al., 1989
	IL-2	Punt et al., 1990
Lymphoeosinophilic cholecystitis	IL-2 + LAK	Chungpark et al., 1990
Myocarditis	IL-2	Samlowski et al., 1989; Kragel et al., 1990; Schuchter et al., 1990
Reactivation of rheumatoid arthritis	IL-2	Lavelle-Jones, 1990

LAK cells developed hypothyroidism, especially those patients whose antithyroid microsomal antibody titers were raised before treatment and who showed partial or complete tumor regressions (Atkins et al., 1988). Among these patients, further minor autoimmune manifestations (cutaneous vaculitis and vitiligo) were observed. Hypothyroidism and goiter concomitant to rises of antithyroid microsomal antibodies were also observed in patients treated with IL-2 alone, i.e., in the absence of LAK cells (Hartmann et al., 1988; van Liessum et al., 1989; Beuzeboc et al., 1989). When combined with IFN-α-2a, which induces thyroid dysfunctions in 17% of patients (10 out of 62) (Burman et al., 1986; Fentimann et al., 1988), the two lymphokines, IL-2 and IFN-α-2a, display an additive effect on thyroiditis induction (Pirchert et al., 1990). In contrast to Hashimoto thyroiditis, the lymphokine-induced autoimmune thyroiditis is self-limited and often starts with a transient hyperthyroid phase. Moreover, it appears intriguing that in most patients IL-2-induced autoimmunity remains confined to the thyroid gland.

Hoffman et al. (1989) reported that two patients receiving rIL-2 and LAK cells developed severe intrahepatic cholestasis with lobular and portal inflammation. Punt et al. (1990) described that during the first cycle of IL-2 therapy some patients developed a transient hyperbilirubinemia and an increase in liver enzyme levels. A liver biopsy revealed features of acute multifocal hepatitis, i.e., periportal hepatocyte necrosis and infiltration of lymphocytes in portal and periportal areas. The lymphocyte phenotype was predominantly $CD2^+$ $CD3^+$ $CD4^+$ $CD45^+$. Whereas these infiltrating cells were predominantly IL-2Rα, Kupffer cells showed marked IL-2Rα expression. It remains elusive whether these changes reflect transient autoimmune phenomena or a nonspecific toxic event, because hyperbilirubinemia did not recur during subsequent rIL-2 cycles. In rabbits, human recombinant IL-2 induces hepatic and myocardial infiltrations by T lymphocytes and mononuclear cells (Marshall et al., 1990). In rats it provokes marked lymphoid infiltration of the liver and eye with hepatocyte necrosis and retinal damage, as well as eosinophilic infiltration of the adrenal medulla accompanied by medulla cell necrosis (Anderson and Hayes, 1989). IL-2 in combination with IFN-β provoked fatal pemphigus vulgaris in one patient (Ramseur et al., 1989). One patient treated with IL-2 alone died from an acute leukoencephalopathy characterized by perivascular infiltration by activated T lymphocytes in foci of demyelination (Vecht et al., 1990). Another patient died from diffuse lymphocytic and eosinophilic myocarditis (Samlowski et al., 1989). Evidence for the capacity of IL-2 to induce systemic autoimmune response is provided by a case report by Lavelle-Jones (1990), according to which IL-2 treatment reactivated latent RA.

Kolb et al. (1986) showed that human recombinant IL-2 induces an acceleration of the development of autoimmune diabetes in a subline of diabetes-prone BB rats with a low spontaneous diabetes incidence. In contrast, the same group (Zielasek et al., 1990) reported that in another subline of BB rats with a high incidence of diabetes the percentage of animals suffering clinically manifest diabetes decreased from 73 to 32% upon IL-2 treatment. However, IL-2 treatment induced a massive interstitial inflammation of the exocrine pancreas, as well as insulitis, in both sublines. Although these two BB lines exhibited differences in thymosin serum levels and in the concentration of an IL-2 antagonist, it is unclear why they differ in their IL-2 response. Possibly, IL-2 may stimulate suppressor cells as well as effector and helper cells. This is suggested by an experiment in which splenocytes from diabetes-prone NOD mice were exposed ex vivo to IL-2 and/or cyclosporin A and reinjected. Whereas all animals receiving autologous cells treated with IL-2 alone developed autoimmunity, the combined treatment with cyclosporin A and IL-2 attenuated autoimmune insulitis. This may be due to selective stimulation of suppressor cells by IL-2, because helper cells are particularly sensitive to the growth-blocking effect of cyclosporin A (Formby et al., 1988).

C. LACK OF AUTOIMMUNE MANIFESTATIONS IN MICE TRANSGENIC
FOR HUMAN IL-2 OR IL-2R COMPONENTS

Massive interventions in the IL-2/IL-2R system by introducing the human IL-2 and/or IL-2Rα genes into the mouse germ line fail to induce fulminant autoimmune phenomena. Mice that carry the intact human genomic IL-2 gene or a mouse metallothionin-I promoter/human IL-2 chimeric gene develop motor ataxic symptoms associated with perivascular lymphocyte accumulation in the cerebellar meninx followed by a destructive nonspecific inflammatory process (Katsuki et al., 1989). The possible autoimmune nature of this phenomenon remains elusive. Ishida et al. (1989a) constructed transgenic mice that carry the human cDNA IL-2 gene under the control of the H-$2K^d$ promoter and express human IL-2 mRNA in lymphoid organs, lung, muscle, and skin. These mice exhibited profound phenotypic alterations, including splenomegaly, focal lymphocyte infiltrations in the lung, alopecia, and increased Thy-1^+ dermal epithelial cells when compared to control mice, but do not develop autoimmune symptoms or autoantibodies. Reciprocal skin transplantations revealed that alopecia was a primary feature of the transgenic skin. Whereas antibody responses and allogeneic mixed lymphocyte culture responses were defective, the transgenic mice exhibited normal T cell subset distributions and normal NK function. When such mice are crossed with mouse H-$2K^d$ promoter/human IL-2Rα transgenic mice, the resulting offspring are characterized by growth retardation, ataxia, premature death before 4 weeks of age, small spleens, a T lymphocyte depletion in spleen and thymus, and a prefential expansion of cells bearing a NK phenotype (Ishida et al. 1989b). This latter observation might be hypothetically explained by the possibility that early class I promoter-triggered expression of IL-2 leads pluripotent lymphoid cell precursors to a NK rather than to a T lymphocyte commitment (de la Hera et al., 1989a). Alternatively, high doses of IL-2 might arrest T cell precursor growth at an immature stage (Waanders and Boyd, 1990) or might induce NK cells that kill thymic stromal cells essential for supporting T cell maturation (Skinner et al., 1987; Plum et al., 1990). Nonetheless, the conclusions drawn from these studies are biased in view of the ectopic expression of both the IL-2 and IL-2Rα transgenes.

More subtle interventions in the IL-2/IL-2R system using a different transgenic approach, however, may induce autoimmune disease. Certain mouse lines transgenic for the transactivator protein Tax of HTLV-I develop autoimmune symptoms resembling those of Sjögren's syndrome (Green et al., 1989a). Tax has been shown to induce nuclear proteins that allow transcription of both the IL-2 and the IL-2Rα genes. Nonetheless, it remains to be established whether Tax really induces autoimmunity via the IL-2/IL-2R pathway, because Tax transactivates further genes (e.g., granulocyte/macrophage colony-stimulating factor), which could also contribute to the inflammatory process (Green et al., 1989b). In addition, Tax transgenic animals have been reported to exhibit nonautoimmune symptoms, e.g., neurofibromatosis-like mesenchymal tumors, which are unlikely to involve IL-2 and its receptor (Hinrichs et al., 1987).

D. Effect of a Recombinant IL-2/Vaccinia Virus Construct on SLE of MRL/Mp-*lpr/lpr* Mice

Lupus-prone inbred strains of the mouse provide excellent models for human systemic lupus erythematosus (SLE) because most of the immunopathological abnormalities fundamental in the human disease appear to be operative (Theofilopoulos and Dixon, 1985; Theofilopoulos et al., 1989). MRL/Mp-*lpr/lpr* mice exhibit an acute disseminated autoimmune disease characterized by hypergammaglobulinemia, anti-DNA and other antinuclear antibodies, circulating immune complexes, interstitial glomerulonephritis, necrotizing polyarteritis, and a rheumatoid arthritis-like pathology accompanied by rheumatoid factors in serum. This lupus-prone inbred strain carries a disease accelerating gene that adds into a genetic background that normally produces late-onset lupus. A single-locus recessive lymphoproliferation (*lpr*) gene causes angioimmunoblastic lymphadenopathy and a massive expansion of an unusual subset of T cells that are phenotypically Thy-1$^+$ CD3$^+$ CD5$^+$ CD4$^-$ CD8$^-$. These cells also display the B cell marker B220 and express rearranged TCRα and TCRβ chains (α/βTCR/CD3^{dull}), among which V$_\beta$8.2 and V$_\beta$8.3 gene products account for 60% of the expressed TCR (Singer et al., 1989). The *lpr* cells are defective in IL-2 production, but transcribe cytokines involved in the inflammatory process (IFN-γ, TNF-α, TNF-β, and IL-6) (Murray et al., 1990). Although the exact pathogenetic contribution of the *lpr* cell remains unknown, it is clear that the *lpr* gene precipitates development of a generalized autoimmune disease when present on a lupus background. Different approaches aimed at reducing the number of *lpr* T cells also lead to an improvement of autoimmunity. Thus, neonatal thymectomy, anti-Thy-1.2 treatment, or treatment with 5-azacytidine prevent lymphoproliferation and delay disease onset in MRL/Mp-*lpr/lpr* mice by several months (Steinberg et al., 1980; Yoshida et al., 1990). T cell vaccination, i.e., injection of a low number of *lpr* cells (2.5 × 10^5) into young MRL/Mp-*lpr/lpr* mice, causes diminution in the development of both lymphoproliferation and autoimmune lesions

(Gutierrez-Ramos et al., 1990b, 1991a). Conversely, injection of MRL/ Mp-*lpr/lpr* stem cells into lethally irradiated MRL/Mp-+/+ recipients provokes a syndrome resembling graft-verus-host disease (Theofilopoulos et al., 1985).

Adoptive bone marrow transfer experiments have demonstrated that hematopoietic stem cells from MRL/Mp-*lpr/lpr* mice carry the *lpr* defect, whereas the genotype of the thymus has no importance (Theofilopoulos et al., 1981). Nonetheless, the presence of a thymus is necessary for the selective expansion of the abnormal *lpr* T cell subset whose exact cellular derivation remains obscure. Because *lpr* cells express a combination of immature ($CD4^-$ $CD8^-$) and mature ($\alpha/\beta TCR/CD3^+$) surface markers, it has been difficult to accommodate them in currently accepted schemes of intrathymic T cell maturation. Whereas some authors speculate that *lpr* cells represent developing T cells that fail to die in the thymus or "branch off" from the main pathway and are exported into the periphery (reviewed by Strominger, 1989), others believe that *lpr* cells represent a relatively mature, postdeletional stage, given that both I-E- and Mls-related V_β clonal deletions may be detected among these cells (Theofilopoulos et al., 1989). RNA protection analysis of *lpr* T cells revealed the identical set of I-E tolerance-related clonal deletions (i.e., $V_\beta 5.1$, 11, 12, and 16) of the non-*lpr* congenic MRL/Mp-+/+ strain (Singer et al., 1989; Mountz et al., 1990). It therefore has been proposed that *lpr* cells derive from a "loss pathway" where the accessory molecules CD4 and/or CD8 are down-regulated on cells that previously have expressed these molecules. Accordingly, *in vivo* treatment of MRL/Mp-*lpr/lpr* mice with anti-CD4 mAb inhibits the accumulation of *lpr* cells (Santoro et al., 1983). The *lpr* cell has an equivalent in mice carrying the *gld* mutation, an abnormality that is nonallelic to *lpr* and also predisposes to the development of autoimmunity (Roths et al., 1984; Davidson et al., 1986). An analogous T lymphocyte population has been found in a patient with features of graft-versus-host disease (Wirt et al., 1989). Also, individuals suffering from SLE have increased numbers of $\alpha/\beta TCR/CD3^+$ $CD4^-$ $CD8^-$ lymphocytes that augment the production of anti-DNA autoantibodies *in vitro* (Marcos et al., 1988; Shivakumar et al., 1989).

To test the effect of IL-2 on the development of lupus of MRL/Mp-*lpr/lpr* mice, 10^7 plaque-forming units of a recombinant IL-2/ vaccinia virus (IL-2. VV; construct vCF13), i.e., a vaccinia containing the human IL-2 gene under the control of the vaccinia p7.5 promoter (Flexner et al., 1987), were injected four times in 14-day intervals into the peritoneal cavity of female MRL/Mp-*lpr/lpr* mice. As controls, either wild-type vaccinia virus (WT.VV) or a recombinant human IL-4/ vaccinia virus (IL-4.VV) was administered. Because human IL-4 does not bind to mouse IL-4 receptors (Mosely et al., 1989), this latter construct may be considered as a negative control. IL-2.VV proved to be a convenient virus vector lymphokine-delivery system that allows maintenance of a protracted course of high systemic IL-2 levels. IL-2 ELISA activity (8.9 ± 0.6 U/ml) was detectable in the serum of IL-2.VV-treated mice from day 10 after the first intraperitoneal injection to 25 days after the last injection.

1. Effect of IL-2.VV on Life Expectancy and Autoimmune Manifestations

In contrast to the antitolerance activity of IL-2 and its putative involvement in mediating autoimmune disease (see above), IL-2.VV treatment has a beneficial effect on the spontaneously developing systemic autoimmunity of MRL/Mp-*lpr/lpr* mice (Gutierrez-Ramos et al., 1990b, 1991a). Whereas the mean survival time for untreated or control *lpr* females observed in our colony is 190–195 days, IL-2.VV mice treated at 2 or 4 months of age displayed a significant increase in life span (408 days; Fig. 3). No effect was observed when IL-2.VV treatment was started at 6 months, indicating that either the autoimmune lesions responsible for premature death have already advanced in an irreversible fashion or that the mechanism of the life-prolonging effect of IL-2 relies on an interference with early events in the pathogenetic cascade.

The prolonged survival rate after IL-2.VV treatment was accompanied by a lack of clinical symptoms such as cutaneous ulcers, arthritis, and general deterioration (Fig. 4). Mononuclear kidney infiltrates, as well as glomerular lesions typical of untreated MRL/Mp-*lpr/lpr* mice (intraglomerular proliferation, enlargement of glomeruli, renal deposits of complement factor 3, and immunocomplexes), do not develop when animals are treated at 2 or 4 months of age with IL-2.VV (Fig. 5). The improvement in the pathohistomorphology of the kidney correlates with a normalization of proteinuria. Moreover, swollen joints, pannus formation, synovial proliferation, and degradation of the articular cartilage were absent in IL-2.VV-infected mice. Kidneys from 6-month-old *lpr* mice treated with IL-2.VV showed detectable scar lesions and sclerosed glomeruli. As the major cause of death in MRL/Mp-*lpr/lpr* mice is glomerulonephritis, this may explain why late IL-2.VV treatment did not augment life expectancy. Concomitant to the amelioration of macro- and microscopic symptoms of autoimmune disease, some but not all autoantibody titers decrease. Serum levels of

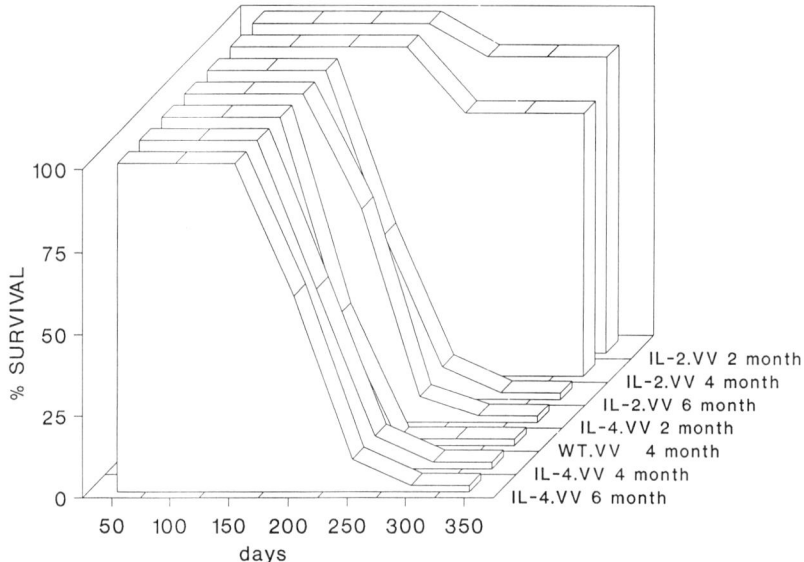

FIG. 3. Survival rates of MRL/Mp-*lpr/lpr* mice treated with a recombinant IL-2/vaccinia virus (IL-2.VV). Plaque-forming units (10^7) were injected four times in 14-day intervals into the peritoneal cavity of female MRL/Mp-*lpr/lpr* mice starting at 2, 4, or 6 months of age. As controls, either wild-type vaccinia virus (WT.VV) or a vaccinia virus construct (IL-4.VV) that releases human IL-4 (which does not bind to mouse IL-4 receptors) was administered. Note that IL-2.VV prolongs the life span when animals are treated at 2 or 4 months of age, but has no effect at 6 months.

IgM (not IgG), antibodies against double-stranded DNA, and IgM rheumatoid factors that combine with IgG_{2a} (not those that bind IgG_1) were significantly reduced (Table VII). This could mean that the humoral component of the autoimmune reaction is less susceptible to IL-2-mediated modulation than is the cellular component.

2. Effect of IL-2.VV on the T Cell Compartment

A common feature of SLE developing in humans and MRL/Mp-*lpr/lpr* mice is a severe thymic atrophy, most accentuated in the cortical area (Theofilopoulos and Dixon, 1985). IL-2.VV, not IL-4.VV or WT.VV, induces a nearly 10-fold increase in thymic cellularity, but does not change the overall architecture of the thymus (Table VII). In contrast with untreated or IL-4.VV-infected mice, in which the majority of thymocytes belong to a selectively expanded double-negative ($CD4^-$ $CD8^-$) subset, the distribution of thymocytes in IL-2-VV infected mice is close to that found in normal nonautoimmune strains.

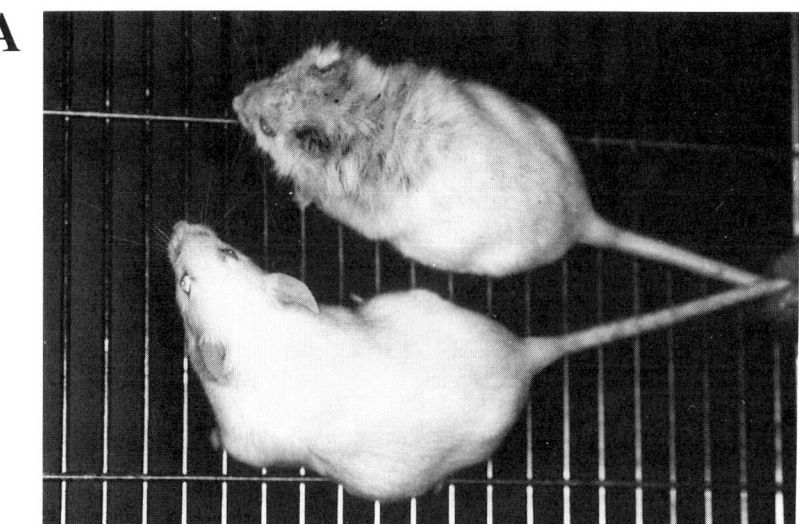

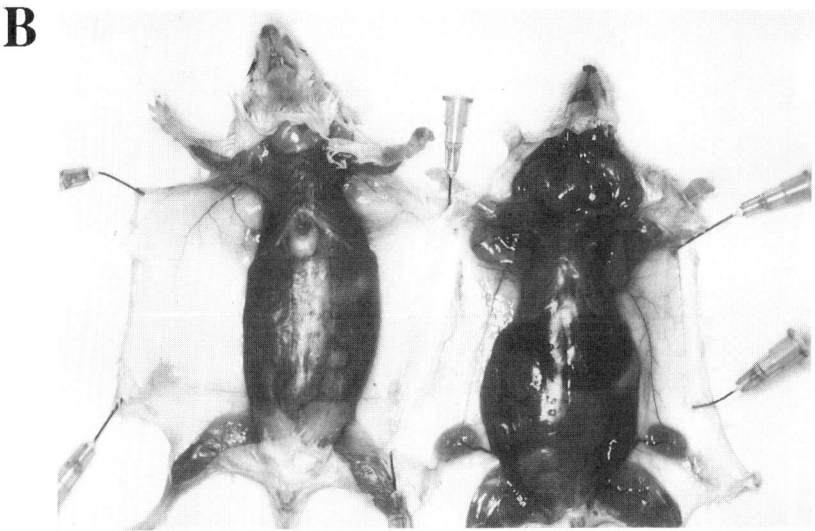

Fig. 4. IL-2.VV improves the clinical picture and reduces lymphadenopathy in MRL/Mp-*lpr/lpr* mice. Female MRL/Mp-*lpr/lpr* mice were either treated with WT.VV or with IL-2.VV starting from 2 months of age. The external aspect (A) and the necropsy (B) of two 9-month-old individuals are depicted. As do untreated controls, mice injected with WT.VV (upper mouse in A and mouse on left in B) develop cutaneous ulcera, acral necrosis (A), hepatomegaly, and a massive lymphadenopathy visible in the cervical and inguinal areas (B). In contrast, IL-2.VV-treated animals (lower mouse in A, mouse on right in B) exhibit a normal phenotype.

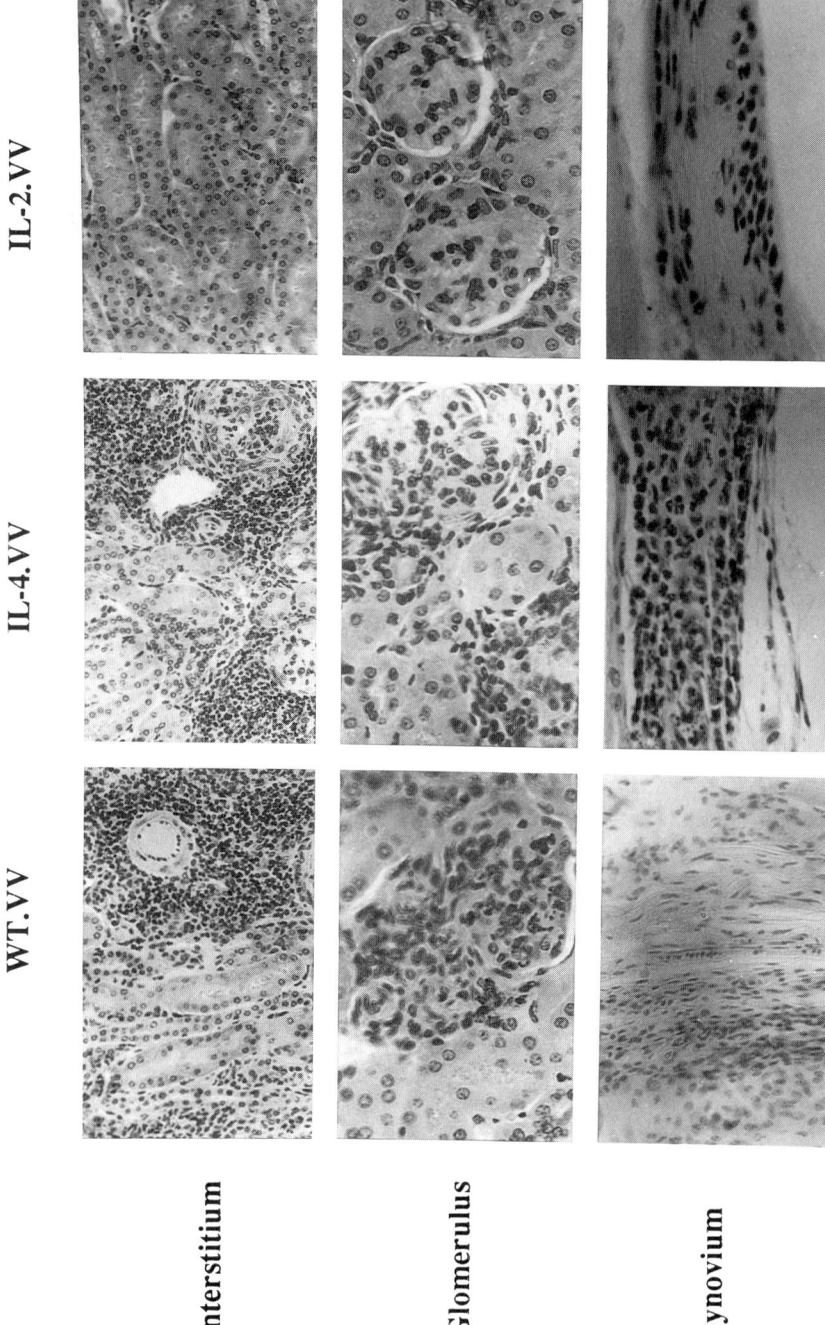

TABLE VII

AMELIORATION OF LYMPHADENOPATHY AND AUTOIMMUNE SYMPTOMS IN MRL/Mp-*lpr/lpr* MICE TREATED AT DIFFERENT AGES WITH IL-2.VV

Parameter[a]	Age at time of treatment (months)					
	2		4		6	
	IL-4.VV	IL-2.VV	IL-4.VV	IL-2.VV	IL-4.VV	IL-2.VV
Clinical findings			Number of cases/total			
Proteinuria	16/16	0/16*[b]	22/22	0/23*	16/16	12/16
Cutaneous ulcers	16/16	0/16*	22/22	0/23*	16/16	2/16*
Lymph node enlargement	16/16	0/16*	22/22	0/23*	16/16	0/16*
Arthritis	11/16	0/16*	17/22	0/23*	10/16	2/16*
Rheumatoid factors			Concentration (μg/ml)			
Anti-IgG2a	73 ± 6	18 ± 7*	70 ± 8	19 ± 8*	75 ± 8	22 ± 8*
Anti-IgG1	21 ± 2	23 ± 2	22 ± 2	20 ± 1	22 ± 2	20 ± 2
Anti-dsDNA autoantibodies			Optical density ($\times 10^{-2}$)			
IgM	51 ± 5	22 ± 2*	44 ± 7	13 ± 3*	62 ± 6	15 ± 1*
IgG	112 ± 13	85 ± 8	106 ± 22	91 ± 22	95 ± 14	108 ± 14
Cellularity			Count ($\times 10^{-6}$)			
Thymus	17 ± 2	165 ± 14*	20 ± 2	148 ± 13*	18 ± 2	25 ± 2
Inguinal lymph nodes	180 ± 17	23 ± 2*	190 ± 16	26 ± 2*	213 ± 21	32 ± 3*

[a] Parameters were evaluated at 7 months of age.
[b] Asterisks mark significant reduction of parameters ($p < 0.01$, χ^2 test, Student t test).

IL-2.VV leads to a decrease in $CD4^- CD8^-$, $CD4^+ CD8^-$, and $CD3^+ CD4^- CD8^-$ cells and entails an increase in $CD4^+ CD8^+$ and $Thy-1^+ CD3^{-/dull}$ thymocytes (Table VIII). Collectively, these data support the assertion that IL-2.VV promotes the intrathymic differentiation of double-negative precursor thymocytes into $Thy-1^+ CD3^{-/dull} CD4^+ CD8^+$ thymocytes, the main intermediate stage in T cell development (Gutierrez-Ramos *et al.*, 1989a).

In the peripheral immune system, IL-2.VV treatment leads to a drastic reduction of the *lpr* ($CD3^+ CD4^- CD8^-$) population in spleen, lymph nodes, and peritoneum (Table VII), as well as a concomitant decrease in lymphadenopathy (Table VIII). To obtain this effect, the

FIG. 5. Representative photomicrographs of kidney interstitia, glomeruli, and synovia from female MRL/Mp-*lpr/lpr* mice infected intraperitoneally in 2-week intervals with 10^7 plaque-forming units (four cycles starting from 4 months of age) of wild-type vaccinia virus (WT.VV) or vaccinia virus constructs that release human IL-4 (IL-4.VV) or human IL-2 (IL-2.VV). The samples were removed from 7-month-old animals and subjected to hematoxylin–eosin histology.

TABLE VIII
PHENOTYPIC ANALYSIS OF LYMPHOID CELLS RECOVERED FROM 7-MONTH-OLD MRL/Mp-*lpr/lpr* MICE TREATED WITH IL-2.VV OR IL-4.VV AT DIFFERENT STAGES OF LIFE[a]

		Age at time of treatment (months)					
		2		4		6	
Organ	Phenotype	IL-4.VV	IL-2.VV	IL-4.VV	IL-2.VV	IL-4.VV	IL-2.VV
		Cell subset (%)					
Thymus	$CD4^+ CD8^+$	16	44	15	42	11	50
	$CD4^+ CD8^-$	44	29	46	26	28	22
	$CD3^+ CD4^- 8^-$	17	5	20	6	31	18
	$CD3^- Thy\text{-}1^+$	10	35	17	39	10	8
Spleen	$CD3^+ CD4^- 8^-$	15	8	17	4	30	13
Lymph node	$CD3^+ CD4^- 8^-$	35	22	58	7	61	59
Peritoneum	$CD3^+ CD4^- 8^-$	20	20	67	4	48	35

[a] Cells were subjected to double staining and two-color cytofluorometric analysis.

optimal initiation point of treatment was 4 months, with 6 months having only a marginal effect. The relative amount of mature $CD3^+$ $CD4^+ CD8^-$ or $CD3^+ CD4^- CD8^+$ subsets, but not of NK cells, augments in spleen and lymph nodes. Among the T lymphocytes, IL-2.VV results in an expansion of cells expressing the $V_\beta 8.1$ gene at the expense of $V_\beta 8.2^+$ and $V_\beta 8.3^+$ cells. Thus, the treatment of MRL/Mp-*lpr/lpr* with IL-2.VV tends to correct the preponderance of $V_\beta 8.2$ and $V_\beta 8.3$. In the bone marrow, a twofold increase in cells bearing the surface markers of lymphoid and myeloid precursors (Joro 37.5, BP1, PC61, and B220) was observed after IL-2.VV (relative to IL-4.VV) inoculation (Gutierrez-Ramos *et al.*, 1990b).

3. *Possible Mechanism of the Beneficial Effect of IL-2 on Lymphoproliferation and Autoimmunity*

IL-2.VV prolongs the life span of MRL/Mp-*lpr/lpr* mice (Gutierrez-Ramos *et al.*, 1989b, 1990a), prevents the development of a series of pathomorphologically or serologically detectable symptoms of autoimmune disease, and reduces the phenotypic manifestation of the *lpr* mutation, i.e., the accumulation of abnormal *lpr* cells in the peripheral immune system. As the *lpr* gene represents the major disease-accelerating factor, it is tempting to speculate that the beneficial effect of IL-2.VV is functionally linked to its capacity to counteract this abnormality. This idea is supported by the abundant functional and genetic evidence that links autoimmune phenomena to the *lpr* defect. It also is sustained by the observation that the beneficial effect of IL-2 on survival rates and autoimmune disease is only obtained when the therapy is initiated early (2 or 4 months) and still reduces the accumulation of *lpr* cells. Furthermore, the same protocol of IL-2.VV treatment that proved to be efficient in MRL/Mp-*lpr/lpr* mice had no effect on the development of lupus in the strains (NZB × NZW)F_1, which lack the *lpr* mutation (our unpublished observation). Long-term treatment with purified human recombinant IL-2 similarly did not alter consistently the SLE parameters of (NZB × NZW)F_1 hybrids (Owen *et al.*, 1989).

As to the mechanism by which IL-2 prevents or attenuates the *lpr* phenomenon, several speculative explanations may be forwarded. The progenitor of the *lpr* cell is known to be bone marrow derived and to pass through an obligatory intrathymic stage. As discussed above, IL-2 has direct effects on the frequency of lymphoid and myeloid progenitors in the bone marrow, and apparently stimulates the intrathymic differentiation of double-negative precursor thymocytes into Thy-1^+ $CD3^{-/dull}$ $CD4^+$ $CD8^+$ thymocytes. It could thus inhibit the phenotypic expression of the *lpr* mutation on the early precursor stage or circumvent a putative block in cell differentiation that accounts for the branching off of the *lpr* cells from the streamline of intrathymic T cell differentiation. However, the fact that IL-2.VV reduces an established lymphadenopathy in 6-month-old animals without affecting the thymic architecture argues against the possibility that IL-2 might interfere with the intrathymic generation of *lpr* cells (Table VII). Fresh *lpr* T cells recovered from lymph nodes lack the IL-2Rβ chain that is essential for IL-2-mediated signal transduction, and possess low, if any, levels of low-affinity IL-2-binding sites. Culture of these cells with Con A and PMA only induces IL-2Rα chains but not IL-2Rβ chains (Rosenberg *et al.*, 1989). In contrast, another recent study demonstrated that a MRL/Mp-*lpr/lpr*-derived cell line with characteristics of *lpr* cells expressed IL-2Rβ but not IL-2Rα, even after induction. This cell line responds to IL-2 by transducing a negative signal resulting in growth inhibition (Gutierrez-Ramos *et al.*, 1989a), a finding that suggests that at some stage of their development *lpr* cells are susceptible to direct IL-2-mediated suppression. This possibility appears even more probable in light of the inhibitory effect of IL-2 on T cell precursor growth reported for transgenic mice bearing the human IL-2 gene (Ishida *et al.*, 1989) or fetal murine thymic lobe cultures incubated in the presence of high IL-2 doses (Waanders and Boyd, 1990). In addition, it is possible that IL-2 activates some cell types that counteract the surge of *lpr* cells. This is suggested by transfer experiments in which lymph node cells from MRL/Mp-*lpr*/*lpr* mice treated with IL-2.VV adoptively transfer protection against lymphadenopathy and premature death to young MRL/Mp-*lpr*/*lpr* animals.

E. INDUCTION OF MANIFEST AUTOIMMUNITY IN ATHYMIC MICE BY IL-2/VACCINIA VIRUS

To evaluate the effect of IL-2 on nondeletional autotolerance, we have chosen to administer IL-2.VV to mice in which clonal deletion is inoperational due to the absence of a thymus. Congenitally athymic (nude, or *nu/nu*) mice, i.e., mutant strains whose thymic primordium is rudimentary and lacks colonization by pro-T cells at any stage of development, are not completely devoid of lymphocytes expressing T cell markers and bearing rearranged TCR genes, apparently due to the existence of extrathymic T cell differentiation pathways (Hünig, 1983). The *nu/nu* mice are characterized by a numeric T cell defect and exhibit an oligoclonal T cell repertoire, a notion that is based on Southern blots that reveal a more restricted pattern of TCRβ chain rearrangement in *nu/nu* T cells than in *nu/+* controls (MacDonald *et al.*, 1987). T lymphocytes isolated from BALB/c *nu/nu*, C3H/HeN *nu/nu*, and B10.D2 *nu/nu* mice comprise a high percentage of cells expressing products of "forbidden" TCR V_β gene families, i.e., T cells that in euthymic *nu/+* littermates or +/+ controls are deleted due to their unwarranted reactivity with self MHC-encoded $E_\alpha E_\beta$ products ($V_\beta 6$, $V_\beta 11$) or self minor lymphocyte-stimulating (Mls^c) antigens ($V_\beta 3$) (Fry *et al.*, 1989; Hodes *et al.*, 1989; Rocha, 1990). The *nu/nu* T cells express decreased amounts of CD3/TCR complexes, as well as CD4 and CD8 (Hodes *et al.*, 1989). Moreover, BALB/c *nu/nu* T cells exhibit comparatively low proliferative responses to Con A, to CD3, and to the bacterial superantigen staphylococcal enterotoxin A (SEA), which in normal mice specifically stimulates $V_\beta 3^+$ and $V_\beta 11^+$ T cells (Lawetzky and Hünig, 1988; Yuuki *et al.*, 1990). Defective SEA responses of *nu/nu* T cells are restored by exogenous IL-2 (Yuuki *et al.*, 1990). Thus, T cells from athymic mice exhibit a behavior reminiscent of anergic T cells (see Section III,C.), a phenomenon that might explain why *nu/nu* mice do not develop manifest autoimmunity, although their T cell repertoire is potentially autoreactive.

Similar to the *nu/nu* mutation, neonatal thymectomy skews the T cell repertoire of adult mice to one enriched in those T cells that are normally deleted in the adult thymus (Smith *et al.*, 1989; Jones *et al.*, 1990). In neonatally thymectomized BALB/c and DBA/J mice, T cells that escape clonal deletion are not functionally competent, i.e., they fail to proliferate in response to clonotypic anti-$V_\beta 11$ (neoTx BALB/c) or $V_\beta 5$ and $V_\beta 6$ (neoTx DBA/2 mice) mAbs (Jones et al., 1990). Functional inactivation of cells bearing a forbidden TCR repertoire, however, is not absolute in neoTx mice, because, for example, neoTx (C57BL/6 × A/J)F_1 develop organ-specific autoimmune disease in the gonads, thyroid, and stomach (Smith et al., 1989).

1. Effect of IL-2.VV on nu/nu Mice

One decade ago, IL-2-containing preparations were reported to induce autoimmune hemolytic anemia in nu/nu mice if administered together with rat erythrocytes (Reimann and Diamantstein, 1981). Since at that time recombinant IL-2 was not yet available and semipurified material probably contaminated by other lymphokines was used, it remains unclear whether this effect was truly due to IL-2. We therefore tested the effect of IL-2.VV on unchallenged nu/nu mice with respect to the possible induction of autoimmune phenomena.

When wild-type vaccinia virus or a recombinant construct expressing E. coli β galactosidase, influenza hemagglutinin, or nucleoprotein and/or herpes simplex kinase are injected into outbred Swiss nu/nu or inbred BALB/c nu/nu mice, the animals succumb to infection in a dose-dependent fashion. If human or murine IL-2 is expressed by the vector, mortality is significantly reduced, a finding that correlates with an accelerated clearance of the virus (Ramshaw et al., 1987), enhanced antibody titers (Flexner et al., 1987), elevated NK cell activities (Karupiah et al., 1990a), IL-2-mediated induction of IFN-γ (Karupiah et al., 1990b), and possibly cytotoxic T lymphocyte responses (Mizuochi et al., 1989).

In accordance to these data, we observed that intraperitoneal injection of 1×10^7 plaque-forming units of wild-type vaccinia virus killed 10- to 12-week-old athymic BALB/c nu/nu and C57BL/6 nu/nu mice during the second week after treatment (Gutierrez-Ramos et al., 1991b). In contrast, the same dose of recombinant human IL-2/vaccinia virus (vCF13, IL-2.VV) allowed survival of all animals. Fourteen days after treatment these animals exhibited significantly elevated serum concentrations of rheumatoid factors, as well as anti-single-stranded DNA antibody titers, accompanied by proteinuria. Repeated injection of IL-2.VV at 14-day intervals resulted in increased proteinuria levels, a progressive leukopenia with a relative increase of $CD3^+$ $Thy-1^+$ cells in spleen, and death several days after the third IL-2.VV injection (Table IX). Moreover, a significant relative increase in $\alpha/\beta TCR^+$ cells (up to 20% of total splenocytes) and cells expressing products of the potentially autoreactive (I-E specific) $V_\beta 11$ gene family

TABLE IX
Induction of Autoimmune Phenomena in Athymic Mice by Recombinant IL-2 Vaccinia Virus Infection

Animals and treatment[a]		Rheumatoid factors		Anti-dsDNA		Proteinuria	Interstitial nephritis	Reduced life span
		anti-IgG1	anti-IgG2a	IgM	IgG			
CBA/H neo Tx	—	—[b]	—	—	—	—	—	—
CBA/H sham	IL-2.VV	—	—	—	—	—	—	—
CBA/H neoTx	WT.VV	—	—	—	—	—	—	—
CBA/H neoTx	IL-2.VV	+	+	+	+	+	+	+
BALB/c nu/nu	PBS	—	—	—	—	—	—	—
BALB/c +/+	IL-2.VV	—	—	—	—	—	—	—
BALB/c nu/nu	IL-2.VV	+	+	+	+	+	ND	+
C57BL/6 nu/nu	PBS	—	—	—	—	—	ND	—
C57V1/6 nu/nu	IL-2.VV	+	+	+	+	+	ND	+

[a] CBA/H mice were neonatally thymectomized (neoTx) at 3 days of age or were sham treated. Mice from each group received 1×10^7 plaque-forming units of recombinant IL-2/vaccinia virus (IL-2.VV) or wild-type vaccinia virus (WT.VV) starting from the sixth week of age. Treatment was repeated three times every 2 weeks and necropsy was performed at 12 weeks of age. BALB/c nu/nu or +/+ and C57BL/6 nu/nu mice were treated according to an analogous protocol with the difference that the treatment was strated at 10 to 12 weeks of age for a maximum of two cycles. Antibody titers were evaluated by standard ELISA, renal histopathology was assessed after hematoxylin-eosin staining, and reductions in life span were considered up to 8 months of age. The nu/nu animals treated with WT.VV die within 14 days after the first cycle of treatment and therefore cannot be used as controls.

[b] Plus and minus symbols indicate significant changes in a given parameter, when compared to untreated controls. Note that autoimmune symptoms are only manifest when two requirements are met: (1) IL-2 treatment and (2) either hereditary or surgical athymia. ND, Not determined.

was found. In contrast to athymic BALB/c *nu/nu* mice, euthymic controls did not develop autoantibodies or proteinuria, and survived several cycles of IL-2.VV treatment. Thus, in this model the induction of autoimmune phenomena by IL-2 is critically dependent on the absence of a thymus.

2. *Effect of IL-2.VV on Neonatally Thymectomized Mice*

In accordance with previous reports (Smith *et al.*, 1989; Jones *et al.*, 1990), neonatal thymectomy of CBA/H mice (H-2^k, I-E$^+$) results in a numeric defect in splenic and lymph node T cells and in a relative increase of cells expressing products of a "forbidden" V_β gene family, specifically a $V_\beta 11$ gene product that is clonally deleted in euthymic CBA/H controls due to its high reactivity with as-yet unidentified non-MHC antigens presented by class II I-E molecules (Tomonari, 1990). $V_\beta 11$ is expressed on up to one-third of CD3$^+$ cells in the spleen and lymph nodes of neoTx CBA/H mice. Interestingly, $V_\beta 11$ expression is highest among peritoneal T cells (up to 68%), which increased in neoTx CBA/H mice relative to non-T cells. Peritoneal T cells exhibit an abnormal, predominantly CD3$^+$ CD4$^-$ CD8$^-$ phenotype during the first 3 months of age. Such a "double-negative" phenotype has been associated with autoimmune phenomena in mice affected by the *lpr, gld*, and *nu* mutations, as well as in patients with systemic autoimmune diseases (reviewed by Gutierrez-Ramos *et al.*, 1990a). "Double-negative" T lymphocytes may either represent an immature stage of T cell maturation preceding acquisition of the accessory molecules CD4 and/or CD8 (Toribio *et al.*, 1988), or, alternatively, derive from a "loss pathway," where CD4 or CD8 is down-regulated on autoreactive T cells that have escaped negative selection (Singer *et al.*, 1989; Teh *et al.*, 1989; Egerton and Scollay, 1990), thus rendering them anergic. In agreement with these considerations, Jones *et al.* (1990) have shown that nondeleted $V_\beta 11^+$ cells from neoTx BALB/c mice lack the capacity to produce IL-2 or to undergo proliferation in response to matrix-bound anti-$V_\beta 11$ mAb, whereas $V_\beta 11^+$ cells from euthymic C57BL/6 mice do respond to this stimulus. To assess the possible antitolerance effect of IL-2, IL-2.VV was repeatedly administered to young (6-week old) neoTx CBA/H mice. In contrast to the athymic model, injections of WT.VV are readily survived and WT.VV may be used as a control. IL-2.VV administration did not alter the V_β repertoire. However, it induced conversion of double-negative peritoneal cells to either CD4$^+$ CD8$^-$ or CD4$^-$ CD8$^+$ single positives and reversed the antigen-nonresponsive state of $V_\beta 11^+$ cells, which acquired the ability to proliferate in response to clonotypic antibodies (Table X). This is in con-

TABLE X
IL-2 Allows Proliferative Responses of Cells Bearing a Potentially Autoaggressive TCR[a]

Stimulus	[^{3}H]TdR uptake[b] of peritoneal cells derived from neonatally thymectomized CBA/H mice treated with	
	WT.VV	IL-2.VV
Anti-CD3	76,500 ± 11,200	97,300 ± 6500
Anti-V$_\beta$17a	−300 ± 200	−200 ± 200
Anti-V$_\beta$11	2200 ± 1300	10,600 ± 1100[c]

[a] Neonatally thymectomized CBA/H mice injected with 10^7 plaque-forming units of IL-2.VV or a wild-type control (WT.VV) at 6, 8, and 10 weeks of age. The proliferation of peritoneal T cells (1×10^5/well) in response to plastic-absorbed mAbs directed against CD3 (positive control), V$_\beta$17a (negative control; CBA/H mice express the V$_\beta$17b allele), and V$_\beta$11 was measured in the presence of 10 U/ml of human recombinant IL-2 during a 3-day culture period followed by an 18-hour pulse label with 1 μCi of [^{3}H]thymidine. Values are presented as mean values ± SEM after subtraction of background values ($<2 \times 10^3$ cpm). Cytofluorometric analysis revealed that 40–50% of peritoneal T cells expressed V$_\beta$11, a TCR V$_\beta$ family that is deleted in euthymic CBA/H mice due to its self-reactivity. The percentage of V$_\beta$11$^+$ cells does not change upon WT.VV or IL-2.VV treatment. However, the percentage of double-negative (CD4$^-$ CD8$^-$) T cells (58% in WT.VV controls) is reduced by IL-2.VV (12%). [Data from Andreu et al. (1991).]

[b] Uptake measured as change in counts per minute (X ± SEM).

[c] Significant difference between the two groups of animals ($p > 0.01$, Student t test).

trast with the findings of Jones et al. (1990), who were not capable of abrogating the anergy of nondeleted V$_\beta$11$^+$ cells by *in vitro* short-term (3 days) exposure to IL-2. Nevertheless, the different duration of IL-2 treatment (6 weeks versus 3 days) and the very different experimental design (*in vitro* versus *in vivo*) may explain this discrepancy. Thus, IL-2.VV is known to induce *in vivo* production of IFN-γ (Karupiah et al., 1990b), i.e., another lymphokine with potent antitolerance properties. As to be expected, the induction of phenotypically mature and functionally competent T cells bearing products of the "forbidden" V$_\beta$11 gene family is accompanied by the development of autoimmune phenomena. IL-2.VV, not WT.VV, provokes the production of rheumatoid factors and DNA antibodies, as well as renal autoimmunity (minimal-change glomerulonephritis and interstitial lymphoid infiltrates in the parenchyma) in neoTx CBA/H mice, but fails to do so in euthymic sham-treated littermates (Table IX; Fig. 6). This finding indicates that IL-2 may precipitate autoimmune diseases mediated by a nondeleted T cell repertoire.

As to the mechanism by which IL-2 induces autoimmune phenomena in athymic but not in euthymic conditions, several nonexclusive possibilities may be considered. The absence of a thymus implies that

NeoTx + WT.VV

Sham + IL-2.VV

NeoTx + IL-2.VV

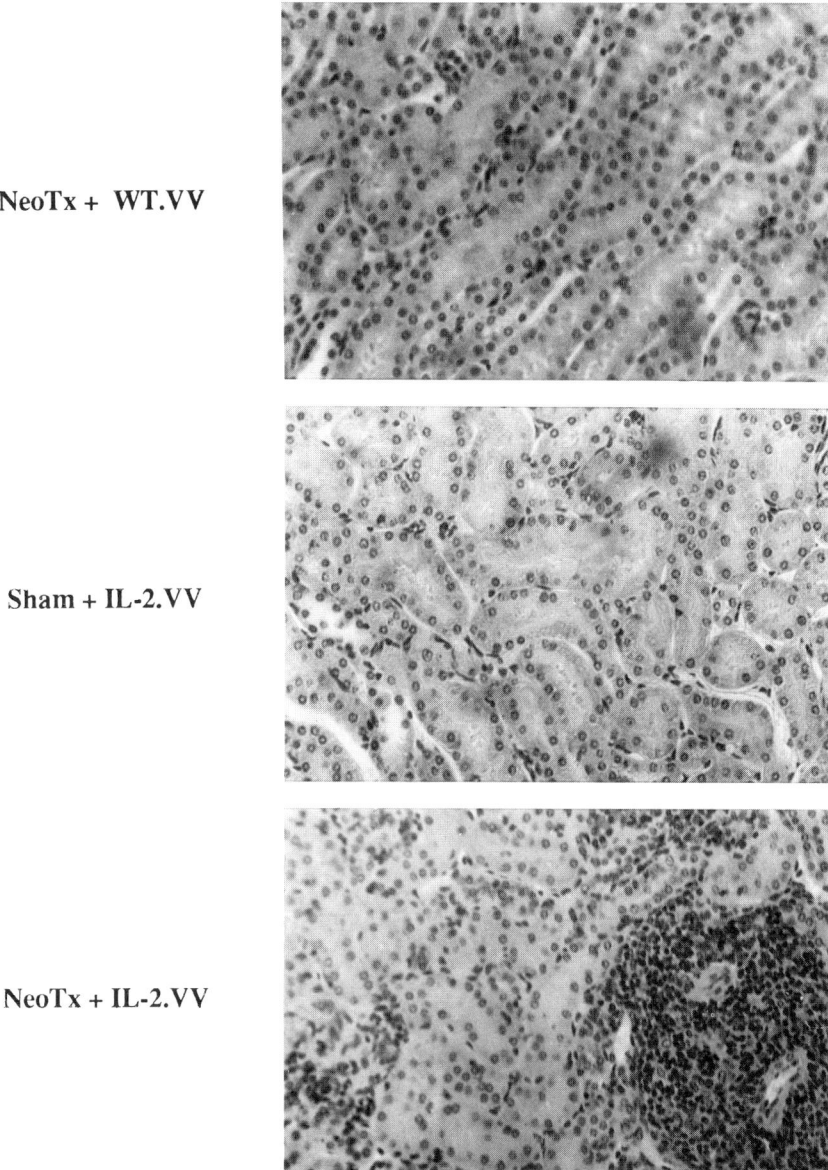

Fig. 6. Interstitial nephritis induced by IL-2 in neonatally thymectomized (neoTx) CBA/H mice. CBA/H mice, either sham-treated or thymectomized 3 days after birth, were treated by intraperitoneal injection of 10^7 plaque-forming unit of wild-type vaccinia virus (WT.VV) or recombinant IL-2/vaccinia virus (IL-2.VV) starting from the sixth week of age. Treatment was repeated three times in 2-week-intervals followed by necropsy at 3 months of age. Note the interstitial infiltrate visible only in animals that have been thymectomized and receive IL-2.VV.

autotolerance of the cellular immune system relies entirely on nondeletional mechanisms. It therefore appears probable that this precarious equilibrium is particularly vulnerable to the antianergy effect of IL-2. The absence of a thymus may generally predispose to the development of autoaggression. Neonatal thymectomy of Obese chickens (Wick et al., 1970) and thymectomy at weaning (3 weeks) of NOD mice (Dardenne et al., 1989) accelerate the development of autoimmune thyroiditis and diabetes, respectively. Neonatal thymectomy alone may induce autoimmune phenomena in several normal mouse strains (Kojima and Prehn, 1981). BALB/c nu/nu mice spontaneously develop systemic and organ-specific autoimmune diseases when transplanted with neonatal BALB/c thymus (Sakaguchi and Sakaguchi, 1990). It remains unknown whether imbalances in distinct T cell populations, altered helper/suppressor ratios, a deficient lymphokine-absorptive capacity ("lymphokine sink") due to lymphopenia, or the nonexistence of an elaborated idiotypic network may contribute to the autoimmunity-predisposing effect of athymia.

F. CYCLOSPORIN A AND AUTOIMMUNITY

One of the best studied effects of cyclosporin A (CsA), a cyclic undecapeptide of fungal origin, is its capacity to inhibit IL-2 and IFN-γ gene expression at the level of mRNA transcription both in vitro and in vivo (Kronke et al., 1984; Granelli-Piperno, 1990). This probably accounts for the well-documented immunosuppressive effects of the drug that is clinically exploited to prevent allograft rejections. In addition, CsA exerts beneficial effects on a series of different human autoimmune diseases including autoimmune uveitis, diabetes mellitus, minimal-change focal and segmental nephropathies, membranous and IgA nephropathies, RA, SLE, and polymyositis (reviewed by Kahan, 1989). In diabetes patients the side effects of CsA, especially its high nephrotoxicity, are too important to warrant its therapeutic use. Nonetheless, this drug increases the rate and the length of remissions in patients with insulin-dependent diabetes of recent onset (Feultren et al., 1986). Treatment of diabetes-prone BB rats with CsA prevents the appearance of diabetes (Like et al., 1983), a finding that has been explained by the capacity of CsA to interfere with IL-2-dependent and IL-2-independent NK cell functions (Baquerizo et al., 1989). CsA can prevent the development of mercury-induced autoimmune glomerulonephritis of the Brown–Norway rat (Baran et al., 1986), as well as spontaneously developing renal lesions of (NZB × NZW)F_1 mice (Jones et al., 1983). In MRL/Mp-lpr/lpr mice, CsA prolongs survival, prevents the development of lymphadenopathy, leads to a 100-fold decrease in $CD4^-$ $CD8^-$ $B220^+$ cells in lymph nodes, restores the capacity of T cells to mount Con A responses, reduces glomerulonephritis and arthritis, but does not suppress autoantibody or immune complex levels (Mountz et al., 1987). Thus, the consequences of both IL-2 and CsA treatment in MRL/Mp-*lpr/lpr* mice appear to be roughly the same. This suggests that the CsA effect is not mediated by down-regulation of IL-2 production, but rather via an independent mechanism. Alternatively, endogenous IL-2 whose production is blocked by CsA might be implicated in local infiltrative processes and thus play a rate-limiting role in the disease.

Despite its potent immunosuppressive activity, CsA paradoxically induces autoimmune phenomena in certain conditions. Its application to patients suffering from severe Grave's ophthalmopathy may have a deteriorating effect (Howlett et al., 1984) and it aggravates Hashimoto-like thyroiditis of Obese strain chickens (Wick et al., 1982). Irradiated hosts transplanted with syngeneic bone marrow and then treated with, and withdrawn from, CsA develop a "syngeneic graft-versus-host reaction" that bears all the pathological and clinical hallmarks of allogeneic graft-versus-host disease including erythroderma, dermatitis, and alopecia (reviewed by Hess and Fischer, 1989). This phenomenon, observed in mice, rats (Glazier et al., 1983), and humans (Jones et al., 1989), only develops in individuals that bear an intact thymus (Sorokin et al., 1986) *after* discontinuation of CsA therapy, and may be transferred to irradiated or cyclophosphamide-treated hosts. Interestingly, certain normal mouse strains (e.g., CBA/N) are particularly susceptible to the induction of a disseminated autoimmune syndrome that develops after sublethal irradiation and CsA application without syngeneic bone marrow transfer (Marcos et al., 1986). Neonatal administration of CsA alone induces organ-specific autoimmunity in BALB/c mice (Sakaguchi and Sakaguchi, 1989).

In synthesis, the effects of CsA on the development of autoimmune phenomena are multifaceted. The paradoxical effects of this drug may reflect the complex pleiotropic involvement of lymphokines in the cascade of events leading to manifest autoimmunity. As to the possible mechanism of autoimmunity induction, CsA has been shown to provoke marked changes in the thymic architecture, including a reduction of I-E expression on medullary epithelial cells and a loss of medullary, not cortical, epithelium (Cheney and Sprent, 1985). CsA does not interfere with early steps in thymocyte maturation nor with the differentiation of γ/δ T cells, but it blocks the differentiation of immature $CD4^+$ $CD8^+$ thymocytes bearing the α/βTCR into $CD4^+$ $CD8^-$ and $CD4^-$ $CD8^+$ cells. CsA-treated mice show incomplete deletion of T cells expressing TCR molecules reactive to self MHC I-E molecules [V$_\beta$11 in irradiated (B6 × CBA/Ca)F$_1$ mice, V$_\beta$17a in C57BR mice] (Jenkins et al., 1988; Gao et al., 1988). It remains to be elucidated whether the CsA-induced changes in thymocyte subset distribution and TCR repertoire are secondary to a direct effect on the developing thymocyte or to a reduction in I-E molecule expression on stromal elements thought to be critical for tolerance induction. In this regard, recent studies have shown that CsA inhibits activation-induced programmed cell death in T cell hybridomas (Shi et al., 1989).

VI. Role of IL-2-Induced Cytokines in Autoimmunity

IL-2 induces *in vivo* as well as *in vitro* production of several lymphokines, specifically TNF-α, IFN-γ, and IL-6 (Heslop et al., 1989; Kasid et al., 1989; Jablons et al., 1989) (Table I), all of which have been implicated in the pathogenesis of autoaggression. Whereas IL-6 appears to be an unequivocal proinflammatory mediator, TNF and IFN-γ have been reported to have ambiguous effects on autoimmune disease development. Thus the per se complex effects of IL-2 are further complicated by the action of additional IL-2-induced pleiotropic substances.

A. INTERLEUKIN-6 AND AUTOIMMUNITY

IL-6 is produced in a nonspecific fashion during inflammatory processes, e.g., experimental autoimmune encephalomyelitis (Gijbels et al., 1990), human RA (Hirano et al., 1988; Firestein et al., 1990), and thyroid autoimmune disease (Grubeck-Loebenstein et al., 1989). This molecule probably mediates the paraneoplastic autoimmunity of patients with IL-6-producing tumors such as cardiac myxoma (Jourdan et al., 1990), uterine carcinoma, and bladder cell carcinoma (Hirano et al., 1987). IL-6 serves as an autocrine mediator of mesangioproliferative glomerulonephritis (Horii et al., 1989). It also appears to be involved in the spontaneous proliferation and differentiation of SLE B cells into immunoglobulin-producing cells (Tanaka et al., 1988). Introduction of the IL-6 gene into hematopoietic stem cells of the mouse and deregulated expression of the gene result in a syndrome resembling Castleman's disease: lymphadenopathy, splenomegaly, polyclonal hypergammaglobulinemia, and extensive plasma cell infiltration of peripheral lymphoid organs, liver, and lung (Brandt et al., 1990). In human Castleman's disease, IL-6 is produced by the hyperplastic lymph nodes. IL-6 transgenic mice develop diffuse plasmacytosis, hyperglobulinemia, and mesangioproliferative glomerulonephritis

(Suematsu et al., 1989). To our knowledge, no beneficial effect of IL-6 on autoimmune diseases has been published.

B. TUMOR NECROSIS FACTOR-α AND AUTOIMMUNITY

Several lines of evidence suggest that TNF-α is present in autoimmune lesions and is involved in the mediation of the disease. TNF-α mRNA is hyperexpressed in an inflammation-dependent fashion in the kidneys of (NZB × W)F_1 mice (Boswell et al., 1988). TNF-α is synthesized by freshly isolated, unstimulated Kupffer cells from female MRL/Mp-*lpr/lpr* mice and aged MRL/Mp-+/+ mice, but not in normal controls (Magilavy and Rothstein, 1988). It is produced by astrocytes and macrophages in multiple sclerosis plaques (Hofman et al., 1989) where it possibly mediates the proliferation of astrocytes responsible for reactive gliosis and exerts cytotoxic effects on oligodendrocytes, thus causing demyelination (Robbins et al., 1987; Lieberman et al., 1989). Together with IL-1, TNF-α is increased in the blood and spinal fluid of patients with multiple sclerosis (Merrill et al., 1989). *In situ* hybridization studies revealed that TNF-α is expressed in the intraislet infiltrate of NOD mice, predominantly in lymphocytes adjacent to insulin-producing β cells (Campbell et al., 1988a; Held et al., 1990). TNF is produced by T cell clones isolated from autoimmune lesions (Mariotti et al., 1989; Ferrarini et al., 1990). TNF α and β production of myelin basic protein-specific T cell lines correlates with their encephalitogenicity. A mAb that recognizes both molecules prevents adoptive transfer of encephalomyelitis by such clones (Powell et al., 1990). TNF-α is a potent class II inducer on human islet β cells. Synergistic with IFN-γ, it disintegrates pancreatic islets *in vitro* and destroys islet cell monolayers on which it induces aberrant expression of class II MHC molecules (Pukel et al., 1988). Moreover, TNF-α is cytotoxic for thyroid epithelial cells (Taverne et al., 1987) and stimulates cartilage breakdown, bone resorption, synovial cell growth, and fibroblast proliferation *in vitro* (Saklatvala, 1986; Dayer et al., 1987; Bunning and Russel, 1989). In rabbits, TNF-α causes transient synovitis after intraarticular injection (Henderson and Pettipher, 1989). Pretreatment with human recombinant TNF-α exacerbates antibody-mediated injury in a passive model of antiglomerular basement membrane antibody-mediated nephritis in rats (Tomosugi et al., 1989). Short-term continuous infusion of TNF induces glomerular damage in rabbits (Bertani et al., 1989). In patients with SLE, spontaneous TNF-α production by peripheral blood mononuclear cells is higher than in healthy controls, but PMA-induced levels are lower (Malavé et al., 1989). Using a radioimmunoassay, Maury and Teppo

(1989) showed that circulating TNF levels are elevated in some patients with SLE and RA.

In sharp contrast, several authors suggest that autoimmune diseases are associated with deficient TNF production. As compared to normal controls, NOD mice display poor elevations in TNF-α serum levels subsequent to injection of LPS or IFN-γ (Satoh *et al.*, 1989). Long-term treatment with recombinant TNF-α prevents the development of type 1 diabetes in NOD mice (Jacob *et al.*, 1990a) and BB rats (Satoh *et al.*, 1990). A polymorphic microsatellite in the TNF-α promoter identifies an allele unique to the NZW mouse strain (Jongeneel *et al.*, 1990). The contribution of the NZW parent to the lupuslike glomerulonephritis of (NZB × NZW)F_1 mice has been traced back to a single dominant allele mapping within or close to the MHC, which could well be the TNF-α/β locus (Kotzin and Palmer, 1987). A note of caution is required with this interpretation of the data because the association of the NZB MHC haplotype and the autoimmune phenotype is imperfect, with a discordance of about 10% (Babcock *et al.*, 1989). Peritoneal macrophages from NZW mice are defective in their ability to secrete TNF in response to LPS, and treatment with high doses of TNF-α can postpone the development of the disease in (NZB × NZW)F_1 hybrids (Jacob and McDevitt, 1988), although this beneficial effect cannot be sustained indefinitely by chronic therapy (Gordon *et al.*, 1989). At lower doses, however, TNF-α has been reported to accelerate renal disease and mortality rate in (NZB × NZW)F_1 animals (Brennan *et al.*, 1989). The effect of high doses of TNF-α may be related to a general nonspecific immunosuppressive effect on accessory cells, T cells, and NK lymphocytes (Gordon *et al.*, 1989; Gordon and Wofsy, 1990). Peripheral blood lymphocytes from HLA-DR3- or DR4-positive donors with SLE exhibit high TNF-α production in response to Con A plus PMA, whereas DR2- and DQw1-positive subjects, i.e., individuals at risk of developing lupus nephritis, show low levels of *in vitro* TNF-α production (Jacob *et al.*, 1990b).

C. INTERFERON-γ AND AUTOIMMUNITY

Several organ-specific autoimmune diseases are associated with a genetically conferred, primary IFN-γ-mediated hyperinducibility of MHC class II antigens on target cells of the autoaggressive attack, i.e., pancreatic β cells in NOD mice (Leiter *et al.*, 1989), thyroid epithelial cells in Obese strain chickens (Wick *et al.*, 1989), and astrocytes in mouse and rat strains susceptible to the induction of experimental allergic encephalomyelitis (Massa *et al.*, 1987). Expression of transgenic IFN-γ under the control of the insulin receptor in mouse pancreatic β cells results in a loss of immune tolerance to islet cells and provokes autoimmune diabetes mediated by cytotoxic T cells (Sarvetnick et al., 1988, 1990).

Similar to TNF, IFN-γ also has parodoxical effects on autoimmune diseases. Intraperitoneal injection of rat recombinant IFN-γ augments both myelin-induced and T cell line-mediated experimental neuritis of Lewis rats, and *in vivo* administration of a mAb to IFN-γ suppresses the disease (Hartung et al., 1990). In contrast, intraventricular administration of IFN-γ completely suppresses clinical signs of experimental allergic encephalitis, and intraperitoneal injection of anti-IFN-γ just prior to the onset of clinical signs results in a more severe disease course in Lewis rats (Voorthuis et al., 1990). Quite comparable results were reported for murine allergic encephalitis, wherein neutralizing mAb against IFN-γ causes an increase in morbidity and mortality rates (Alfons et al., 1988). This finding was interpreted in the sense that endogenous IFN-γ produced during the course of the pathogenetic process has a disease-limiting role (Biliau et al., 1988). In patients with multiple sclerosis, however, systemic administration of IFN-γ precipitates disease exacerbations (Panitch et al., 1987). IFN-γ treated (NZB × NZW)F_1 mice display an accelerated development of fatal immune complex glomerulonephritis relative to sham-treated controls (Jacob et al., 1987). Murine streptozotocin-induced diabetes (Campbell et al., 1988b) and collagen-induced arthritis are augmented with IFN-γ (Mauritz et al., 1988). Diabetes of BB rats is prevented by anti-IFN-γ mAbs (Nicoletti et al., 1990). Dependent on the timing of anti-IFN-γ mAb injection, rat adjuvant arthritis is suppressed or augmented (Wiesenberg et al., 1989). Pernice et al. (1989) published a study in which recombinant IFN-γ caused partial improvements in clinical symptoms in seven out of nine severely ill patients with systemic juvenile rheumatoid arthritis.

In synthesis, both TNF-α and IFN-γ exert ambiguous effects on autoaggressive processes. Depending on the administration schedule and the experimental system, opposite effects may be obtained with both substances when they are applied *in vivo*. Although their implication in the local disease process is probable, systemic administration of these substances with the consequent aphysiological sequels entails unforseen effects. It remains to be established which conditions determine whether negative or positive effects on ongoing disease processes follow from the application of TNF-α and IFN-γ, as well as their respective antagonists.

VII. Concluding Remarks

The system composed of the IL-2 and its receptor constitutes a dual pivot in T lymphocyte biology. On the one hand, the IL-2/IL-2R system is involved in the antigen-nonspecific phase of T cell precursor proliferation and differentiation. On the other hand, IL-2 is a lymphocytotropic hormone released by a defined subset of T helper cells subsequent to antigen-specific stimulation, provided the T cell receives costimulatory signals via monomorphic receptors. In the peripheral immune system, IL-2 is a pleiotropic mediator whose effects are rigorously controlled by different compartmentalization mechanisms. Under physiological conditions, the topographical and chronological restriction of IL-2 production, bioavailability, and responsiveness guarantees that this mediator, which plays a fundamental role in the regulation of peripheral tolerance, does not endanger the specificity of the immune response. IL-2 is a transient signal that communicates across a short gap between cooperating pairs of cells (T–B, T–APC, and T–T) and/or functions as an autocrine growth factor for T helper cells. Endogenous IL-2 has a predominantly immunostimulatory function, a notion that is based on the fact that disruption of the IL-2/IL-2R system by anti-IL-2R mAbs and drugs such as cyclosporin A has immunosuppressive effects. Nonetheless, the systemic administration of high doses of recombinant IL-2, submerging the whole organism in IL-2 in a nonphysiological way, has ambiguous effects on the immune system. While reestablishing antigen responsiveness in a variety of tolerant states, IL-2 may also exert inhibitory effects in certain experimental systems. These paradoxical effects may be related to the multiple actions of IL-2 on suppressor, helper, and effector cells of the immune system, as well as to feedback inhibition phenomena. Numerous *in vivo* and *in vitro* studies indicate that IL-2 interferes with at least one of the fundamental mechanisms ensuring autotolerance—peripheral anergization. T cells whose IL-2 gene has been silenced due to TCR triggering in the absence of costimulatory signals, as well as T lymphocytes that have down-regulated the cell surface expression of TCR/CD3 or CD8 during the anergization process, respond to IL-2 by reacquiring antigen-responsiveness. Whether IL-2 abrogates clonal deletion in the thymus remains a matter of debate.

A common denominator of the vast majority of diseases of probable autoaggressive etiology is their association with abnormalities in the IL-2/IL-2R system. In organ-specific and systemic autoimmune diseases either developing in animals or in humans, several indications point to an endogenous activation of the system composed of IL-2 and its receptor components. Elevated concentrations of IL-2 and the soluble IL-2Rα chain are encountered during active disease phases in various body fluids, whereas the *in vitro* production of IL-2 by peripheral lymphocytes in response to polyclonal stimuli is deficient, most probably due to a transient exhaustion of *in vivo*-activated IL-2-producing cells. Elevated sIL-2R may be secondary to an overproduction of IL-2, because the lymphokine stimulates expression and secretion of its own receptor. Alternatively, high serum IL-2 and sIL-2R may reflect abnormal autoantigenic and/or polyclonal stimulation of T cells expressing both IL-2 and IL-2Rα chains. Abnormalities in the IL-2/IL-2R system may have a genetic basis. Moreover, they may be acquired during the process of autoimmunization or result from viral infections, e.g., with HTLV-I, which transactivates both the IL-2 and the IL-2Rα genes. IL-2-producing and IL-2-receptive cells are present in autoimmune lesions and exert an important effector function. Antagonizing the IL-2/IL-2R system by administration of IL-2Rα-targeted drugs prevents, postpones, or mitigates the development of autoimmune diseases in virtually any animal model investigated in this respect, but has limited effects on already established autoimmune diseases. Therefore, IL-2 appears to play a crucial role in the initial stage of the pathogenesis, but is of limited importance once the destructive process has entered into the self-perpetuating phase.

In spite of the well-documented antitolerance effects of IL-2, the majority of patients receiving recombinant IL-2 do not develop fulminant autoimmune diseases, but rather display transient and organ-specific autoimmune symptoms that, in most cases, become manifest in the thyroid gland. Similarly, mice transgenic for human IL-2 do not exhibit signs of overt autoimmunity. Administration of a recombinant vaccinia virus construct that serves as an autonomously replicating device releasing human IL-2 (IL-2.VV) does not induce autoimmune symptoms in normal mice, nor does it change the disease course of spontaneous lupus erythematosus in (NZB × NZW)F_1 mice. In contrast, IL-2.VV provokes the production of antinuclear and rheumatoid factors, as well as lymphoid kidney infiltrations in athymic mice, irrespective of whether athymia is congenital or acquired during the neonatal period. Athymia is known to result in numerous alterations in the cellular immune system, above all in an abrogation of clonal deletion that results in the surge of T cells bearing a "forbidden," potentially autoreactive T cell repertoire. Whereas such T cells from untreated neonatally thymectomized mice do not proliferate in response to TCR-specific triggers, T cells from IL-2.VV-treated mice acquire antigen responsiveness. Thus, IL-2 appears to stimulate the development of autoimmune disease mediated by a nondeleted T lymphocyte repertoire. In this context, it is tempting to speculate that patients who develop autoimmune lesions upon IL-2 administration do so because they bear—for whatever reason—forbidden T cell clones. IL-2 would fail to induce autoimmune lesions in individuals bearing a "normal," deleted repertoire, but would reactivate latent autoimmunity in persons whose autotolerance relies on peripheral postdeletional mechanisms.

In marked contrast to its capacity to induce autoimmune phenomena in undiseased individuals or animals, IL-2.VV exerts a paradoxical, beneficial effect on the systemic autoimmune disease developing in MRL/Mp-*lpr/lpr* mice: (1) it augments the life expectancy from 7 to 13 months when treatment is started at 2 or 4 months of age, (2) it prevents the development of pathomorphologically detectable symptoms of autoimmune disease, and (3) it normalizes some parameters of humoral autoimmunity. This therapeutic effect of IL-2 probably is due to its capacity of counteracting the phenotypic manifestation of the *lpr* mutation, which represents the major disease-accelerating factor thus far characterized in the MRL/Mp-*lpr/lpr* strain. IL-2.VV treatment results in the almost complete disappearance of cells bearing the *lpr* phenotype ($CD3^+$ $CD4^-$ $CD8^-$ $B220^+$), it reduces the predominance of cells expressing products of the TCR $V_\beta 8.2$ and $V_\beta 8.3$ families, and it abolishes the lymphadenopathy and thymic atrophy characteristic of untreated or control MRL/Mp-*lpr/lpr* mice. It remains elusive whether the effect of IL-2.VV on the manifestation of the *lpr* genotype is due to effects on the bone marrow, the intrathymic *lpr* precursor, the peripheral *lpr* cell, or the three compartments together.

It thus emerges that the effect of exogenous IL-2 on autoimmune disease development is less straightforward than might have been expected. The contradictory consequences of lymphokine administration also become manifest at the level of two lymphocyte products whose production is induced by IL-2, tumor necrosis factor-α and interferon-γ. The involvement of these two mediators in the pathogenesis of autoimmune disease is extremely controversial. Depending on the experimental system, exogenous TNF-α and IFN-γ both are able to accelerate or prevent autoimmunity. Along the same line, it is intriguing that cyclosporin A, a drug whose capacity to suppress IL-2 production is well documented, mitigates many autoimmune diseases—including systemic autoimmunity and lymphoproliferation of MRL/Mp-*lpr/lpr* mice—but, once withdrawn, may entail autoimmune manifestations by preventing clonal deletion, especially in conditions of high intrathymic T cell turnover.

At first glance, it appears an impossible task to integrate these conflicting data into a simple model. One possibility might be that high doses of exogenous lymphokine simultaneously influence different mechanisms implicated in the maintenance of autotolerance. Depending on the defect predominantly responsible for the development of a given disease, IL-2 might reequilibrate or aggravate an imbalance in the tolerance-preserving processes. Whereas IL-2 is capable of breaking peripheral (auto)tolerance, it equally may reduce the generation of double-negative T cells ($CD4^- CD8^-$) that precipitate systemic autoimmunity, at least in the *lpr* model. Moreover, IL-2 could activate suppressor cell circuits or initiate immunosuppressive feedback mechanisms, e.g., by virtue of its capacity of down-regulating TCR-mediated signaling in T_{H1} cells or the elicitation of high systemic glucocorticoid levels. Because systemic IL-2 injections provoke the secretion of mediators such as IL-6, TNF-α, and IFN-γ that rise to detectable serum levels, additional pleiotropic effects may superimpose themselves over the *per se* complex biology of IL-2. A future challenge will be to decipher how individual lymphokine effects that have been characterized on the cellular level sum up to a determined *in vivo* outcome. Nonetheless, we feel that systemic application of lymphokines will not necessarily reproduce the physiology of local—paracrine or autocrine—mediators. In contrast, administration of substances that preclude cytokine effects such as specific receptor antagonists, soluble receptor derivatives, or mAbs, may furnish us with a more precise picture of what their principal role might be under normal conditions. If this premise is true, then it may be postulated that IL-2 has one predominant effect, that of stimulating the initiation of (auto)immune reactions.

Acknowledgments

We are indebted to Drs. I. Moreno de Alboran and R. de Cid for allowing us to comment on their unpublished data; Drs. J. E. Alés Martínez, G. Wick, J. Penninger, and B. Fedecka-Bruner for stimulating discussions; E. Leonardo for skilled technical assistance; K. M. Sweeting for editorial assistance; and B. Campos Egozcue for intellectual stimuli. This work was supported in part by grants from FISS, EC, and CICyT.

References

Adachi, K., Kumamoto, T., and Araki, S. (1989). *Lancet* 1, 559–560.
Akoglu, T. F., Direskeneli, H., Yazici, H., and Lawrence, R. (1990). *J. Rheumatol.* 17, 1107–1108.
Akuffo, H., Kaplan, G., Kiessling, R., Teklemariam, S., Dietz, M., McElrath, J., and Cohn, Z. A. (1990). *J. Infect. Dis.* 161, 775–780.

Alarcón-Riquelme, M. E., Alarcón-Segovia, D., Loredo-Abdala, A., and Alcocer-Varela, J. (1990). *Clin. Immunol. Immunopathol.* **55**, 120–128.
Alcocer-Varela, J., and Alarcón-Segovia, D. (1982). *J. Clin. Invest.* **69**, 1388–1392.
Alcocer-Varela, J., Laffon, A. C., and Alarcón-Segovia, D. (1983). *Clin. Exp. Immunol.* **54**, 125–132.
Alex-Martínez, J. E., Alvarez-Mon, M., Merino, F., Bonilla, F.,. Martínez-A., C., Durantez, A., and de la Hera, A. (1988). *Eur. J. Immunol.* **18**, 1827–1832.
Alfons, B., Heremans, H., Vandekerckhove, F., Dijkmans, R., Sobis, H., Meulepas, E., and Carton, H. (1988). *J. Immunol.* **140**, 1506–1510.
Allen, J. B., McCartney-Francis, N., Smith, P. D., Simon, G., Gartner, S., Wahl, L. M., Popovic, M., and Wahl, S. M. (1990). *J. Clin. Invest.* **85**, 192–199.
Altman, A., Theofilopoulos, A. N., Weiner, R., Katz, D. H., and Dixon, F. J. (1981). *J. Exp. Med.* **154**, 791–808.
Altman, A., Mustelin, T., and Coggeshall, K. M. (1990). *Crit. Rev. Immunol.* **10**, 347–391.
Anderson, T. D., and Hayes, T. J. (1989). *Lab. Invest.* **60**, 331–346.
Andreu, J. L., Moreno de Alborán, I., Marcos, M. A. R., Sánchez-Movilla, A., Martínez-A, C., and Kroemer, G. (1991). *J. Exp. Med.* (in press).
Arai, K. I., Lee, F., Miyajima, A., Miyatake, S., Arai, N., and Yokota, T. (1990). *Annu. Rev. Biochem.* **59**, 783–836.
Asao, H., Takeshita, T., Nakamura, M., Nagata, K., and Sugamura, K. (1990a). *Int. Immunol.* **2**, 469–472.
Asao, H., Takeshita, T., Nakamura, M., Nagata, K., and Sugamura, K. (1990b). *J. Exp. Med.* **171**, 637–644.
Ashwell, J. D., and Klausner, R. D. (1990). *Annu. Rev. Immunol.* **8**, 139–167.
Atkins, M. B., Mier, J. W., Parkinson, D. R., Gould, J. A., Berkman, E. M., and Kaplan, M. M. (1988). *N. Engl. J. Med.* **318**, 1557–1563.
Auffray, C., Piatier-Tonneau, D., and Kroemer, G. (1991). *Trends Biotechnol.* **9**, 24–30.
Babcock, S. K., Appel, V. B., Schiff, M., Palmer, E., and Kotzin, B. L. (1989). *Proc. Natl. Acad. Sci. U.S.A.* **86**, 7552–7555.
Balderas, R. S., Josimovic-Alasevic, O., Diamantstein, T., Dixon, F. J., and Theofilopoulos, A. N. (1987). *J. Immunol.* **139**, 1496–1500.
Bandeira, A., Larsson, E.-L., Forni, L., Pereira, P., and Coutinho, A. (1987). *Eur. J. Immunol.* **17**, 901–908.
Banerjee, S., Wei, N.-Y., Hillman, K., Luthra, H. S., and David, C. S. (1988). *J. Immunol.* **141**, 1150–1154.
Baquerizo, H., Leone, J., Pukel, C., Wood, P., and Rabinovitch, A. (1989). *J. Autoimmun.* **2**, 133–150.
Baran, D., Vendeville, B., Vial, M. C., Cosson, C., Bascou, C., Teychenne, P., and Druet, P. (1986). *Clin. Nephrol.* **25** (Suppl. 1), S175.
Baran, D., Lantz, O., Dosquet, P., Sfaksi, A., and Druet, P. (1988). *Clin. Exp. Immunol.* **73**, 401–405.
Bárcena, A., Toribio, M. L., Pezzi, L., and Martínez-A, C. (1990). *J. Exp. Med.* **172**, 439–446.
Bárcena, A., Sánchez, M. J., de la Pompa, J. L., Toribio, M. L., Kroemer, G., and Martínez-A, C. (1991a). *Proc. Natl. Acad. Sci. U.S.A.* (in press).
Bárcena, A., Toribio, M. L., Kroemer, G., and Martínez-A., C. (1991b). *Int. Immunol.* **3**, (in press).
Barron, K. S., Montalvo, J. F., Joseph, A. K., Hilario, M. O., Saadeh, C., Giannini, E. H., and Orson, F. M. (1990). *Arthritis Rheum.* **33**, 1371–1377.
Bass, H., and Strober, S. (1990). *Cell. Immunol.* **126**, 129–142.
Bazan, J. F. (1990). *Proc. Natl. Acad. Sci. U.S.A.* **87**, 6934–6938.

Ben-Nun, A., Wekerle, H., and Cohen, J. R. (1981). *Eur. J. Immunol.* **11**, 195–199.
Bermudez, L. E. M., Stevens, P., Kolonoski, P., Wu, M., and Young, L. S. (1989). *J. Immunol.* **143**, 2996–3000.
Berridge, M. J., and Irvine, R. F. (1989). *Nature (London)* **341**, 197–203.
Berrih-Aknin, S., Cohen-Kaminsky, S., Neumann, D., Safar, D., Eymard, B., Gaud, C., Levasseur, P., Fuchs, S., and Bach, J. F. (1988). *Immunol. Res.* **7**, 189–199.
Bertani, T., Abbate, M., Zoja, C., Corna, D., Perico, N., Ghezzi, P., and Remuzzi, G. (1989). *Am. J. Pathol.* **134**, 45.
Beuzeboc, P., Escourolle, H., Dorval, T., Dieras, V., Mastorakos, G., Luton, J. P., and Pouillard, P. (1989). *La Presse Médicale* **18**, 727.
Billiau, A., Heremans, H., and Vandekerckhove, F. (1988). *J. Immunol.* **140**, 1506–1510.
Blackman, M., Kappler, J., and Marrack, P. (1990). *Science* **248**, 1335–1341.
Blaise, D., Maraninchi, D., Mawas, C., Stoppa, A. M., Hirn, M., Guyotat, D., Attal, M., and Reiffers, J. (1989). *Lancet* **1**, 1333–1334.
Blaise, D., Olive, D., Stoppa, A. M., Viens, P., Pourreau, C., Lopez, M., Attal, M., Jasmin, C., Monges, G., Mawas, C., Mannoni, P., Palmer, P., Franks, C., Philip, T., and Maraninchi, D. (1990). *Blood* **76**, 1092–1097.
Bonneville, M., Ishida, I., Itohara, S., Verbeek, S., Berns, A., Kanagawa, O., Haas, W., and Tonegawa, S. (1990). *Nature (London)* **344**, 163–165.
Boswell, J. M., Yui, M. A., Burt, D. W., and Kelley, V. E. (1988). *J. Immunol.* **141**, 3050–3054.
Boyer, P. D., Diamond, R. A., and Rothenberg, E. V. (1989). *J. Immunol.* **142**, 4121–4130.
Brandhuber, B. J., Boone, T., Kenney, W. C., and McKay, D. B. (1987). *Science* **238**, 1707–1709.
Brandt, S. J., Bodine, D. M., Dunbar, C. E., and Nienhuis, A. W. (1990). *J. Clin. Invest.* **86**, 592–599.
Brennan, D. C., Yui, M. A., Wuthrich, R. P., and Kelley, V. E. (1989). *J. Immunol.* **143**, 3470–3475.
Buchan, G., Barrett, K., Turner, M., Chantry, D., Maini, R. N., and Feldmann, M. (1988). *Clin. Exp. Immunol.* **73**, 449–455.
Bunning, R. A. D., and Russell, R. G. (1989). *Arthritis Rheum.* **32**, 780–784.
Burkly, L. C., Lo, D., Kanagawa, O., Brinster, R. L., and Flavell, R. A. (1989). *Nature* **342**, 564–566.
Burkly, L. C., Lo, D., and Flavell, R. A. (1990). *Science* **248**, 1364–1368.
Burman, P., Tötterman, T. H., Öberg, K., and Karlsson, F. A. (1986). *J. Clin. Endocrinol. Metab.* **63**, 1086.
Burton, J., Goldman, C. K., Moos, M., and Waldmann, T. A. (1990). *Proc. Natl. Acad. Sci. U.S.A.* **87**, 7329–7333.
Caligiuri, M. A., Zmudzinas, A., Maneley, T. J., Levine, H., Smith, K. A., and Ritz, J. (1990). *J. Exp. Med.* **171**, 1509–1526.
Calvino, I. (1983). In "Palomar," pp. 110–114. Einaudi, Turin.
Campbell, I. L., Oxbrow, L., Koulmanda, M., and Harrison, L. C. (1988a). *J. Immunol.* **140**, 1111–1116.
Campbell, I. L., Iscaro, A., and Harrison, L. C. (1988b). *J. Immunol.* **141**, 2325–2329.
Carding, S. L., Jenkinson, E. J., Kingston, R., Hayday, A. C., Bottomly, K., and Owen, J. T. (1989). *Proc. Natl. Acad. Sci. U.S.A.* **86**, 3342.
Case, J. P., Lorberbioum-Galski, H., Lafyatis, R., FitzGerald, S., Wilder, R. L., and Pastan, I. (1989). *Proc. Natl. Acad. Sci. U.S.A.* **86**, 287–291.
Caspi, R. R., Roberge, F. G., McAllister, C. G., El-Saied, M., Kuwabara, T., Gery, I., Hanna, E., and Nussenblatt, R. B. (1986). *J. Immunol.* **136**, 928–933.
Cathely, G., Amor, B., and Fournier, C. (1986). *Clin. Rheumatol.* **5**, 482–492.

Chatenoud, L., Dugas, B., Bearurain, G., Tovem, M., Drveke, T., Vasquez, A., Galanaud, P., Bach, J.-F., and Delfraissy, J.-F. (1986). *Proc. Natl. Acad. Sci. U.S.A.* **83**, 7457–7461.
Cheney, R. T., and Sprent, J. (1985). *Transplant Proc.* **17**, 528–537.
Cheney, P. R., Dorman, S. E., and Bell, D. S. (1989). *Ann. Int. Med.* **110**, 321.
Chungpark, M., Kim, B., Marmolya, G., Karlins, N., and Wojcik, E. (1990). *Arch. Pathol. Lab. Med.* **114**, 1073–1075.
Cicardi, M., Gardinali, M., Bisiani, G., Rosti, A., Allavena, P., and Agostoni, A. (1990). *Ann. Int. Med.* **113**, 475–477.
Cohen-Kaminsky, S., Levasseur, P., Binet, J. P., and Berrih-Aknin, S. (1989). *J. Neuroimmunol.* **24**, 75–85.
Collins, L., Tsien, W. H., Seals, C., Hakimi, J., Weber, D., Bailon, P., Hoskings, J., Greene, W. C., Toome, V., and Ju, G. (1988). *Proc. Natl. Acad. Sci. U.S.A.* **85**, 7709–7713.
Colvin, R. B., Fuller, T. C., and Kung, P. C. (1987). *Clin. Immunol. Immunopathol.* **43**, 273–276.
Cornaby, A. J., Simpson, M. A., Madras, P. N., Dempsey, R. A., Clowes, G. H. A., and Monaco, A. P. (1989). *Transplant. Proc.* **21**, 1861–1862.
Crabtree, G. R. (1989). *Science* **243**, 355–361.
Dardenne, M., Lepault, F., Bendelac, A., and Bach, J.-F. (1989). *Eur. J. Immunol.* **19**, 889–895.
Dauphinée, M. J., Kipper, S. B., Wofsy, D., and Talal, N. (1981). *J. Immunol.* **127**, 2483–2487.
Dautry, F., Weil, D., Yu, J., and Dautry-Varsat, A. (1988). *J. Biol. Chem.* **263**, 17615–17620.
Davidson, W. F., Dumont, F. J., Bedigian, H. G., Fowlkes, B. J., and Morse, H. C. (1986). *J. Immunol.* **136**, 4057–4084.
Dayer, J.-M., Beutler, B., and Cerami, A. (1987). *J. Exp. Med.* **162**, 2163–2168.
Daynes, R. A., and Araneo, B. A. (1989). *Eur. J. Immunol.* **19**, 2319–2325.
de Cid, R., Kroemer, G., Moreno de Alboran, I., von Boehmer, H., and Martínez-A, C. (1991). Submitted.
deFreitas, E. C., Sandberg-Wollheim, M., Schonely, K., Boufal, M., and Koprowsky, H. (1986). *Proc. Natl. Acad. Sci. U.S.A.* **83**, 2637–2641.
Degiannis, D., Seibold, J. R., Czarnecki, M., Raskova, J., and Raska, K. Jr. (1988). *Arthritis Rheum.* **33**, 375–380.
de la Hera, A., Toribio, M.-L., and Martínez-A, C. (1989a). *Int. Immunol.* **1**, 471–478.
de la Hera, A., Toribio, M.-L., and Martínez-A, C. (1989b). *Int. Immunol.* **1**, 496–502.
Diamantstein, T., and Osawa, H. (1986). *Immunol. Rev.* **92**, 5–27.
Diamantstein, T., Eckert, R., Volk, H.-D., and Kupiec-Weglinski, J. W. (1988). *Eur. J. Immunol.* **18**, 2101–2103.
Disanto, J. P., Keever, C. A., Small, T. N., Nichols, G. L., O'Reilly, R. J., and Flomenberg, N. (1990). *J. Exp. Med.* **171**, 1697–1704.
Djeu, J. Y., Kasahara, T., Balow, J. E., and Tsokos, G. C. (1986). *Clin. Exp. Immunol.* **65**, 279–285.
Doi, T., Hatakeyama, M., Minamoto, S., Kono, T., Mori, H., and Taniguchi, T. (1989). *Eur. J. Immunol.* **19**, 2375–2378.
Donnelly, R. P., Levine, J., Hartwell, D. W., Frendl, G., Fenton, M. J., and Beller, D. I. (1990). *J. Immunol.* **145**, 3231–3239.
Donohue, J. H., and Rosenberg, S. A. (1983). *J. Immunol.* **130**, 2203–2208.
Durand, D. B., Shaw, J. P., Bush, M. R., Replogle, R. E., Belageje, R., and Carbtree, G. R. (1988). *Mol. Cell. Biol.* **8**, 1715–1724.

Egerton, M., and Scollay, R. (1990). *Int. Immunol.* **2**, 157–164.
Egerton, M., Shortman, K., and Scollay, R. (1990). *Int. Immunol.* **2**, 501–507.
Eisenstein, Z., Engelsman, E., Weiss, M., Kalechman, Y., and Sredni, B. (1988). *J. Clin. Immunol.* **8**, 349–355.
Elder, M. E., and MacLaren, N. K. (1983). *J. Immunol.* **130**, 1723–1731.
Emery, P., Panayi, G. S., and Nouri, A. M. E. (1984). *Clin. Exp. Immunol.* **57**, 123–129.
Endres, S., Cannon, J. G., Ghorbani, R., Dempsey, R. A., Sisson, S. D., Lonnemann, G., Van der Meer, J. W. M., Wolff, S. M., and Dinarello, C. A. (1989). *Eur. J. Immunol.* **19**, 2327–2333.
Espinoza-Delgado, I., Ortaldo, J. R., Winkler-Pickett, W., Sugamura, K., Varesio, L., and Longo, D. L. (1990). *J. Exp. Med.* **171**, 1821–1826.
Essery, G., Feldmann, M., and Lamb, J. R. (1988). *Immunology* **64**, 413–417.
Famularo, G., Procopio, A., Giacomelli, R., Danese, C., Sacchetti, S., and Perego, M. A. (1990). *Clin. Exp. Pathol.* **81**, 368–372
Favier, R., Edelman, P., Mary, J.-Y., Sadouil, G., and Douay, L. (1990). *N. Engl. J. Med.* **321**, 270.
Fearon, E. R., Pardoll, D. M., Itaya, T., Golumbek, P., Levitsky, H. I., Simons, J. W., Karasuyama, H., Vogelstein, B., and Frost, P. (1990). *Cell* **60**, 397–403.
Fedecka-Bruner, B., Penninger, J., Vaigot, P., Lehmann, A., Martinez-A., C., and Kroemer, G. (1991). *Dev. Immunol.* **2** (in press).
Fentimann, I. S., Balkwill, F. R., Thomas, B. S., Russell, M. J., Todd, I., and Bottazzo, G. F. (1988). *Eur. J. Cancer Clin. Oncol.* **24**, 1299.
Ferrarini, M., Steen, V., Medsger Jr., A., and Whiteside, T. L. (1990). *Clin. Exp. Pathol.* **79**, 346–352.
Ferris, D. K., Willette-Brown, J., Ortaldo, J. R., and Farrar, W. L. (1989). *J. Immunol.* **143**, 870–876.
Feultren, G., Assen, R., Karsenty, G., DuRostu, H., Sirmai, J., Papoz, J., Vialettes, B., Rodier, M., Lallemand, A., and Bach, J.-F. (1986). *Lancet* **1**, 119–124.
Firestein, G. S., Alvaro-Garcia, J. M., and Maki, R. (1990). *J. Immunol.* **144**, 3347–3353.
Flexner, C., Hügin, A., and Moss, B. (1987). *Nature (London)* **330**, 259–262.
Flomenberg, N., Welte, K., Mertelsmann, R., Kernan, N., Ciobanu, N., Venuta, S., Feldman, C., Kruger, G., Kirkpatrick, D., Dupont, B., and O'Reilly, R. (1983). *J. Immunol.* **130**, 2644–2649.
Formby, B., Miller, N., and Peterson, C. M. (1988). *Diabetes* **37**, 1305–1309.
Fox, R. I., Theofilopoulos, A. N., and Altman, A. (1985). *J. Immunol.* **135**, 3109–3115.
Fraker, D. L., Langstein, H. N., and Norton, J. A. (1989). *J. Exp. Med.* **170**, 1015–1020.
Freundlich, B., and Jimenez, S. A. (1987). *Clin. Exp. Pathol.* **69**, 375–380.
Froelich, C. J., Guiffaut, S., Sosenko, M., and Muth, K. (1989). *Clin. Immunol. Immunopathol.* **50**, 132–145.
Fry, A. M., Jones, L. A., Kruisbeek, A. M., and Matis, L. A. (1989). *Science* **246**, 1044–1046.
Gahring, L. C., and Weigle, W. O. (1990). *J. Immunol.* **145**, 1318–1323.
Gajewski, T. F., Schell, S. R., Nau, G., and Fitch, F. W. (1989). *Immunol. Rev.* **111**, 79–110.
Gao, E.-K., Lo, D., Cheney, R., Kanagawa, O., and Sprent, J. (1988). *Nature (London)* **136**, 176–179.
Gardner, P. (1989). *Cell* **59**, 15–20.
Gaston, J. S. H. (1989). *Autoimmunity* **4**, 143–149.
Geppert, T. D., Davis, L. S., Gur, H., Wacholitz, M. C., and Lipsky, D. (1990). *Immunol. Rev.* **117**, 5–66.

Gijbels, K., van Damme, J., Proost, P., Put, W., Carton, H., and Billiau, A. (1990). *Eur. J. Immunol.* **20,** 233–235.
Gillis, S., Ferm, M. M., Ou, W., and Smith, K. A. (1978). *J. Immunol.* **120,** 2027–2032.
Glazier, A., Tutschka, P. J., Farmer, E. R., and Santos, G. W. (1983). *J. Exp. Med.* **158,** 1–14.
Gordon, C., and Wofsy, D. (1990). *J. Immunol.* **144,** 1753–1758.
Gordon, C., Ranges, G. E., Greenspan, J. S., and Wofsy, D. (1989). *Clin. Immunol. Immunopathol.* **52,** 421–434.
Goronzy, J., Weyand, C., Quan, J., Fathman, C. G., and O'Haneley, P. (1989). *J. Immunol.* **142,** 1134–1138.
Granelli-Piperno, A. (1990). *J. Exp. Med.* **171,** 533–544.
Granstein, R. D., Goulston, C., and Gaulton, G. N. (1986). *J. Immunol.* **136,** 898–903.
Green, J. E., Hinrichs, S. H., Vogel, J., and Jay, G. (1989a). *Nature (London)* **341,** 72–74.
Green, J. E., Begley, C. G., Wagner, D. K., Waldmann, T. A., and Jay, G. (1989b). *Mol. Cell. Biol.* **9,** 4731–4737.
Greene, W. C., Leonard, W. J., Depper, J. M., Nelson, D. L., and Waldmann, T. A. (1986). *Ann. Intern. Med.* **105,** 560–572.
Gromkowski, S. H., Hepler, K. M., and Janeway Jr., C. A. (1988). *Eur. J. Immunol.* **18,** 1385–1389.
Grubeck-Loebenstein, B., Buchan, G., Chantry, D., Kassal, H., Londei, M., Pirich, K., Barrett, K., Tunre, M., Waldhausl, W., and Feldmann, M. (1989). *Clin. Exp. Pathol.* **77,** 324–330.
Gruschwitz, M., Kroemer, G., Moormann, S., Faessler, R., Dietrich, H., Boeck, G., Gershwin, M. E., Abplanalp, H., Boyd, R., and Wick, G. (1991). *J. Autoimmun.* (in press).
Gutierrez-Ramos, J. C., Pezzi, L., Palacios, R., and Martínez-A, C. (1989a). *Eur. J. Immunol.* **19,** 201–204.
Gutierrez-Ramos, J. C., Martínez-A., C., Köhler, G., and Iglesias, A. (1989b). *Res. Immunol.* **140,** 661–667.
Gutierrez-Ramos, J. C., Andreu, J. L., Moreno de Alboran, I., Rodriguez, J., Leonard, E., Kroemer, G., Marcos, M. A. R., and Martínez-A., C. (1990a). *Immunol. Rev.* **118,** 73–101.
Gutierrez-Ramos, J. C., Andreu, J. L., Revilla, Y., Viñuela, E., and Martínez-A., C. (1990b). *Nature (London)* **346,** 271–274.
Gutierrez-Ramos, J. C., Andreu, J. L., Marcos, M. A. R., Vagazo, I. R., and Martínez-A., C. (1991a). *Autoimmunity* (in press).
Gutierrez-Ramos, J. C., Moreno de Alboran, I., Kroemer, G., and Martínez-A., C. (1991b). Submitted.
Hahn, H. J., Lucke, S., Klöting, I., Volk, H. D., Baehr, R. V., and Diamantstein, T. (1987). *Eur. J. Immunol.* **17,** 1075–1078.
Hahn, H. J., Gerdes, J., Lucke, S., Liepe, L., Kauert, C., Volk, H. D., Wacker, H. W., Brocke, S., and Diamantstein, T. (1988). *Eur. J. Immunol.* **18,** 2037–2042.
Hancock, W. W., Lord, R. H., Colby, A. J., Diamantstein, T., Rickles, F. R., Dijkstra, C., Hogg, N., and Tilney, N. L. (1987). *J. Immunol.* **138,** 164.
Hardt, C., Rollsinghof, M., Pfizenmaier, K., Mosmann, H., and Wagner, H. (1981). *J. Exp. Med.* **154,** 262–273.
Hardt, C., Diamantstein, T., and Wagner, H. (1985). *J. Immunol.* **16,** 1092–1097.
Harel-Bellan, A., Joskowicz, M., Fradelizi, D., and Eisen, H. (1983). *Proc. Natl. Acad. Sci. U.S.A.* **80,** 3466–3470.
Hartmann, L. C., Urba, W. J., Steis, R. G., Smith II, J. W., vanderMolen, L., Creekmore, S. P., and Longo, D. L. (1989). *J. Clin. Oncol.* **7,** 686–687.

Hartung, H.-P., Schäfer, B., Diamantstein, T., Fierz, W., Heininger, K., and Toyka, K. V. (1989). *Brain Res.* **489**, 120–125.
Hartung, H. P., Hughes, R. A. C., Taylor, W. A., Heininger, K., Reiners, K., and Toyka, K. V. (1990). *Neurology* **40**, 215–218.
Hasegawa, K., Abe, M., Okada, T., Hirose, S., Sato, H., and Shirai, T. (1989). *Int. Immunol.* **1**, 99–103.
Hashimoto, S., McCombs, C. C., and Michalski, J. P. (1989). *Clin. Exp. Immunol.* **76**, 317–324.
Hashimoto, S., Michalski, J. P., Berman, M. A., and McCoombs, C. (1990). *Clin. Exp. Immunol.* **79**, 227–232.
Haskins, K., Portas, M., Bergman, B., Lafferty, K., and Bradley, B. (1989). *Proc. Natl. Acad. Sci. U.S.A.* **86**, 8000–8004.
Hatakeyama, M., Tsuod, M., Minamoto, S., Kono, T., Doi, T., Miyata, T., Miyasaka, M., and Taniguchi, T. (1987). *Science* **244**, 551–556.
Hatakeyama, M., Mori, H., Doi, T., and Taniguchi, T. (1989a). *Cell* **59**, 837–845.
Hatakeyama, M., Tsudo, M., Minamoto, S., Kono, T., Doi, T., Miyata, T., Miyasaka, M., and Taniguchi, T. (1989b). *Science* **244**, 551–556.
Held, W., MacDonald, H. R., Weissmann, I. L., Hess, M. W., and Mueller, C. (1990). *Proc. Natl. Acad. Sci. U.S.A.* **87**, 2239–2243.
Henderson, B., and Pettipher, E. R. (1989). *Clin. Exp. Pathol.* **75**, 306–310.
Hervé, P., Wijdenes, J., Bergerat, J. P., Bordigoni, P., Milpied, N., Cahn, J. Y., Clément, C., Beliard, R., Morel-Fourrier, B., Racadot, E., Troussard, X., Benz-Lemoine, E., Gaud, C., Legros, M., Attal, M., Kloft, M., and Peters, A. (1990). *Blood* **75**, 1017–1023.
Heslop, H. E., Gottlieb, D. J., Bianchi, A. C. M., Meager, A., Prentice, H. G., Mehta, A. B., Hoffbrand, A. V., and Brenner, M. K. (1989). *Blood* **74**, 1374–1380.
Hess, A. D., and Fischer, A. C. (1989). *Transplantation* **48**, 895–900.
Hinrichs, S. H., Nernberg, M., Reynolds, R. K., Khoury, G., and Jay, G. (1987). *Science* **237**, 1340–1343.
Hirano, T., Taga, T., Yasukawa, K., Nakajima, K., Nakano, N., Takatsku, F., Shimizu, M., Murashima, A., Tsunasawa, S., Sakiyama, F., and Kishimoto, T. (1987). *Proc. Natl. Acad. Sci. U.S.A.* **84**, 228–231.
Hirano, T., Matsuda, T., Turner, M., Miyasaka, M., Buchan, G., Tang, B., Karzuto, S., Shimizu, M., Maini, R., Feldmann, M., and Kishimoto, T. (1988). *Eur. J. Immunol.* **18**, 1797–1801.
Hirohata, S., Davis, L. S., and Lipsky, P. E. (1989). *J. Immunol.* **142**, 3104–3112.
Hodes, R. J., Sharrow, S. O., and Salomon, A. (1989). *Science* **246**, 1041–1044.
Hoffman, M., Mittelman, A., Dworkin, B., Rosenthal, W., Beneck, W., Gafney, E., Arlin, Z., Levitt, D., and Podack, E. (1989). *J. Cancer Res. Clin. Oncol.* **115**, 175–178.
Hoffman, R., Briddellm, R. A., van Besien, K., Srour, E. F., Guscar, T., Hudson, N. W., and Ganser, A. (1989). *N. Engl. J. Med.* **321**, 97–102.
Hofman, F. M., Hinton, D. R., Johnson, K., and Merril, J. E. (1989). *J. Exp. Med.* **170**, 607–612.
Hohlfeld, R., Toyka, K. V., Heininger, K., Grosse-Wilde, H., and Kalies, I. (1984). *Nature (London)* **310**, 244–246.
Holda, J. H., Maier, T., and Claman, H. N. (1986). *J. Immunol.* **137**, 3538–3543.
Holmdahl, R., Klareskog, L., Rubin, K., Larsson, E., and Wigzell, H. (1985). *Scand. J. Immunol.* **22**, 295–306.
Holoshitz, J., Naparstek, Y., Ben-Nun, A., and Cohen, I. R. (1983). *Science* **219**, 56–58.
Hooks, J. J., Chan, C. C., and Detrick, B. (1988). *Investig. Ophthalmol. Visual Sci.* **29**, 1444–1451.

Horii, Y., Muraguchi, A., Iwano, M., Matsuda, T., Hirayama, T., Yamada, H., Fujii, Y., Dohi, K., Ishikawa, H., Ohmoto, Y., Yoshizaki, K., Hirano, T., and Kishimoto, T. (1989). *J. Immunol.* **143**, 3949–3955.
Howlett, T. A., Lawton, N. F., Fells, P., and Besser, G. M. (1984). *Lancet* **2**, 1101.
Hoyos, B., Ballard, D. W., Böhnlein, E., Siekevitz, M., and Greene, W. C. (1989). *Science* **244**, 457–460.
Huang, Y.-P., Miescher, P. A., and Zubler, R. H. (1986). *J. Immunol.* **137**, 3515–3520.
Huang, Y.-P., Perrin, L. H., Miescher, P. A., and Zubler, R. H. (1988). *J. Immunol.* **141**, 827–833.
Huang, C. M., Elin, R. J., Ruddel, M., Silva, C., Lotze, M. T., and Rosenberg, S. A. (1990). *Clin. Chem.* **36**, 431–434.
Hünig, T. (1983). *Immunol. Today* **4**, 84–87.
Iizawa, Y., Nishi, T., Kondo, M., Tsuchiya, K., and Imada, A. (1988). *Infection Immunity* **56**, 45–50.
Ikehara, S., Yasumizu, R., Inaba, M., Izui, S., Hayakawa, K., Sekita, K., Toki, J., Sugiura, K., Iwai, H., Nakamura, T., Muso, E., Hamashima, Y., and Good, R. A. (1989). *Proc. Natl. Acad. Sci. U.S.A.* **86**, 3306–3310.
Ishida, H., Kumagia, S., Umehara, H., Sano, H., Yagaya, Y., Yodoi, J., and Imura, H. (1987). *J. Immunol.* **139**, 1070–1074.
Ishida, Y., Nishi, M., Taguchi, O., Inaba, K., Minato, N., Kaiwaichi, M., and Honjo, T. (1989a). *Int. Immunol.* **1**, 113–120.
Ishida, Y., Nishi, M., Taguchi, O., Inaba, K., Hattori, M., Minato, N., Kaiwaichi, M., and Honjo, T. (1989b). *J. Exp. Med.* **170**, 1103–1115.
Ishizaka, S., and Tsuji, T. (1989). *Cell. Immunol.* **118**, 100–107.
Jablons, D. M., Mule, C. J., McIntosh, J. K., Sehgal, P. B., May, L. T., Huang, C. M., Rosenberg, S. A., and Lotze, M. T. (1989). *J. Immunol.* **142**, 1542–1547.
Jacob, C. O., and McDevitt, H. O. (1988). *Nature (London)* **331**, 356–358.
Jacob, C. O., van der Meide, P. H., and McDevitt, H. O. (1987). *J. Exp. Med.* **166**, 798–803.
Jacob, C. O., Aiso, S., Michie, S. A., McDevitt, H. O., and Acha-Orbea, H. (1990a). *Proc. Natl. Acad. Sci. U.S.A.* **87**, 968–972.
Jacob, C. O., Froner, Z., Lewis, G. D., Koo, M., Hansen, J. A., and McDevitt, H. O. (1990b). *Proc. Natl. Acad. Sci. U.S.A.* **87**, 1233–1237.
Jenkins, M. K., Pardoll, D. M., Mizuguchi, J., Quill, H., and Schwartz, R. H. (1987). *Immunol. Rev.* **95**, 113–135.
Jenkins, M. K., Schwartz, R. H., and Pardoll, D. M. (1988). *Science* **241**, 1655–1658.
Jenkinson, E. J., Kingston, R., and Owen, J. J. T. (1987). *Nature (London)* **329**, 160.
Jones, M. G., Harris, G., and Cowing, G. (1983). *Transplant. Proc.* **15** (Suppl. 1), 2904–2908.
Jones, R. J., Vogelsang, G. B., and Hess, A. D., Farmer, E. R., Mann, R. B., Geller, R. B., Piantadosi, S., and Santos, G. W. (1989). *Lancet* **1**, 754.
Jones, L. A., Chin, L. T., Merriam, G. R., Nelsom, L. M., and Kruisbeck, A. M. (1990). *J. Immunol.* **172**, 1277–1285.
Jongeneel, C. V., Acha-Orbea, H., and Blankenstein, T. (1990). *J. Exp. Med.* **171**, 2141–2146.
Jönsson, J., Boyxe, N. W., and Eichmann, K. (1989). *Eur. J. Immunol.* **19**, 253–256.
Jourdan, M., Bataille, R., Seguin, J., Zhang, X. G., Chaptal, P. A., and Klein, B. (1990). *Arthritis Rheum.* **33**, 398–402.
June, C. H., Ledbetter, J. A., Lindsten, T., and Thompson, C. B. (1989). *J. Immunol.* **143**, 153–161.

Kahaleh, M. B., and LeRoy, E. C. (1989). *Ann. Intern. Med.* **110**, 446–450.
Kahan, B. D. (1989). *N. Engl. J. Med.* **321**, 1725–1729.
Kamio, K., Uchiyama, T., Arima, N., Itoh, K., Ishikawa, T., Hori, T., and Uchino, H. (1990). *Int. Immunol.* **2**, 521–529.
Kanamori, H., Suzuki, N., Siomi, H., Nosaka, T., Sato, A., Sabe, H., Hatanaka, M., and Honjo, T. (1990). *EMBO J.* **9**, 4161–4166.
Kaplan, G., Kiessling, R., Teklmariam, S., Hancock, G., Sheftel, G., Job, C. K., Converse, P., Ottenhoff, T. H. M., Becx-Bleumink, M., Dietz, M., and Cohn, Z. A. (1989). *J. Exp. Med.* **169**, 893–907.
Karupiah, G., Blanden, R. V., and Ramshaw, I. A. (1990a). *J. Exp. Med.* **172**, 1495–1503.
Karupiah, G., Coupar, B. E. H., Andrew, M. E., Boyle, D. B., Phillips, S. M., Mühlbacher, A., Blanden, R. V., and Ramshaw, I. A. (1990b). *J. Immunol.* **144**, 290–298.
Kashima, N., Nishi-Takaoka, C., Fujita, T., Taki, S., Yamada, G., Hamuro, J., and Taniguchi, T. (1985). *Nature (London)* **313**, 402–404.
Kasid, A., Director, E. P., and Rosenberg, S. A. (1989). *J. Immunol.* **143**, 736–739.
Katre, N. (1990). *J. Immunol.* **144**, 209–213.
Katsuki, M., Kimura, M., Ohta, M., Otani, H., Tanaka, O., Yamamoto, T., Nozawa-Kimura, S., Yokoyama, M., Nomura, T., and Habu, S. (1989). *Int. Immunol.* **1**, 214–218.
Kawabe, Y., and Ochi, A. (1990). *J. Exp. Med.* **172**, 1065–1070.
Kawamura, H., Rosenberg, S. A., and Berzofsky, J. A. (1985). *J. Exp. Med.* **162**, 381–386.
Kay, N. E., and Kaplan, M. E. (1986). *Blood* **67**, 578–584.
Kay, T. W. H., Campbell, I. L., Malcolm, L., and Harrison, L. C. (1989). *Cell. Immunol.* **120**, 341–350.
Kaye, W. A., Adri, M. N. S., Soeldner, J. S., Rainowe, S. L., Kaldany, A., Kahn, C. R., Bistrian, B., Srikanta, S., Ganda, O. P., and Eisnebarth, G. S. (1986). *N. Engl. J. Med.* **315**, 920–924.
Kehrl, J. H., Wakefield, L. M., Roberts, A. B., Jakowlew, S., Alvarez-Mon, M., Derynck, R., Sporn, M. B., and Fauci, A. S. (1986). *J. Exp. Med.* **163**, 1037.
Keller, R. J., and Jackson, R. A. (1989). *J. Pediatrics* **May 1989**, 816–819.
Kelley, V. E., Gaulton, G. N., Hattori, M., Ikegami, H., Eisenbarth, G., and Strom, T. B. (1988). *J. Immunol.* **140**, 59–61.
Kennedy, M., Wassmer, P., and Erb, P. (1987). *J. Immunol.* **139**, 110–113.
Kern, D. E., Gillis, S., Okada, M., and Henney, C. S. (1981). *J. Immunol.* **127**, 1323–1331.
Keystone, E. C., Snow, K. M., Bombardier, C., Chang, C.-H., Nelson, D. L., and Rubin, L. A. (1988). *Arthritis Rheum.* **31**, 844–849.
Kirkman, R. L., Barrett, L. V., Gaulton, G. N., Kelley, V. E., Ythier, A., and Strom, T. B. (1985). *J. Exp. Med.* **162**, 358–369.
Kitas, G. D., Salmon, M., Farr, M., Gaston, J. S. H., and Bacon, P. A. (1988). *Clin. Exp. Immunol.* **73**, 242–249.
Klarnet, J. P., Kern, D. E., Dower, S. K., Matis, L. A., Cheever, M. A., and Greenberg, P. D. (1989). *J. Immunol.* **142**, 2187–2191.
Klempner, M. S., Noring, R. N., Mier, J. W., and Atkins, M. B. (1990). *N. Engl. J. Med.* **322**, 959–965.
Koch, B., Regnat, W., Solbach, W., Lanz, R., Hermanek, P., and Kalden, J. R. (1984). *J. Clin. Lab. Immunol.* **13**, 171.
Koevari, S., Rossini, A., Stoller, W., and Chick, W. (1983). *Science* **220**, 727–728.
Kohl, S., Loo, L. S., Drath, D. B., and Cox, P. (1989). *J. Infect. Dis.* **1591**, 239–247.
Koizumi, T., Nakao, Y., Matsui, T., Katakami, Y., Nakagawa, T., and Fujita, T. (1986). *Immunology* **59**, 43–49.
Kojima, A., and Prehn, R. T. (1981). *Immunogenetics* **14**, 15–24.

Kolb, H., Nen-Nun, A., Cohen, I. R., Baraberena, I., and Kiesel, U. (1985). *Immunol. Lett.* **9,** 29–32.
Kolb, H., Zielasek, J., Treichel, U., Freytag, G., Wrann, M., and Kiesel, U. (1986). *Eur. J. Immunol.* **16,** 209–214.
Konishi, K., Moler, D. R., Saltini, C., Kirby, M., and Crystal, R. G. (1988). *J. Clin. Invest.* **82,** 775–781.
Kotzin, B. L., and Palmer, E. (1987). *J. Exp. Med.* **165,** 1237.
Kradin, R. L., Turnick, J. T., Pfeffer, F. J., Dubinett, S. M., Dickersin, G. R., and Pinto, C. (1989). *Clin. Immunol. Immunopathol.* **50,** 184–195.
Kragel, A. H., Travis, W. D., Steis, R. G., Rosenberg, S. A., and Roberts, W. C. (1990). *Cancer* **66,** 1513–1516.
Kroemer, G., and Wick, G. (1989). *Immunol. Today* **10,** 246–251.
Kroemer, H., Sundick, R. S., Schauenstein, K., Hála, K., and Wick, G. (1985a). *J. Immunol.* **135,** 2452–2457.
Kroemer, G., Schauenstein, K., Neu, N., Stricker, K., and Wick, G. (1985b). *J. Immunol.* **135,** 2458–2463.
Kroemer, G., Schauenstein, K., Dietrich, H., Faessler, R., and Wick, G. (1987). *J. Immunol.* **138,** 2104–2110.
Kroemer, G., Faessler, R., Hala, K., Boeck, G., Schauenstein, K., Brezinschek, H.-P., Neu, N., Dietrich, H., Jakober, R., and Wick, G. (1988). *Eur. J. Immunol.* **18,** 1499–1505.
Kroemer, G., Neu, N., Faessler, R., Kuehr, T., Dietrich, H., Hála, K., and Wick, G. (1989). *Clin. Immunol. Immunopathol.* **52,** 202–213.
Kroemer, G., Gastinel, L. N., Neu, N., Auffray, C., and Wick, G. (1990). *Autoimmunity* **6,** 215–233.
Kroemer, G., Toribio, M. L., and Martínez-A., C. (1991). *N. Biolog.* (in press).
Kronke, M., Leonard, W. J., Depper, J. M., Arya, J. M., Wong-Staal, F., Gallo, R. C., Waldmann, T. A., and Greene, W. C. (1984). *Proc. Natl. Acad. Sci. U.S.A.* **81,** 5214–5218.
Ksander, B. R., and Streilin, J. W. (1990). *J. Immunol.* **145,** 2057–2063.
Kucharz, E. J., and Goodwin, J. S. (1988). *Life Sci.* **42,** 1485–1491.
Kyle, V., Coughlan, R. J., Tighe, H., Waldmann, H., and Hazleman, B. J. (1989). *Ann. Rheumatic Dis.* **48,** 428–429.
Lamb, J. R., Zanders, E. D., Sewell, W., Crumpton, M. J., Feldmann, M., and Owen, M. J. (1987). *Eur. J. Immunol.* **17,** 1641–1644.
Lavelle-Jones, M., Hadrani, A. A., Spiers, E. M., Campbell, F. C., and Cuschieri, A. (1990). *Brit. Med. J.* **301,** 97.
Lawetzky, A., and Hünig, T. (1988). *Eur. J. Immunol.* **18,** 409–416.
Lawrence, E. C., Brousseau, K. P., Berger, M. B., Kurman, C. C., Marcon, L., and Neslon, D. L. (1988). *Am. Rev. Resp. Dis.* **137,** 759–764.
Lehmann, K. R., Kotzin, B. L., Portanova, J. P., and Santro, T. J. (1986). *Eur. J. Immunol.* **16,** 1105–1110.
Leiter, E. H., Christianson, G. J., Serreze, D. V., Ting, A. T., and Worthen, S. M. (1989). *J. Exp. Med.* **170,** 1243–1262.
Lemm, G., and Warnatz, H. (1986). *Clin. Exp. Immunol.* **64,** 71–79.
Lenardo, M. J., and Baltimore, D. (1989). *Cell* **58,** 227–229.
Lewis, C. E., McCarthy, S. P., Richards, P. S., Lorenzen, J., Horak, E., and McGee, J. O'D. (1990). *J. Immunol. Methods* **127,** 51–59.
Lieberman, A. P., Pitha, P. M., Shin, H. S., and Shin, M. L. (1989). *Proc. Natl. Acad. Sci. U.S.A.* **86,** 6348–6352.

Like, A. A., Anthony, M., Guberski,D. I., and Rossini, A. (1983). *Diabetes* **32**, 326–330.
Linker-Israeli, M., Bakke, A. C., Kitridou, R. C., Gendler, S., Gillis, S., and Horwitz, D. A. (1983). *J. Immunol.* **130**, 2651–2655.
Linker-Israeli, M., Bakke, A. C. Quismorio, F. P. Jr., and Horwitz, D. A. (1985). *J. Clin. Invest.* **75**, 762–768.
Lo, D., Burkly, L. C., Flavell, R. A., Palmiter, R. D., and Brinster, R. L. (1989). *J. Exp. Med.* **170**, 87–104.
Lobo-Yeo, A., Mieli-Vergagni, G., Mowat, A. P., and Vergagni, D. (1990). *Gut* **31**, 690–693.
Lotze, M. T., Chang, A. E., Seipp, C. A., Simpson, C., Vetto, J. T., and Rosenberg, S. A. (1986). *JAMA (J. Am. Med. Assoc.)* **256**, 3117–3124.
Lotze, M. T., Custer, M. C., Sharrow, S. O., Robin, A. L., Nelson, D. L., and Rosenberg, S. A. (1987). *Cancer Res.* **47**, 2188–2195.
Lowenthal, J. W., Ballard, D. W., Böhnlein, D. W., and Greene, W. C. (1989). *Proc. Natl. Acad. Sci. U.S.A.* **86**, 2331–2335.
MacDonald, H. R., Lees, R. K., Bron, C., Sordat, B., and Miescher, G. (1987).*J. Exp. Med.* **166**, 195–209.
Macdonald, D., Gordon, A. A., Kajitani, H., Enokihara, H., and Barrett, A. J. (1990). *Br. J. Haematol.* **76**, 168–173.
Magilavy, D. B., and Rothstein, J. L. (1988). *J. Exp. Med.* **168**, 789–794.
Mahida, Y. R., Gallagher, A., Kurlak, L., and Hawkey, C. J. (1990). *Clin. Exp. Immunol.* **82**, 75–80.
Malavé, I., Searles, R. P., Montano, J., and Williams, R. C., Jr., (1989). *Int. Arch. Allergy Appl. Immunol.* **89**, 355–361.
Malkovsky, M., and Medawar, P. B. (1984). *Immunol. Today* **5**, 340–342.
Malkovsky, M., Medawar, P. B., Thatcher, D. R., Toy, J., Hunt, R., Rayfield, L. S., and Doré, C. (1985). *Proc. Natl. Acad. Sci. U.S.A.* **82**, 536–538.
Malkovsky, M., Loveland, B., North, M., Asherson, G. L., Gao, L., Ward, P., and Fiers, W. (1987). *Nature (London)* **325**, 262–265.
Manoussakis, M. N., Papadopoulos, G. K., Drosos, A. A., and Moutsopoulos, H. M. (1989). *Clin. Immunol. Immunopathol.* **50**, 321–332.
Marcos, M. A. R., de la Hera, A., Gaspar, M. L., Márquez, C., Bellas, C., Mampaso, F., Toribio, M. L., and Martínez-A., C. (1986). *Immunol. Rev.* **94**, 51–74.
Marcos, M. A. R., de la Hera, A., Gaspar, M. L., Márquez, C., Bellas, C., Mamposo, F., Toribio, M. L., and Martínez-A., C. (1988). *Immunol. Today* **9**, 204–207.
Mariotti, S., Chiovato, L., Vitti, P., Marcocci, C., Fenzi, G. F., del Prete, G. F., Tiri, A., Romagnani, S., Ricci, M., and Pinchera, A. (1989). *Clin. Immunol. Immunopathol.* **50**, S73–S84.
Maron, R., Zerubavel, R., Friedman, A., and Cohen, I. R. (1983). *J. Immunol.* **131**, 2316–2322.
Marshall, M. E., Cibull, M. L., Pearson, T., Hall, C., and Goldblum, S. E. (1990).*J. Biol. Response Mod.* **9**, 279–287.
Martinez, O. M., Gibbons, R. S., Garovoy, M. R., and Aronson, F. R. (1990).*J. Immunol.* **144**, 2211–2215.
Martínez-A., C. (1990). *Res. Immunol.* **141**, 263–312.
Massa, P. T., ter Meulen, V., and Fontana, A. (1987). *Proc.Natl. Acad. Sci. U.S.A.* **84**, 4219–4223.
Massaia, M., Bianchi, A., Dianzani, U., Camponi, A., Attisano, C., Boccadoro, M., and Pileri, A. (1990). *Clin. Exp. Immunol.* **79**, 100–104.

Maurtiz, N. J., Holmdahl, R., Jonsson, R., vand der Meide, P. H., Scheynius, A., and Klareskog, L. (1988). *Arthritis Rheum.* **32**, 1297.
Maury, C. P. J., and Teppo, A.-M. (1989). *Arthritis Rheum.* **32**, 146–150.
McCombs, C. C., Michalski, J. P., deShazo, R. D., Bozelka, B., and Lane, J. T. L. (1986). *Clin. Immunol. Immunopathol.* **39**, 283–293.
McElrath, M. J., Kaplan, G., Burkhardt, R. A., and Cohn, Z. A. (1990). *Proc. Natl. Acad. Sci. U.S.A.* **87**, 5783–5787.
McGuire, K. L., and Rothenberg, E. V. (1987). *EMBO. J.* **6**, 939–946.
McGuire, K. L., Yang, J. A., and Rothenberg, E. V. (1988). *Proc. Natl. Acad. Sci. U.S.A.* **85**, 6503–6507.
Merluzzi, V. J., Welk, K., Savage, D. M., Last-Barney, K., and Mertelsmann, R. (1983). *J. Immunol.* **131**, 806–810.
Merrill, J. E., Mohlstrom, C., Uittenbogaart, C., Kermaniarab, V., Ellison, G. W., and Myers, L. W. (1984). *J. Immunol.* **133**, 1931–1937.
Merrill, J. E., Strom, S. R., Ellison, G. W., and Meyers, L. W. (1989). *J. Clin. Immunol.* **9**, 84–89.
Meuer, S. C., Hussey, R. E., Cantrell, D. A., Hodgdon, J. C., Schlossman, S. F., Smith, K. A., and Reinherz, E. L. (1984). *Proc. Natl. Acad. Sci. U.S.A.* **81**, 1509–1513.
Meuer, S. C., Dumann, H., Meyer zum Büschenfelde, K.-H., and Köhler, H. (1989). *Lancet* **1**, 15–18.
Michie, H. R., Eberlein, T. J., Spriggs, D. R., Manogue, K. R., Cerami, A., and Wilmore, D. W. (1988). *Ann. Surgery* **208**, 493–501.
Mier, J. W., and Gallo, R. C. (1982). *J. Immunol.* **128**, 1122–1126.
Migita, K., Eguchi, K., Tezuka, H., Otsubo, S., Kawakami, A., Nakao, H., Ueki, Y., Shimomura, C., Matsunaga, M., Ishikawa, N., Ito, K., and Nagataki, S. (1989). *Clin. Exp. Immunol.* **77**, 196–201.
Mills, G., Stewart, D. J., Mellors, A., and Gelfand, E. W. (1986). *J. Immunol.* **136**, 3019–3024.
Mills, G. B., Girard, P., Grinstein, S., and Gelfand, E. W. (1988). *Cell* **55**, 91–100.
Minakuchi, R., Wacholtz, M. C., Davis, L. S., and Lipsky, P. E. (1990). *J. Immunol.* **145**, 2616–2625.
Miossec, P., Kashiwado, T., and Ziff, M. (1987). *Arthritis Rheum.* **30**, 121–129.
Miyagi, J., Minato, N., Sumiya, M., Kasahara, T., and Kano, S. (1989). *Arthritis Rheum.* **32**, 1356–1364.
Miyasaka, N., Nakamaura, T., Russell, I. J., and Talal, N. (1984). *Clin. Immunol. Immunopathol.* **31**, 109–117.
Miyasaka, N., Murota, N., Yamaoka, K., Sato, K., Yamada, T., Nishido, T., and Okuda, M. (1986). *Clin Exp. Immunol.* **65**, 497–505.
Mizuochi, T., Hügin, A. W., Morse III, H. C., Singer, A., and Buller, R. M. L. (1989). *J. Immunol.* **142**, 270–273.
Möller, G., ed. (1986). *Immunol. Rev.* **92**.
Möller, G., ed. (1987). *Immunol. Rev.* **95**.
Möller, G., ed. (1990). *Immunol. Rev.* **118**.
Moore, K. W., Vierira, P., Fiorentino, D. F., Trounstine, M. L., Khan, T. A., and Mosmann, T. R. (1990). *Science* **248**, 1230–1234.
Morahan, G., Allison, J., and Miller, J. F. A. P. (1989). *Nature (London)* **339**, 622–624.
Moriyama, T., Suzuki, G., Nakao, I., Aizawa, S., Okumura, K., Nishimura, T., and Takaku, F. (1988). *Cell. Immunol.* **111**, 482–491.
Mosely, B., Beckman, O., March, C. J., Idzerda, I., Gimpel, S. D., Alpert, A., Anderson, J., Smith, C., Urdal, D., Cosman, D., and Park, L. S. (1989). *Cell* **59**, 335–344.

Mosmann, T. (1983). *J. Immunol. Methods* **65**, 55–63.
Mosmann, T. R., and Coffman, R. L. (1989). *Annu. Rev. Immunol.* **7**, 145–173.
Mountz, J. D., Smith, H. R., Wilder, R. L., Reeves, J. P., and Steinberg, A. D. (1987). *J. Immunol.* **138**, 157–163.
Mountz, J. D., Smith, T. M., and Toth, K. S. (1990). *J. Immunol.* **144**, 2159–2166.
Mouzaki, A., Volk, H., Osawa, H., and Diamantstein, T. (1987). *Eur. J. Immunol.* **17**, 335–342.
Muegge, K., and Durum, S. K. (1989). *N. Biologist* **1**, 239–247.
Muegge, K., Williams, T. M., Kant, J., Karin, M., Chiu, R., Schmidt, A., Siebenlist, U., Young, H. A., and Durum, S. K. (1989). *Science* **246**, 249–251.
Mueller, D. L., Jenkins, M. K., and Schwartz, R. H. (1989). *Annu. Rev. Immunol.* **7**, 445–480.
Muñoz, E., Zubiaga, A., Olson, D., and Huber, B. T. (1989). *Proc. Natl. Acad. Sci. U.S.A.* **86**, 9461–9464.
Murakawa, Y., and Sakane, T. (1988). *Arthritis Rheum.* **31**, 826–833.
Murakawa, Y., Takada, S., Ueda, Y., Suzuki, N., Hoshino, T., and Sakane, T. (1985). *J. Immunol.* **134**, 187–195.
Murray, L. J., Lee, R., and Martens, C. (1990). *Eur. J. Immunol.* **20**, 163–170.
Nagler, A., Lanier, L. L., and Phillips, J. H. (1988). *J. Immunol.* **141**, 2349.
Naides, S. J. (1986). *J. Immunol.* **136**, 4113–4117.
Nakanishi, K., Malek, T. P., Smith, K. A., Hamaoka, T., Shevach, E. M., and Paul, W. E. (1984). *J. Exp. Med.* **160**, 1605–1617.
Nau, G. J., Moldwin, R. L., Lancki, D. W., Kim, D.-K., and Fitch, F. W. (1987). *J. Immunol.* **139**, 114–122.
Nelson, D. L., Rubin, L. A., Kurman, C. C., Fritz, M. E., and Boutin, B. (1986). *J. Clin. Immunol.* **6**, 114–120.
Nicoletti, F., Meroni, P. L., Landolfo, S., Gariglio, M., Guzzardi, S., Barcellini, W., Lunetta, M., Mughini, L., and Zanussi, C. (1990). *Lancet* **2**, 319.
Niederkorn, J. Y. (1987). *Transplantation* **43**, 523–528.
Nieto, M. A., and López-Rivas, A. (1989). *J. Immunol.* **143**, 4166–4170.
Nieto, M. A., González, A., López-Rivas, A., Diaz-Espada, F., and Gambón, F. (1990). *J. Immunol.* **145**, 1364–1368.
Nishi, M., Ishida, Y., and Honjo, T. (1988). *Nature (London)* **331**, 267–269.
Nishismura, T., Takeuchi, Y., Ichimura, Y., Gao, X., Akatsuka, A., Tamaoki, N., Yagita, H., Okamura, K., and Habu, S. (1990). *J. Immunol.* **145**, 4012–4017.
Nishizuka, Y. (1986). *Science* **233**, 305–312.
Odaka, C., Kizaki, H., and Tadakuma, T. (1990). *J. Immunol.* **144**, 2096–2101.
Oen, K., Warrington, R., Rosenberg, A. M., and Krzekotowska, D. (1988). *Clin. Exp. Immunol.* **74**, 87–93.
Ognibene, F. P., Rosenberg, S. A., Lotze, M., Skibber, J., Parker, M. M., Shelhamer, J. H., and Parrillo, J. E. (1988). *Chest* **94**, 750–754.
Owen, K. L., Shibata, T., Izui, S., and Walker, S. E. (1989). *J. Biol. Resp. Med.* **8**, 366–374.
Pahwa, R., Chatila, T., Pahwa, S., Paradise, C., Day, N. K., Geha, R., Schwartz, S. A., Slade, H., Oyaizu, N., and Good, R. A. (1989). *Proc. Natl. Acad. Sci. U.S.A.* **86**, 5069–5073.
Panitch, H. S., Hirsch, R. L., Schindler, J., and Johnson, K. P. (1987). *Neurology* **37**, 1097.
Papiernik, M., and Pontoux, C. (1990). *Int. Immunol.* **2**, 407–412.
Parsitol, H. S., Hirsch, R. L., Haley, A. S., and Johnson, K. P. (1986). *Lancet* **1**, 893–894.
Paul, W. E. (1989). *Cell* **57**, 521–524.

Pernice, W., Schuchmann, L., Dippell, J., Suschke, J., Vogel, P., Truckenbrodt, H., Schindera, F., Humburg, C., and Brzoska, J. (1989). *Arthritis Rheum.* **32**, 643-646.
Phadke, K., Carlson, D. G., Gitter, B. D., and Buttler, L. D.(1986). *J. Immunol.* **136**, 4085-4091.
Pirchert, G., Jost, L. M., Zöbeli, L., Odermatt, B., Pedio, G., and Stahel, R. A. (1990). *Br.J. Cancer* **62**, 100-104.
Plaetinck, G., Combe, M.-C., Corthésy, P., Sperisen, P., Kanamori, H., Honjo, T., and Nabholz, M. (1990). *J. Immunol.* **145**, 3340-3347.
Plearse, M., Wu, L., Egerton, M., Wilson, A., Shortman, K., and Scollay, R. (1989). *Proc. Natl. Acad. Sci. U.S.A.* **86**, 1614-1618.
Plum, J., Koning, F., Leclercq, G., Tison, B., and de Smedt, M. (1990). *J. Immunol.* **144**, 3710-3717.
Powell, M. B., Mitchell, D., Lederman, J., Buckmeier, J., Zamvil, S. S., Graham, M., Ruddle, N. H., and Steinman, L. (1990). *Int. Immunol.* **2**, 539-544.
Prince, H. E., Kleinman, S., and Williams, A. E. (1988). *J. Immunol.* **140**, 1139-1141.
Prud'homme, G. J., Fuks, A., Colle, E., Seemayer, T. A., and Guttmann, R. D. (1984). *J. Exp. Med.* **159**, 463-478.
Pukel, C., Baquerizo, H., and Rabinovitch, A. (1987). *Diabetes* **36**, 1217-1222.
Pukel, C., Baquerizo, H., and Rabinovitch, A. (1988). *Diabetes* **37**, 133-135.
Punt, C. J. A., Henzen-Logmans, S. C., Bolhuis, R. L. H., and Stoter, G. (1990). *Br. J. Cancer* **61**, 491.
Ramarli, D., Fox, D. A., and Reinherz, E. L. (1987). *Proc. Natl. Acad. Sci. U.S.A.* **84**, 8598-8602.
Rammensee, H.-G., Kroschewski, R., and Frangoulis, B. (1989). *Nature (London)* **339**, 541-544.
Ramsdell, F., and Fowlkes, B. J. (1990). *Science* **248**, 1342-1348.
Ramsdell, F., Lantz, T., and Fowlkes, B. J. (1989). *Science* **246**, 1038-1041.
Ramseur, W. L., Richards, F., and Duggan, D. B. (1989). *Cancer* **63**, 2005-2007.
Ramshaw, I. A., Andrew, M. E., Philips, S. M., Boyle, D. B., and Coupar, B. E. H. (1987). *Nature (London)* **329**, 545-547.
Reem, G. H., and Yeh, N. H. (1984). *Science* **225**, 429-430.
Reimann, J., and Diamantstein, T. (1981). *Clin. Exp. Pathol.* **43**, 641-644.
Reske-Kunz, A. B., Osawa, H., Josimovic-Alasevic, O., Rüde, E., and Diamantstein, T. (1987). *J. Immunol.* **138**, 192-196.
Ridge, S. C., Zabriskie, J. B., Osawa, H., Diamantstein, T., Oronsky, A. L., and Kerwar, S. S. (1986). *J. Exp. Med.* **164**, 327-339.
Robb, R. J., and Greene, W. C. (1987). *J. Exp. Med.* **165**, 1201-1206.
Robb, R. J., and Kutny, R. M. (1987). *J. Immunol.* **139**, 855-862.
Robbins, D. S., Sirazi, Y., Drysdale, B.-E., Lieberman, A., Shin, H. S., and Shin, M. L. (1987). *J. Immunol.* **139**, 2593-2597.
Roberge, F. G., Lorberboum-Galski, H., Hoang, P. L., de Smet, M., Chan, C. C., Fitzgerald, D., and Pastan, I. (1989). *J. Immunol.* **143**, 3498-3502.
Rocha, B. (1990). *Eur. J. Immunol.* **20**, 919-925.
Roncarlo, M. G., Zoppo, M., Bacchetta, R., Gabiano, C., Sacchetti, C., Cerutti, F., and Tovo, P. A. (1988). *Clin. Immunol. Immunopathol.* **49**, 53-62.
Rosenberg, Y. J., Steinberg, A. D., and Santoro, T. J. (1984).*J. Immunol.* **133**, 2545-2550.
Rosenberg, S. A., Motze, M. T., Muul, L. M., Chang, A. E., Avis, F. P., Leitman, S., Linehan, W. M., Robertson, G. N., Lee, R. E., Rubin, J. T., Seipp, C. A., Simpson, C. G., and White, D. E. (1987). *N. Engl. J. Med.* **316**, 889-897.
Rosenberg, Y. J., Nurse, F., and Begley, C. G. (1989). *J. Immunol.* **143**, 2216-2222.

Rothenberg, E. V., McGuire, K. L., and Boyer, P. D. (1988). *Immunol. Rev.* **104**, 29–54.
Roths, J. B., Murphy, E. D., and Eicher, E. N. (1984). *J. Exp. Med.* **159**, 1–20.
Rubin, L. A., Kurman, C. C., Fritz, M. E., Yarchoan, R., and Nelson, D. L. (1985). *J. Immunol.* **135**, 3172–3177.
Rubin, L. A., Jay, G., and Nelson, D. L. (1986). *J. Immunol.* **137**, 3841–3844.
Rubin, L. A., Snow, K. M., Kurman, C. C., Nelson, D. L., and Keystone, E. C. (1990). *J. Rheumatol.* **17**, 597–602.
Sabath, D. E., Podolin, P. L., Comber, P. G., and Prystowsky, M. B. (1990). *J. Biol. Chem.* **265**, 12671–12678.
Sakaguchi, S., and Sakaguchi, N. (1989). *J. Immunol.* **142**, 471–480.
Sakaguchi, S., and Sakaguchi, N. (1990). *J. Exp. Med.* **172**, 537–545.
Sakane, T., Murakawa, Y., Suzuki, N., Ueda, Y., Tsuchida, T., Takada, S., Yamauchi, Y., and Tsunematsu, T. (1989). *Am. J. Med.* **86**, 385–390.
Saklatvala, J. (1986). *Nature (London)* **322**, 547–549.
Samlowski, W. E., Ward, J. H., Craven, C. M., and Freedman, R. A. (1989). *Arch. Pathol. Lab. Med.* **113**, 838–841.
Sano, H., Kumagai, S., Namiuchi, S., Uchiyama, T., Yodoi, J., Maeda, M., Takatsuki, K., Suginoshita, T., and Imura, H. (1986). *Clin. Exp. Immunol.* **63**, 8–16.
Santoro, T. J., Luger, T. A., Raveche, E. S., Smolen, J. S., Oppenheim, J. J., and Steinberg, A. D. (1983). *Eur. J. Immunol.* **13**, 601–604.
Saragovi, M., and Malek, T. R. (1990). *Proc. Natl. Acad. Sci. U.S.A.* **87**, 11–15.
Sarvetnick, N., Liggitt, D., Pitts, S. L., Hansen, S. E., and Stewart, T. A. (1988). *Cell* **52**, 773–782.
Sarvetnick, N., Shizuru, J., Liggitt, D., Martin, L., McIntyre, B., Gregory, A., Parslow, T., and Stewart, T. (1990). *Nature (London)* **346**, 844–847.
Satoh, J., Seino, H., Abo, T., Tanaka, S.-I., Shintani, S., Ohta, S., Tamura, K., Sawai, T., Nobunaga, T., Oteki, T., Kumagai, K., and Toyota, T. (1989). *J. Clin. Invest.* **84**, 1345–1348.
Satoh, J., Seino, H., Shintani, S., Tanaka, S.-I., Ohteki, T., Masuda, T., Nobunaga, T., and Toyota, T. (1990). *J. Immunol.* **145**, 1395–1399.
Saxena, S., Nouri-Aria, K., Anderson, M. G., Eddleston, A. L. W. F., and Williams, R. (1986). *Clin. Exp. Immunol.* **63**, 541–548.
Schauenstein, K., and Kroemer, G. (1987). In "Avian Immunology: Basis and Practice," Vol. I. (P. Toivanen and A. Toivanen, eds.), pp. 213–227. CRC Press, Boca Raton, Florida.
Schell, S. R., and Fitch, F. W. (1989). *J. Immunol.* **143**, 1499–1505.
Schena, F. P., Mastrolitti, G., Jirillo, E., Munno, I., Pellegrino, N., Fracasso, R., and Aventaggiato, L. (1989). *Kidney Int.* **35**, 875–879.
Schneider, T. M., Kupiec-Weglinski, J. W., Towpik, E., Padberg, W., Araneda, D., Daimantstein, T., Strom, T. B., and Tilney, N. C. (1986). *Transplantation* **42**, 191–195.
Schuchter, L. M., Hendricks, C. B., Holland, K. H., Shelton, B. K., Hutchins, G. M., Baugham, K. L., and Ettinger, D. S. (1990). *Am. J. Med.* **88**, 439–440.
Schwartz, R. H. (1990). *Science* **248**, 1349–1356.
Sekigawa, I., Noguchi, K., Hasegawa, K., Hirose, S., Sato, H., and Shirai, T. (1989). *Clin. Immunol. Immunopathol.* **51**, 172–184.
Selmaj, K., Plater-Zyberg, C., and Rockett, K. A. (1986). *Neurology* **36**, 1392–1395.
Selmaj, K., Nowak, Z., and Tchórzewski, H. (1988). *J. Neurol. Sci.* **85**, 67–76.
Semenzato, G., Bambara, L. M., Biasi, D., Frigo, A., Vinante, F., Zuppini, B., Trentin, L., Feruglio, C., Chilosi, M., and Pizzolo, G. (1988). *J. Clin. Immunol.* **8**, 447–452.
Serreze, D. V., and Leiter, E. H. (1988). *J. Immunol.* **140**, 3801–3807.

Sharon, M., Gnarra, J. R., and Leonard, W. J. (1990). *Proc. Natl. Acad. Sci. U.S.A.* **87**, 4869–4873.
Shaw, J. P., Utz, P., Durand, D. B., Toole, J. J., Emmel, E. A., and Crabtree, G. R. (1988). *Science* **241**, 202–205.
Sherblom, A. P., Sathyamoorthy, N., Decker, J. M., and Muchmore, A. V. (1989). *J. Immunol.* **143**, 939–944.
Shi, Y., Sahai, B. M., and Green, D. R. (1989). *Nature (London)* **339**, 625–627.
Shimonkevitz, R. P., Husmann, L. A., Bevan, M. J., and Crispe, I. N. (1987). *Nature (London)* **329**, 157–159.
Shivakumar, S., Tsokos, G. C., and Datta, S. K. (1989). *J. Immunol.* **143**, 103–112.
Shoenfeld, Y., and Schwarts, R. S. (1984). *N. Engl. J. Med.* **311**, 1019–1029.
Siegel, J. P., Sharon, M., Smith, P. L., and Leonhard, W. J. (1987). *Science* **238**, 75–78.
Sierakowski, S., Kucharz, E. J., Lightfoot, R. W., and Goodwin, J. S. (1989). *J. Clin. Immunol.* **9**, 496–500.
Simon, M. M., Moller, D. R., Saltini, C., Kirby, M., and Crystal, R. G. (1984). *Clin. Immunol. Immunopathol.* **33**, 39–53.
Singer, P. A., Balderas, R. S., McEvilly, R. J., Bobardt, M., and Theofilopoulos, A. N. (1989). *J. Exp. Med.* **170**, 1869–1877.
Sitkovsky, M. V., and Paul, W. (1988). *Nature (London)* **332**, 306–308.
Skidmore, B. J., Stamnes, S. A., Townsend, K., Glasbrook, A. L., Sheehan, K. C. F., Schreiber, R. D., and Chiller, J. M. (1989). *Eur. J. Immunol.* **19**, 1591–1597.
Skinner, M., Le Gros, G., Marbrook, J., and Watson, J. D. (1987). *J. Exp. Med.* **165**, 1481–1490.
Smith, K. A. (1988). *Science* **240**, 1169–1176.
Smith, K. A. (1989). *Annu. Rev. Biochem.* **5**, 397–425.
Smith, J. B., and Talal, N. (1982). *Scand. J. Immunol.* **16**, 269–278.
Smith, H., Chen, I.-M., Kobo, R., and Tung, K. S. K. (1989). *Science* **245**, 749–752.
Smith, L. R., Brown, S. L., and Blalock, J. E. (1989). *J. Neuroimmunol.* **21**, 249–254.
Sorokin, R., Kimura, H., Schroeder, K., Wilson, D. H., and Wilson, D. B. (1986). *J. Exp. Med.* **164**, 1615–1624.
Soulillou, J. P., Peyronnet, P., LeMauff, B., Olive, D., Delaage, M., Peyronnet,P., Hourmant, M., Mawas, C., and Hirn, M. (1987). *Lancet* **1**, 1339–1342.
Steinberg, A. D., Roths, J. B., Murphy, E. D., Steinberg, R. T., and Raveche, E. S. (1980). *J. Immunol.* **125**, 871–873.
Stötter, H., Rüde, E., and Wagner, H. (1980). *Eur. J. Immunol.* **18**, 295–300.
Streck, R. J., Helinski, E. H., Ovak, G. M., and Pauly, J. L. (1990). *J. Leukocyte Biol.* **48**, 237–246.
Strominger, J. L. (1989). *Science* **244**, 943–950.
Stünkel, K. G., Theisen, P., Mouzaki, A., Diamantstein, T., and Schlumberger, H. D. (1988). *Immunology* **64**, 683–689.
Suematsu, S., Matsuda, T., Aozasa, K., Akira, S., Nakano, N., Ohno, S., Miyazaki, J.-L., Yamamura, K. I., Hirano, T., and Kishimoto, T. (1989). *Proc. Natl. Acad. Sci. U.S.A.* **86**, 7547–7551.
Sugamura, K., Nakai, S., Fujii, M., and Hinuma, Y. (1985). *J. Exp. Med.* **161**, 1243–1254.
Sugie, K., Nakamura, Y., Tagaya, Y., Koyasu, S., Yahara, I., Takakura, K., Kumagai, S., Imura, H., and Yodoi, J. (1990). *Int. Immunol.* **2**, 391–397.
Suzuki, J., Takeshita, T., Ohbo, K., Tada, K., Asao, H., and Sugamura, K. (1989). *Int. Immunol.* **1**, 373–377.
Sykes, M., Romick, M. L., Hoyles, K. A., and Sachs, D. A. (1990). *J. Exp. Med.* **171**, 645–658.

Symons, J. A., Wood, N. C., DiGiovine, F. S., and Duff, G. W. (1988). *J. Immunol.* **141**, 2612–2618.
Tadakuma, T., Kizaki, H., Odaka, C., Kubota, R., Ishimura, Y., Yagita, H., and Okumura, K. (1990). *Eur. J. Immunol.* **20**, 779–784.
Taira, S., Matsui, M., Hayakawa, K., Tokoyama, T., and Nariuchi, H. (1987). *J. Immunol.* **139**, 2957–2964.
Takeshita, T., Goto, Y., Tada, K., Nagata, K., Asao, H., and Sugamura, K. (1989). *J. Exp. Med.* **169**, 1323–1335.
Takeshita, T., Asao, H., Suzuki, J., and Sugamura, K. (1990). *Int. Immunol.* **2**, 477–480.
Tanaka, K., Tozawa, H., Hayami, M., Sugamura, K., and Hinuma, Y. (1985). *Microbiol. Immunol.* **29**, 959.
Tanaka, Y., Saito, K., Shirakawa, F., Ota, T., Suzuki, H., Eto, S., and Yamashita, U. (1988). *J. Immunol.* **141**, 3043–3049.
Tanaka, T., Saiki, O., Negoro, S., Igarashi, T., Kuritani, T., Hara, H., Suemara, M., and Kishimoto, S. (1989). *Arthritis Rheum.* **32**, 552–559.
Taniguchi, T., Matsui, H., Fujita, T., Takaoka, C., Kashima, N., Yoshimoto, R., and Hamuro, J. (1983). *Nature (London)* **302**, 305–310.
Taverne, J., Rayner, D. C., Van der Meide, P. H., Lydyard, P. M., and Cooke, A. (1987). *Eur. J. Immunol.* **17**, 1855–1858.
Teh, H. S., Kishi, H., Scott, B., and von Boehmer, H. (1989). *J. Exp. Med.* **169**, 795–806.
Tendler, C. L., Greenberg, S. J., Blattner, W. A., Manns, A., Murphy, E., Fleisher, T., Hanchard, B., Morgan, O., Burton, J. D., Nelson, D. L., and Waldmann, T. A. (1990). *Proc. Natl. Acad. Sci. U.S.A.* **87**, 5218–5222.
Tentori, L., Longo, D. L., Zuñiga-Pflucker, J. C., Wing, C., and Kruisbeek, A. M. (1988). *J. Exp. Med.* **168**, 1741–1747.
Teodorczyk-Injeyan, J. A., Sparkes, B. G., Mills, G. B., Peters, W. J., and Falk, R. E. (1986). *Clin. Exp. Immunol.* **65**, 570–581.
Tezabwala, B. U., Johnson, P. M., and Rees, R. C. (1989). *Immunology* **67**, 115–119.
Theofilopoulos, A., and Dixon, F. J. (1985). *Adv. Immunol.* **37**, 269–390.
Theofilopoulos, A. N., Balderas, R. S., Shawler, D. L., Lee, S., and Dixon, F. J. (1981). *J. Exp. Med.* **153**, 1405–1420.
Theofilopoulos, A. N., Balderas, R. S., Gozes, Y., Aguado, M. T., Hang, L. H., Morrow, P. R., and Dixon, F. J. (1985). *J. Exp. Med.* **162**, 1–18.
Theofilopoulos, A. N., Kofler, R., Singer, P. A., and Dixon, F. J. (1989). *Adv. Immunol.* **46**, 61–109.
Thijs, L. G., Hack, C. E., van Schijndel, J. M. S., Nuijens, J. H., Wolbink, G. J., Eerenberg-Belmer, A. J. M., van der Vall, H., and Wagstaff, J. (1990). *J. Immunol.* **144**, 2419–2424.
Thomen, M L., and Weigle, W. O. (1981). *J. Immunol.* **127**, 2102–2107.
Tigges, M. A., Casey, L. S., and Koshland, M. E. (1989). *Science* **243**, 781–786.
Tilden, A. B., and Balch, C. M. (1982). *J. Immunol.* **129**, 2469–2475.
Tomonari, K. (1990). *Immunol. Rev.* **116**, 139–157.
Tomosugi, N. I., Cashman, S. J., Hay, H., Pusey, C. D., Evans, D. J., Shaw, A., and Rees, A. J. (1989). *J. Immunol.* **142**, 3083–3090.
Toribio, M. L., de la Hera, A., Borst, J., Marcos, M. A. R., Márquez, C., Alonso, J. M., Bárcena, A., and Mártinez-A., C. (1988). *J. Exp. Med.* **168**, 2231–2249.
Toribio, M. L., Gutierrez-Ramos, J. C., Pezzi, L., Marcos, M. A. R., and Martínez-A., C. (1989). *Nature (London)* **342**, 82–85.
Treiger, B. F., Leonard, W. J., Svetlik, P., Rubin, L. A., Nelson, D. L., and Greene, W. C. (1986). *J. Immunol.* **136**, 4099–4105.

Trinchieri, G., Matsumoto-Kobayashi, M., Clark, S. C., Sechra, J., London, L., and Perussia, B. (1984). *J. Exp. Med.* **160**, 1147–1159.
Trotter, J. L., Clifford, D. B., Anderson, C. B., van der Veen, R. C., Hicks, B. C., and Banks, G. (1988). *N. Engl. J. Med.* **318**, 1206.
Tsudo, M., Karasuyama, H., Kitamura, F., Nagasaka, Y., Tanaka, T., and Miyasaka, M. (1989). *J. Immunol.* **143**, 4039–4043.
Tung, S. K., Umland, E., Matzinger, P., Nelson, K., Schauf, V., Rubin, L. A., Wagner, D., Scollard, D., Prakong, V., Vithayasai, V., Worobec, S., Smith, T., and Suriyanond, V. (1987). *Clin. Exp. Immunol.* **68**, 10–15.
Ullman, K. S., Northrop, J. P., Verweij, C. L., and Crabtree, G. R. (1990). *Annu. Rev. Immunol.* **8**, 421–452.
Umehara, H., Kumagai, S., Ishida, H., Suginoshita, T., Maeda, M., and Imura, H. (1988). *Arthritis Rheum.* **31**, 401–407.
van de Water, J., Wilson, T. J., Haapanen, L. A., Boyd, R. L., Abplanalp, H., and Gershwin, M. E. (1990). *Clin. Immunol. Immunopathol.* **56**, 169–184.
van Liessum, P. A., de Mulder, P. H. M., Mattijssen, E. J. M., Corstens, F. H. M., and Wagener, D. J. T. (1989). *Lancet* **1**, 224.
Vecht, C. J., Keohane, C., Menon, R. S., Henzenlogmans, S. C., Punt, C. J. A., and Stoter, G. (1990). *N. Engl. J. Med.* **323**, 1146–1147.
Via, C. S., and Shearer, G. M. (1988a). *J. Exp. Med.* **168**, 2165–2181.
Via, C. S., and Shearer, G. M. (1988b). *Immunol. Today* **9**, 207–212.
Vink, A., Uyttenhove, C., Wauters, P., and van Snick, J. (1990). *Eur. J. Immunol.* **20**, 1–6.
Volk, H. D., and Diamantstein, T. (1986). *Clin. Exp. Immunol.* **66**, 525–531.
Volk, H.-D., Brocke, S., Osawa, H., and Diamantstein, T. (1986). *Eur. J. Immunol.* **16**, 1309–1312.
von Boehmer, H. (1990). *Annu. Rev. Immunol.* **8**, 531–556.
Voothuis, J. A. C., Uitdehaag, B. M. J., de Groot, C. J. A., Goede, P. H., van der Meide, P. H., and Dijkstra, J. B. (1990). *Clin. Exp. Immunol.* **1990**, 183–188.
Waanders, G. A., and Boyd, R. L. (1990). *Int. Immunol.* **2**, 461–468.
Wainberg, M. A., Vydelingum, S., and Margolese, R. G. (1983). *J. Immunol.* **130**, 2372–2378.
Wakasugi, N., Tagaya, Y., Wakasugi, H., Mitsui, A., Maeda, M., Yodoi, J., and Tursz, T. (1990). *Proc. Natl. Acad. Sci. U.S.A.* **87**, 8282–8286.
Waldmann, T. A. (1989). *J. Natl. Cancer Institute* **81**, 914–923.
Wang, A. M., Doyle, M. V., and Mark, D. F. (1989). *Proc. Natl. Acad. Sci. U.S.A.* **86**, 9727–9721.
Warrington, R. J., Sauder, P. J., Homik, J., and Ofosu-Appiah, W. (1989). *Clin. Exp. Immunol.* **77**, 163–167.
Watkins, D., and Cohen, N. (1985). *Dev. Comp. Immunol.* **9**, 819–830.
Weinberg, K., and Parkman, R. (1990). *N. Engl. J. Med.* **322**, 1718–1724.
Wekerle, H., and Diamantstein, T. (1986). *Ann. N. Y. Acad. Sci.* **475**, 401–403.
Welte, K., Ciobanu, N., Moore, M. A., Gulati, S., and O'Reilly, R. J. (1984). *Blood* **64**, 380–385.
Weston, K. M., Yeh, E. T. H., and Man-Sun, S. (1987). *J. Immunol.* **139**, 734–742.
Weston, K. M., Ju, S.-T., Lu, C. Y., and Sy, M.-S. (1988). *J. Immunol.* **141**, 1941–1948.
Wick, G., Kite, J. H., and Witebsky, E. (1970). *J. Immunol.* **101**, 54–59.
Wick, G. Müller, P., and Schwartz, S. (1982). *Eur. J. Immunol.* **12**, 877–879.
Wick, G., Hàla, K., Brezinschek, H.-P., Dietrich, H., Wolf, H., and Kroemer, G. (1989). *Adv. Immunol.* **47**, 433–500.
Wiesenberg, I., van der Meide, P. H., Schellekens, H., and Alkan, S. S. (1989). *Clin. Exp. Immunol.* **78**, 245–249.

Wilde, D. B., Prystowsky, M. B., Ely, J. E., Vogel, S. N., Dialynas, D. P., and Fitch, F. W. (1984). *J. Immunol.* **133**, 636–641.
Wilkins, J. A., Olivier, S. L., and Warrington, R. J. (1984). *Clin. Exp. Immunol.* **58**, 1–6.
Williams, M. E., Lichtman, A. H., and Abbas, A. K. (1990). *J. Immunol.* **144**, 1208–1214.
Wirt, D. P., Brooks, E. G., Vaidya, S., Klimpel, G. R., Waldmann, T. A., and Goldblum, R. M. (1989). *N. Engl. J. Med.* **321**, 370–374.
Wofsy, D., Murphy, E. D., Roths, J. B., Dauphine, M. J., Kipper, S. B., and Talal, N. (1981). *J. Exp. Med.* **154**, 1671–1680.
Wolf, R. E., and Bathge, B. A. (1990). *Arthritis Rheum.* **33**, 1007–1015.
Wolf, R. E., and Brelsford, W. G. (1988). *Arthritis Rheum.* **31**, 729–735.
Xiao, G. X., Chopra, R. K., and Adler, W. H. (1988). *J. Trauma* **28**, 1669–1672.
Yagita, H., Nakata, M., Azuma, A., Nitta, T., Takeshita, T., Sugamura, K., and Okumura, K. (1989). *J. Exp. Med.* **170**, 1445–1450.
Yamada, T., Fujishima, A., Kawahara, K, Kato, K., and Nishimura, O. (1987). *Arch. Biochem. Biophys.* **257**, 194.
Yamagata, N., Kobayashi, K., Kasama, T., Fukushima, T., Tabata, M., Yoneya, I., Shikama, Y., Kaga, S., Hashimoto, M., Yoshida, K., Sekine, F., Negishi, M., Ide, H., Mori, Y., and Takahashi, T. (1988). *J. Rheumatol.* **15**, 1623–1627.
Yamaguchi, A., Ide, T., Hatakeyama, M., Doi, T., Kono, T., Uchiyama, T., Kikuchi, K., Taniguchi, T., and Uede, T. (1989). *Int. Immunol.* **1**, 160–168.
Yamaguchi, Y., Suda, T., Shiosaki, H., Miura, Y., Hitoshi, Y., Tominaga, A., Takatsu, K., and Kasahara, T. (1990). *J. Immunol.* **145**, 873–877.
Yokono, K., Kawase, Y., Nagat, M., Hatamori, N., and Baba, S. (1989). *Diabetologia* **32**, 67–73.
Yoshida, H., Yoshida, M., Merino, R., Shibata, T., and Izui, S. (1990). *Eur. J. Immunol.* **20**, 1989–1993.
Yuuki, H., Yoskijai, Y., Kishihara, K., Iwasaka, A., Matsuzaki, G., Takimoto, H., and Nomoto, K. (1990). *Eur. J. Immunol.* **20**, 1475–1482.
Zielasek, J., Burkart, V., Naylor, P., Goldstein, A., Kiesel, U., and Kolb, H. (1990). *Immunology* **69**, 209–214.
Zier, K. S., Leo, M. M., Spielman, R. S., and Baker, L. (1984). *Diabetes* **33**, 552–555.
Zuñiga-Pflücker, J. C., and Kruisbeek, A. M. (1990). *J. Immunol.* **144**, 3736–3740.
Zuñiga-Pflücker, J. C., Smith, K. A., Tentori, L., Pardoll, D. M., Longo, D. L., and Kruisbeek, A. M. (1990). *Dev. Immunol.* **1**, 59–66.
Zurawski, S. M., Imler, J.-L., and Zurawski, G. (1990). *EMBO J.* **9**, 3899–3905.

This article was accepted for publication on 4 March 1991.

Histamine Releasing Factors and Cytokine-Dependent Activation of Basophils and Mast Cells

ALLEN P. KAPLAN, SESHA REDDIGARI, MARIA BAEZA, AND PIOTR KUNA

Division of Allergy, Rheumatology, and Clinical Immunology, Department of Medicine, SUNY—Stony Brook, Health Sciences Center, Stony Brook, New York 11794

I. Introduction

The release of histamine from human basophils or mast cells is typically initiated by the interaction of antigen with surface-bound IgE antibody. This has physiologic/pathologic significance for that 20% of the population who are atopic (allergic) individuals with a genetic predisposition to produce IgE antibody to allergens (antigens) that are either inhaled or ingested. A second immunologic pathway that causes a rapid release of histamine from basophils or mast cells is the activation of complement so as to cleave the third, fourth, and fifth components of complement. The peptides, C3a, C4a, and C5a, are released (1–3) and interact with specific cell receptors (C3a/4a receptor and C5a receptor) to initiate cell activation (1,4). These are the three known anaphylatoxins derived from the complement pathway and are generated during immune complex-mediated activation of complement or by complement-dependent injury to cells or tissues.

Histamine releasing factors are defined as products of activated cells (cytokine-like molecules) that interact with basophils and/or mast cells to cause the release of histamine. This concept was originally defined by Thueson *et al.* (5,6), who reported the presence of such a factor in streptokinase/streptodornase (SK/SD)-activated human mononuclear cells. It had a molecular weight of approximately 15,000 to 30,000, appeared to be a small protein, and was distinguished from C5a. Subsequently, a variety of reports confirmed that lymphoid cells produce such a factor and HRF-like activities were described as products of T lymphocytes (7–9), mixed mononuclear cells (10), alveolar macrophages (11), platelets (11), B lymphocytes (12), and neutrophils (13–15). Such activity also appeared to be present in nasal washings of allergic subjects (16) and in antigen-challenged cutaneous late-phase reactions (17). In this review, I will describe the various types of histamine releasing factors (HRFs), discuss their interaction with basophils or mast cells, and speculate regarding their role in inflammation in general, and in allergic diseases in particular.

II. Purification of Histamine Releasing Factors

Since histamine releasing factors have been described as products of a wide variety of cells, it has become clear that until such factors are purified, it is not possible to know whether different gene products are secreted by different cell types or even whether the multiplicity of species derived from a single cell type represents molecular heterogeneity of a single gene product. The earliest attempts to characterize histamine releasing factors indicated that the mononuclear cell product is protein in nature (5) and that at least two forms could be resolved by gel filtration (18), reverse-phase high-performance liquid chromatography (HPLC) (19), or by isoelectric focusing (10). These procedures separate molecules by their size, charge distribution, or hydrophobicity, and the molecular weights reported for HRF varied widely from as low as 1300 (19) to as high as 90,000 (9). Lett-Brown et al. (18,19) reported finding molecular-weight species of 1300, 10,000 to 20,000, and 50,000, and were able to characterize these molecules by either ion-exchange chromatography or high-pressure liquid chromatography. Our own early studies using quaternary aminoethyl (QAE)–Sephadex A-50 chromatography or isoelectric focusing (10) confirmed the presence of at least two major forms of HRF, whereas gel filtration gave little resolution, as a single broad peak of HRF activity extending from 15,000 to 45,000 kDa is obtained with the peak activity at 30,000–35,000 kDa. Goetzl et al. (9) reported a "basophil-activating factor" that was derived from human T lymphocytes. This was an acidic protein with an isoelectric point of 4.5 and had an estimated molecular weight of 70,000–90,000. The addition of monocytes appeared to degrade it to a 15,000-Da form. Thus, comparison of observations from different laboratories gave disparate findings.

More recently, an isolation procedure for further characterization of human HRFs was reported that utilized gel filtration, Accell QMA anion-exchange HPLC, and preparative elution after electrophoresis in sodium dodecyl sulfate (SDS) gels. This isolation scheme is shown in Fig. 1 and indicates the various species of HRFs found. The initial step utilized concentrated crude supernatants of SK/SD-stimulated mononuclear cells applied to a 0.75 × 60-cm TSK G3000 SW gel filtration column in phosphate-buffered saline (PBS) containing 0.3% polyethylene glycol 8000. A single peak containing HRF activity was found (Fig. 1) and this was next fractionated by Accell QMA anion-exchange HPLC (20) in 0.02 M ammonium acetate, pH 8. After the flow-through peak was obtained, the column was washed with 10 bed volumes of buffer and a gradient was applied with 0.5 M ammonium

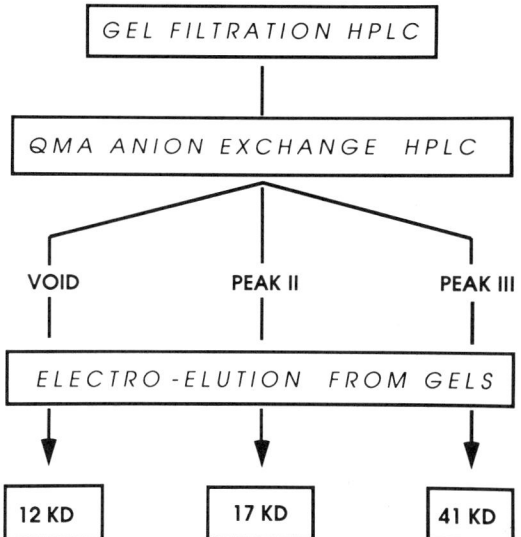

FIG. 1. HRF purification protocol. Schematic diagram for the purification of three molecular species of human histamine releasing factor. (From Ref. 20.)

acetate as the limit buffer. After washing with this latter buffer, a second gradient was applied with 1.0 M ammonium acetate as limit buffer. Three peaks containing HRF activity were obtained. The first was obtained in the void volume and presumably represented HRFs with an alkaline isoelectric point that would not adhere to a positively charged ion exchanger. A second peak containing HRF activity was found at 0.5 M ammonium acetate and eluted during the wash after the initial gradient was run. The third peak was obtained during the latter half of the second gradient at approximately 0.8 M ammonium acetate (Fig. 2). Each peak is then concentrated, lyophilized, and further fractionated by SDS and polyacrylamide gel electrophoresis (SDS–PAGE) using 1.5- or 3.0-mm slab gels and 12% acrylamide. At the end of the run, the gel lanes containing samples were sliced and each slice dialyzed (3500-Da cutoff) with 3 ml 25 mM phosphate buffer, pH 9. Each dialysis bag was placed in an electroblotting tank and 15–17 V were applied for 18 hours at room temperature with three buffer changes. Samples were then dialyzed, filtered, lyophilized, and resuspended in 250 μl PBS, and a portion was passed over a Sephadex G-25 minicolumn prior to assay. In this fashion, all traces of SDS are removed. We obtained peaks of HRF activity at molecular weights of 10,000–12,000 (void peak), 15,000–17,000 (peak 2 at 0.5 M ammonium

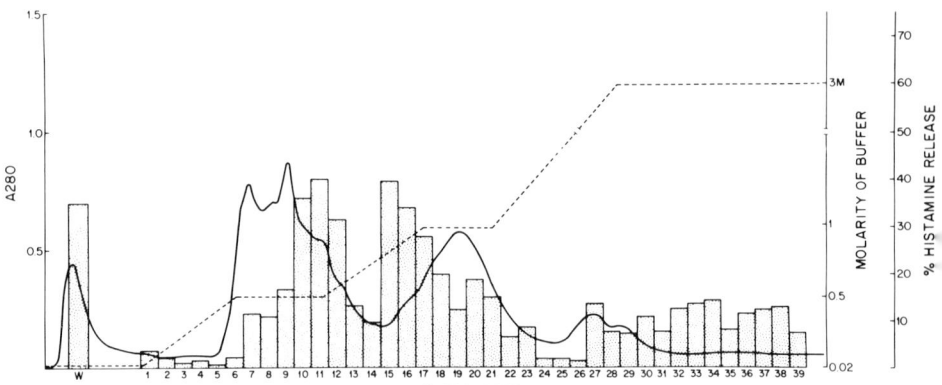

Fig. 2. Accell QMA anion-exchange HPLC of mononuclear cell-derived supernatant containing HRF activity. Molarity of the buffer in gradients and wash phase are indicated by the dashed line. The optical density at 280 nm is shown by the solid line; present histamine release is shown by the vertical bars. Major areas of activity are in the void volume, during the 0.5 M wash phase, and at the end of the 1.0 M gradient. (From Ref. 20.)

acetate), and 40,000 (peak 3 at 0.8 M ammonium acetate). Because our initial purification attempts yielded only small amounts of proteins, we radiolabeled a portion of each peak and subjected it to SDS gel electrophoresis and radioautography. Peaks 2 and 3 are shown in Fig. 3a and b and each revealed a single sharp band at the indicated molecular weight.

This material was next subjected to amino acid sequencing to determine whether HRF corresponded to any previously described protein (21). The 10,000- to 12,000-kDa form was first subjected to SDS gel electrophoresis using a 15% gel in order to obtain better resolution. Stained gels now revealed a broad band rather than the narrow one seen with a 12% gel. This was electroblotted onto nitrocellulose paper, the band was divided into an upper and lower half, and each half was subjected to sequence analysis. As seen in Table I, there was close homology in the upper half (HRF-1) and in the amino acid sequence of connective tissue-activating peptide (CTAP III), a derivative of the platelet α granules that cause activation and proliferation of fibroblasts (22). CTAP III stimulates synthesis of DNA and sulfated glycosaminoglycans, formation of hyaluronic acid, augments glycolysis, and induces secretion of PGE_2. When the amino acid sequence of the lower half of the band was determined (HRF-2), it also was homologous to CTAP III if the alignment began with the alanine at residue 16. Thus, the broad band observed appeared to represent microheterogeneity due to cleavage of the N-terminal 15 amino acids from the native

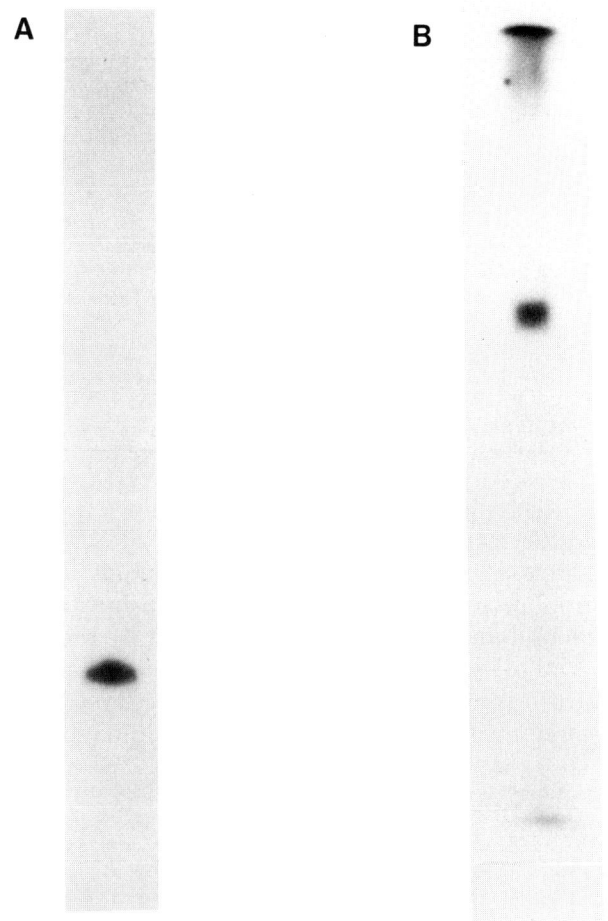

FIG. 3. Autoradiogram of purified histamine releasing factor; 15–17 kDa (A) and 41 kDa (B). (From Ref. 20.)

protein (21). Such a cleavage has been described for CTAP III in which it is converted to a molecule that is a potent activator of neutrophils, designated neutrophil-activating peptide-2 (NAP-2) (23). CTAP III possesses little or no activity on neutrophils (24), and conversion of CTAP III to NAP-2 can be accomplished using neutrophil elastase or supernatants derived from mononuclear cells (25). We have prepared NAP-2 by cleavage of CTAP III with elastase (Fig. 4) and, after isolating NAP-2 by affinity chromotography, compared the ability of CTAP III and NAP-2 to activate basophils (26). We found them to be equipotent. This is similar to studies of their comparative activity on fibroblasts and

TABLE I
N-Terminal Sequence Comparison of CTAP III and HRF[a]

	1	2	3	4	5	6	7	8	9	10	11	12	13
CTAP III:	Asn	Leu	Ala	Lys	Gly	Lys	Glu	Glu	Ser	Leu	Asp	Ser	Asp
HRF-1:	Asn	Leu	Ala	Lys	Gly	Lys	Glu	Glu	Ser	Leu	Asp	Ser	Asp
HRF-2:													

	14	15	16	17	18	19	20	21	22	23	24	25	26
CTAP III:	Leu	Tyr	Ala	Glu	Leu	Arg	Cys	Met	Cys	Ile	Lys	Thr	Thr
HRF-1:	Leu	Tyr	Ala	Glu	Leu			Met					
HRF-2:			Ala	Glu	Leu	Arg			Met	Cys	Ile	Lys	Thr	Thr

	27	28	29	30	31	32	33	34	35	36	37	38	39	40
CTAP III:	Ser	Gly	Ile	His	Pro	Lys	Asn	Ile	Gln	Ser	Leu	Glu	Val	Ile
HRF-1:														
HRF-2:	Ser	Gly	Ile	His	Pro	Lys	Asn	Ile	Gln	Ser	Leu	Glu	Val	Ile

[a] From Ref. 21.

in contrast to the markedly augmented activity of NAP-2 on neutrophils. The range of protein used is in micrograms and the dose–response cause is essentially the same as is seen using fibroblasts. This is a plentiful protein—5 to 35 µg are secreted by the activation of platelets in 1 ml of blood (27).

Our initial preparations of this form of HRF were derived from a mixture of mononuclear cells and platelets. Although we included

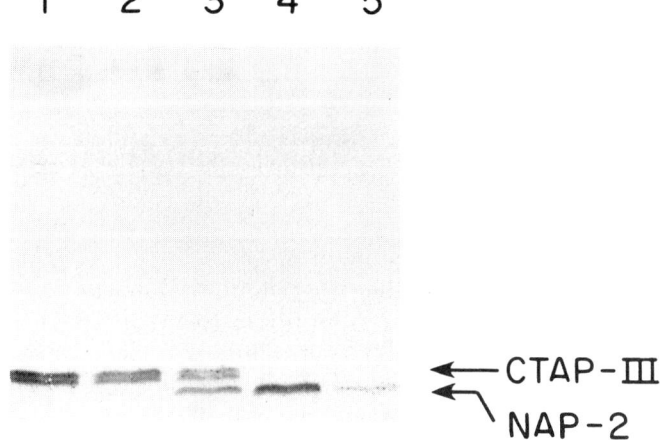

Fig. 4. Conversion of CTAP III to NAP-2 by digestion of crude platelet releasates with increasing concentrations of human neutrophil elastase for 60 minutes at 37°C. Amounts of elastase in units/ml are 0 (lane 1), 0.104 (lane 2), 0.208 (lane 3), 0.415 (lane 4), and 0.83 (lane 5).

steps to diminish platelet contamination, it was found that removal of 99% of platelets still leaves one platelet per mononuclear cell. Thus, it was possible that CTAP III/NAP-2 was derived from platelets and represented one or more forms of platelet-derived HRF. We also considered the possibility that mononuclear cells are capable of producing a similar or identical molecule. Using Percoll gradients, we are now able to reduce platelet contamination to 1–2% relative to mononuclear cells and this may allow us to clarify the latter issue. Purity beyond this may not be possible because platelets and monocytes rosette together (28). On the other hand, it is simple to purify platelets free of mononuclear cells using platelet-phoresis packs, allowing isolation of large quantities of CTAP III. We prepared a monoclonal antibody to the 10- to 12-kDa form of HRF; this recognized purified or recombinant CTAP III (21) and could be used to isolate this form of HRF from platelet or mononuclear cell/platelet supernatants in a single step by affinity chromatography. This monoclonal antibody does not recognize the HRF present in peak 2 or 3 and it is assumed that these represent different gene products. Further, when these latter proteins were subjected to N-terminal sequence analysis, no N-terminal amino acid was identified and both appeared to be blocked. When the protein was cleaved with cyanogen bromide or trypsin, sequence analysis was possible (with a mixture of N-terminal residues), indicating that sufficient protein was tested and confirming the likelihood of a blocked N-terminus.

Several proteins have now been isolated that are structurally related to CTAP III; these include β-thromboglobulin and NAP-2, which are derived from platelet basic protein (29); a monocyte-dependent chemotactic factor for neutrophils (NAP-1 or interleukin-8) (30); a derivative of Rous sarcoma virus-transformed chick embryo fibroblasts (31); a growth-regulated gene in transformed Chinese hamster fibroblasts and human cells (32); a melanoma growth-stimulating factor (33); and gene products derived from T and B lymphocytes, whose function is not yet known (34). These appear to comprise a family of proteins with significant homology to CTAP III. However, it is unclear which, if any, are sufficiently related to share epitopes and cross-react immunologically. It is known, however, that NAP-1 or interleukin-8 affects basophil histamine release, as is discussed in a later section.

III. The Cell Source(s) of HRF

HRF, defined functionally as a cell product that causes basophil degranulation and histamine release, has been described as being secreted by human neutrophils, platelets, alveolar macrophages, and mononuclear cells (7–15). HRF has also been detected in nasal washings (16) and in induced blisters over cutaneous late-phase reactions (17). Neutrophil-derived HRF appears to be quite different from each of the other types thus far characterized. It is constitutively produced, thus no cell stimulation is required to purify it from supernatants. It is heat stable, has a molecular weight of 1400–2300, activates basophils regardless of atopic status of the donor, can degranulate rat basophil leukemia (RBL) cells and therefore can cross species barriers, and is active on human cutaneous mast cells (14–16). The HRF derived from nasal washings (16), platelets (12), and alveolar macrophages (11) is described as a polypeptide of 15,000 kDa that is primarily active on basophils of atopic (allergic) subjects as opposed to nonallergics (35). However, among allergic subjects, there was a 50% response rate. Data to suggest a requirement for cell surface IgE for the functioning of this form(s) of HRF have been described (see Section IV). Mononuclear cell-derived HRF is quite heterogeneous and contains species of 8–10, 15–17, and 40 kDa, as described in Section III. One or more of these forms of HRF appear to be synthesized by T lymphocytes (7), B lymphocytes (36), and monocytes/macrophages (11), but when cell mixtures are tested for HRF production, B cells appear to predominate (37). However, the specific molecular species of HRF that is synthesized by any particular cell type is unknown. Clearly, it is possible that the aforementioned three forms of HRF may each be produced by a different cell type or that individual cells can produce more than one form of HRF. With the exception of neutrophils, other cells capable of producing HRF (mononuclear cells, alveolar macrophages, and platelets) require cell activation for HRF production, and only traces are found in control cell suspensions. When purified T cells, B cells, and monocytes were examined, B cells secreted substantial HRF activity in the culture supernatant in the absence of stimulation whereas T cells and monocytes produced none without activation. HRF found in nasal washings of subjects was found in unstimulated washings of allergic or nonallergic donors and did *not* require nasal antigen challenge (in atopics) to produce the activity. The cell source of this form of HRF is not known.

IV. Mechanism of Action of Human HRFs

Although receptors for the various forms of HRF have not been defined in molecular terms, two general types of HRFs have been described, one whose function appears to be dependent upon cell surface IgE and others that are IgE independent. This distinction has been an area of some controversy; however, it is difficult to compare results of different studies because the cell source and degree of purity of the HRF preparations has been different and therefore not comparable. The responsiveness of subjects to HRF has also varied considerably regardless of which form of HRF is examined, and this variability is a separate issue from the requirement for any interaction with IgE.

Our own studies using crude mononuclear cell supernatants indicated that there was considerable variability in response to the preparations; however, basophils obtained from allergic persons appeared to be more responsive than basophils of nonallergic subjects, i.e., a greater number of subjects with more than 5% histamine release in response to a fixed dose of HRF and, as a group, the percentage histamine release was greater. There were, however, very good responders within the nonallergic subjects, as well as the atopic, and there were nonresponders in both groups (38). The same result was obtained when purified CTAP III/NAP-2 was tested (21). This response does not appear to require interaction with cell surface IgE and the HRF derived from neutrophils has also been reported to be independent of IgE (15). For example, neutrophil-derived HRF was shown to release histamine from cultured rat basophil leukemia cells, which lack any cell surface IgE.

There is also considerable evidence to suggest that one or more forms of HRF are IgE dependent, and such studies have been performed with HRF derived from platelets (12), alveolar macrophages (39), and nasal washings (16). In each case, the material tested was either a concentrate of crude supernatants or a gel-filtered supernatant. When the latter procedure was performed, the estimated molecular weight of the active fraction was 15,000 regardless of source. The means used to demonstrate IgE dependence include the following criteria: (1) Cells desensitized to anti-IgE appeared to be much less responsive to HRF but normally responsive to other secretagogues. (2) Cells would be stripped of their surface-bound IgE by treatment with dilute acid and eliminate responsiveness to HRF. (3) Cells stripped of their IgE could be again rendered responsive if incubated with IgE derived from some patients' basophils and not others (Fig. 5). Donors of IgE that reconstituted such cells were designated IgE(+) and donors that failed to reconstitute were deemed IgE(−). (4) Blocking IgE receptors of stripped cells by an IgE myeloma would render them nonresponsive to IgE(+). (5) Responsiveness to lentil lectin paralleled responsiveness to HRF and identification of basophils coated with IgE(+), suggesting an interaction of HRF with carbohydrate (40). Based on these studies, it was postulated that posttranslational modi-

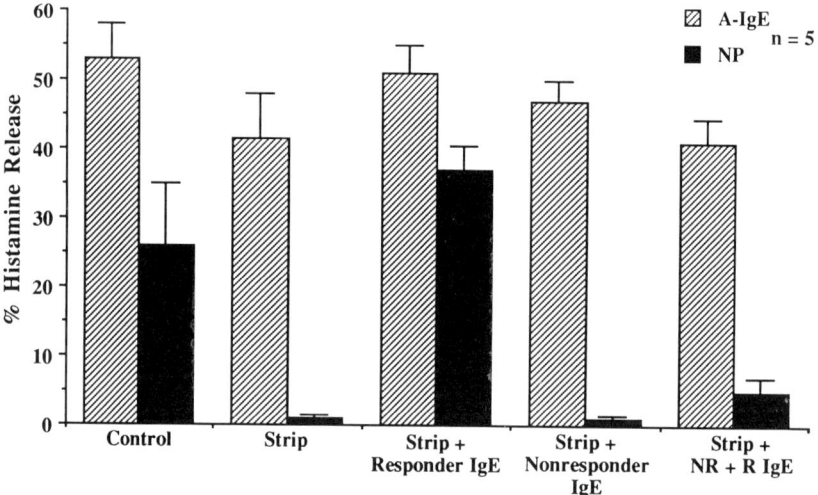

FIG. 5. Effect of saturating IgE receptors with nonresponder IgE on the ability of responder IgE to interact with nasal HRF. The results are expressed as the mean ± SEM of duplicate determinations from the designated number of experiments. Cells were incubated with buffer (control) or lactic acid (strip). Lactic acid-treated cells were sensitized with responder IgE (300 ng/ml) or nonresponder IgE (25 μg/ml when myeloma was used or 2500 ng/ml when highly purified penicillin-specific IgE was employed). Lactic acid-treated cells were also sensitized with a mixture of nonresponder and responder IgE (strip + NR + R IgE). The cells were washed and were challenged with either anti-IgE (A-IgE, 0.1 μg/ml) or nasal HRF (NP, 1:21 dilution). (From Ref. 16.)

fications in IgE synthesis create molecules that differ in carbohydrate content and that the difference can reflect responsiveness to HRF. In general, IgE(+) was found to be present in about 50% of atopic individuals and the other half and nonatopics possessed IgE(−).

The aforementioned studies not only suggest an interaction of HRF with cell surface IgE, but postulate that IgE is heterogeneous based on carbohydrate content. Further, because about 20% of the population at large consists of allergic individuals, responsiveness to this form of HRF would, in general, be limited to only about 10% of the population. Recently, Alam et al. have confirmed this apparent requirement for cell surface IgE using HRF derived from mononuclear cells, and the bulk of evidence suggests that either the 15- to 17-kDa and/or the 40-kDa species we have purified may correspond to this type of HRF. McDonald has recently succeeded in culturing human B cells in the presence of mitogens and appropriate cytokines so that IgE synthesis can be demonstrated *in vitro*. By use of selective inhibitors of carbohydrate synthesis, he has demonstrated IgE(+) and IgE(−) in their cultures based on the aforementioned criteria using HRF responsiveness of stripped and then reconstituted basophils (41).

One difficulty with such studies is that there has been no direct evidence of binding of any form of HRF to IgE and the data are suggestive of such an interaction but do not prove it directly. We have considered the possibility that some myeloma proteins might possess carbohydrate variance that would reflect IgE(+) and IgE(−) and could be used to demonstrate binding of HRF, and perhaps be used to purify HRF. We tested many IgE myeloma proteins and all were negative but one (IgE-AZD), which weakly bound HRF and retarded it when used as an affinity column (38). However, the binding appeared insufficient to be used as a purification step, and this approach was not pursued further.

A more recent study has reproduced the experiments indicating that lactic acid stripping of basophils diminishes responsiveness to HRF and that the response can be repleted by sera from allergic asthmatics but not normal sera or myeloma IgE (42), again suggesting an IgE-dependent mechanism. Responsiveness was, however, not limited to allergic subjects, although as a group their response was significantly greater than in normal subjects.

V. Cytokines and Histamine Release

Although the isolated forms of HRF appear to be different from any known interleukins or colony-stimulating factors, many of them were tested for HRF-like activity. Interleukin-3 and GM-CSF are capable of inducing histamine release upon incubation with basophils. The dose response for interleukin-3 was between 40 and 400 pg/ml in the initial study (43), and these observations were confirmed (44,45), although the dose–response range reported was somewhat higher. The dose–response seen in responding subjects to GM-CSF was between 8 and 800 pg/ml. The rate of histamine release with these cytokines was far slower than any known secretagogue, with a half-maximum secretion at 15–20 minutes and a plateau at 30–40 minutes (43). When atopic and nonatopic subjects' responsiveness was compared, there were responders and nonresponders in each group; however, the magnitude of release was clearly greater in atopic subjects (43). Addition of deuterium oxide in the incubation mixture increased histamine release by three- to fourfold, thus affecting the dose–response and effective concentrations reported (42). Neither of these cytokines correspond to the histamine releasing factors described herein (43,45), and, although present within supernatants tested, the histamine release observed with HRF is so much more pronounced, the contribution of IL-3 and GM-CSF to the total histamine release observed is small. Thus, the various forms of HRF are either much more potent on a molar basis, or are much more plentiful. CTAP III/NAP-2 likely corresponds to the latter circumstance. Other cytokines reported to cause histamine release from basophils are IL-1 (46,47) and IL-8 (48); however, it is clear that the required concentrations of these cytokines are too high to be physiologically significant, and IL-8, in fact, at very low concentrations, acts as an inhibitor of histamine release. Specific receptors for IL-3 on human basophils have been demonstrated and it is likely that histamine release is a receptor-mediated event. GM-CSF may have a unique receptor, or, as has been observed in other tissues (49), may interact with the IL-3 receptor. Both cytokines likely act by a similar mechanism because they share the same unusually slow release of histamine. Interleukin-2, -4, -5, -6, -7, and -9 have been tested for histamine release on human basophils, and none was observed. Tumor necrosis factor-α (TNF-α), TNF-β (lymphotoxin), interferon-γ, G-CSF, and M-CSF are also negative (43) as is the soluble IgE-binding factor derived from CD23 (36), and an IgE-binding factor with lectinlike activity (50) derived from rat basophil leukemia cells (P. Kuna and M. Liu, unpublished observation).

VI. Modulation of Histamine Release by Cytokines

It has been known for years that basophils of some subjects release more histamine when stimulated by a variety of secretagogues and in some instance (for example, basophils of asthmatics) will spontaneously secrete histamine upon incubation with D_2O (51,52). These cells are said to be primed, i.e., possess an augmented stimulus–response setting. One possible explanation of this phenomenon is that exposure to low doses of certain cytokines can augment secretion of basophils. The first such example was the observation that selected viruses, such as influenza A (53,54), or interferons (α/γ) are capable of augmenting basophil histamine release to anti-IgE, or to IgE plus allergen (54–56). Direct histamine release by these agents was not observed. Priming of basophils has been found to be a property of a variety of cytokines and IL-3 is the best studied. When basophils are incubated with concentrations of IL-3 that do not cause histamine release, a subsequent incubation with anti-IgE, F-met-Leu-Phe, or C5a leads to augmented histamine release (57,58). Low concentration of interleukin-1 (59) and interleukin-5 (60) may also prime basophils and can affect production and secretion of leukotriene C4 even more than histamine secretion (60). Dahinden *et al.* have presented data to suggest that interleukin-8 can cause histamine release from basophils if the cells have been preincubated with interleukin-3 (61). Concentrations of each agent were chosen such that neither alone caused any detectable release. The authors cautioned that the phenomenon ascribed to HRF when studied as a crude supernatant may have histamine release that is the result of a combination of factors rather than the activity of a single factor. Nevertheless, isolation of HRF as a distinct molecular species suggests that secretion due to a combination of IL-3 plus IL-8 is a separate phenomenon. During the aforementioned isolation of HRF, IL-8 was found in the void peak with CTAP III/NAP-2. Thus, the peaks at 15–17 and 35–41 kDa have no contaminating IL-8, but IL-3 is present until later fractionation steps are completed.

Alam *et al.* have described an inhibitor of histamine secretion found in mononuclear cell supernatants; they called the activity histamine release inhibiting factor, or HRIF (62). When supernatants were fractionated and assayed for inhibiting activity, two peaks were found at 8000 kDa and less than 1000 kDa. The latter appears to be a prostaglandin (likely PGE_1), which raises cyclic AMP in basophils and inhibits release (63). The 8000-kDa moiety is a small protein but it has not been purified directly from supernatants. Because of the earlier report that IL-8 can cause basophil histamine release between 10^{-4} and 10^{-6} M (48) or at 10^{-7}–10^9 M if the cells are pretreated with IL-3 (61), we reassessed the capability of IL-8 to affect basophil histamine release. We found few direct responders to IL-8 and as reported (48) these were only observed at concentrations of 10^{-6} M or higher—levels that are not physiologic. When we attempted to reproduce the data using preincubation with IL-3, although synergism was obtained in a small number of subjects, our data were not confirmatory in most, i.e., either no effect or an additive effect (P. Kuna and A. P. Kaplan, unpublished observations). We also preincubated basophils with IL-8 to test the possibility that it acts as an inhibitor of HRF-induced histamine release. We found that IL-8 is inhibitory between 10^{-11} and $10^{-8} M$, with peak inhibition at about 10^{-9} M (64) (Fig. 6). Like the original description of HRIF (they are both 8 kDa), IL-8 does *not* inhibit release to anti-IgE, F-met-Leu-Phe, or concanavalin A, but does inhibit release to HRF. Thus, there is relative specificity to its inhibitory spectrum. Further, we have shown that IL-8 inhibits the response to crude supernatants containing HRF and it inhibits purified CTAP III/NAP-2, as well as partially purified 15- to 17-kDa and 40-kDa fractions of HRF.

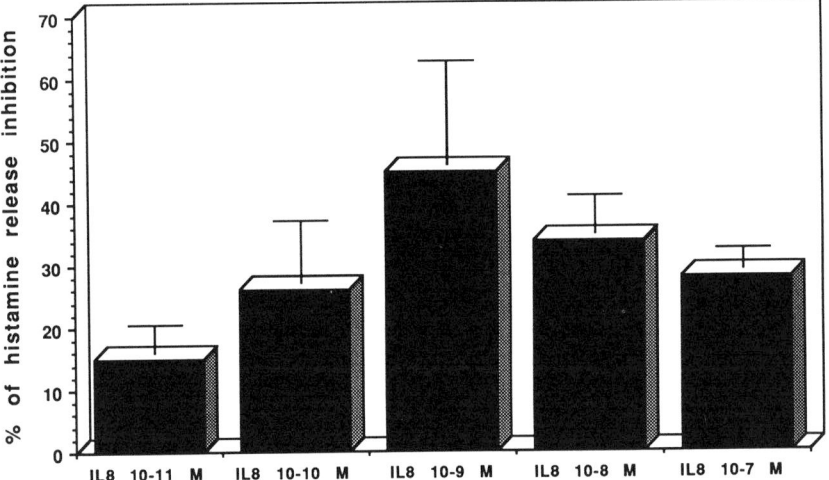

FIG. 6. Dose-dependent inhibition of histamine release by IL-8. Basophils were preincubated without or with IL-8 at various concentrations (10^{-11}–10^{-7} M.). The cells were then exposed to HRF for 40 minutes. Results are mean ± SEM of two experiments in four different subjects. (From Ref. 93.)

Finally, IL-8 inhibited histamine release to interleukin-3. The data thus far suggest, but do not prove, that IL-8 corresponds to HRIF and therefore provides a critical control for cell responsiveness to HRF. Table II summarizes the known effects of cytokines upon histamine release and distinguishes those that are secretagogues, those that prime basophils for secretion, and those that have no effect.

VII. Clinical Considerations

Histamine releasing factors cause a noncytotoxic granule exocytosis from basophils when assessed by electron microscopy (65) and cause histamine release as well as release of lipid mediators such as leukotriene C_4 (66). These same functions have been reported to act on mast

TABLE II
THE EFFECTS OF CYTOKINES ON HISTAMINE RELEASE

Secretion	Priming	Inhibition	No effect
IL-3	IL-3	IL-8 (NAP-1)	IL-2, -4, -6, -7, -9
GM-CSF	IL-5		G-CSF, M-CSF
(IL-1, IL-8)	Interferon-γ		TGF-β

cells derived from the lung (11,67) and synovial tissue (68), although the sensitivity of mast cells to these agents seems less than basophils based on the limited data obtained thus far. HRFs may therefore have a pathologic role in allergic and rheumatic disorders characterized by a subacute or chronic inflammatory process in which mononuclear cells (and perhaps platelets) are in proximity to basophils and tissue mast cells. Specific disorders one might consider are the late-phase allergic reaction relevant to chronic rhinitis, sinusitis, asthma, chronic urticaria, rheumatoid arthritis, and scleroderma.

Late-phase allergic reactions are a consequence of a prior IgE-dependent acute allergic reaction. The phenomenon was first described in cutaneous reactions in which the acute wheal and flare response to allergen was followed by localized swelling 6–12 hours after the reaction subsided, which gradually subsided and resolved by 12–24 hours. When biopsied, these lesions revealed a mixture of neutrophils, eosinophils, basophils, and mononuclear cells (69,70). A similar phenomenon has been described in the nasal mucosa after allergen stimulation to produce rhinitis (71), and a late asthmatic reaction is seen hours after inhalation challenge in allergic asthmatics (72,73). In the last circumstance, bronchospasm induced by allergen occurs within a few minutes, subsides, and then reoccurs hours later in the absence of additional allergen. Thus, in all instances, this late-phase reaction is initiated by a prior IgE-dependent event but occurs in the absence of additional antigen and is no longer dependent on IgE antibody (74). It is assumed that the acute allergic reaction leads to later cell accumulation by release of chemotactic factors as well as a variety of vasodilators, and that the infiltrating inflammatory cells become activated. Among the vasoactive substances revealed during the late phase are neutrophil chemotactic factor (NCF) (73), prostaglandin D_2 (in skin) (75), platelet activating factor (PAF) (76,77), leukotriene C_4 (78), and kallikrein and bradykinin (79). All except the kallikrein/bradykinin appearance are products of mast cells and/or basophils. The cellular infiltrate has been studied in greater detail recently (80) and includes basophils (particularly in nasal mucosa) (81), eosinophils (82), and mononuclear cells (74).

Studies of histamine release throughout such reactions indicate that elevated levels persist over a 12-hour period (75,78,83) or, in some instances, a second peak is seen between 4 and 8 hours (79). When mast cell tryptase levels are examined, they are elevated along with histamine within the first hour and become undetectable thereafter (83). Tryptase is contained in mast cells but not basophils (84), whereas histamine is contained in both. Thus, it appears that basophils infiltrating the area become activated and continue to secrete histamine over many hours. Cytokine-like molecules such as HRF are prime candidates as activators of basophils in such reactions. HRF has been reported to be present in cutaneous late-phase reactions (17) and in bronchoalveolar lavage fluids of asthmatics (85).

Although virtually all studies thus far have examined basophils as effector cells responsive to HRF, it appears likely that mast cells in some tissues are also responsive. Lung mast cells appear to respond to alveolar macrophage-derived HRF (11), although the response was less than that seen with basophils, and preliminary results suggest that cutaneous mast cells also respond to HRF (unpublished observations). This latter observation is, perhaps, related to observations regarding the pathogenesis of chronic idiopathic urticaria. The etiology of this form of chronic hives is unknown and consists of a nonnecrotizing mononuclear cell infiltrate of superficial venules in the skin (86). There is no immunoglobulin or complement deposition. When the cellular infiltrate was further examined, it consisted of 50% T lymphocytes, no B cells, 20% monocytes, 11% mast cells, and 20% lymphocyte-like cells that could not be identified (87). These may represent immature cells that are not identifiable, or perhaps γ/δ T cells. When compared to normal skin, there was increased T lymphocytes and monocytes, and the number of mast cells were increased ninefold. We postulated that these mononuclear cells secrete a mast cell activator and HRF appears to be a likely candidate. Studies to examine this possibility are in progress.

Recent studies of the inflammatory response in a variety of autoimmune diseases suggest a role for mast cells that is unanticipated. For example, mast cells are prominent in the synovium of patients with rheumatoid arthritis (88) and appear in aggregates about areas of bone resorption (89). Degranulated mast cells have also been identified in sclerodermatous skin using reagents that detect surface-bound IgE, because the degranulated cells are not identifiable using the usual staining methods (90,91). For each of these disorders, mast cells are surrounded by mononuclear cells (T cells, B cells, monocytes) and synovial mast cell activation by HRF or analogous factors represents one possibility (68).

Some preliminary studies of asthma have examined the possibility that HRF as an inflammatory agonist and HRIF as a control mechanism are important contributors to symptoms and that the relative ratio of the two might determine the clinical course. Alam and Rozniecki (92) demonstrated that stimulation of patient lymphocytes with allergen or with a nonspecific mitogen such as phytohemagglutinin augmented

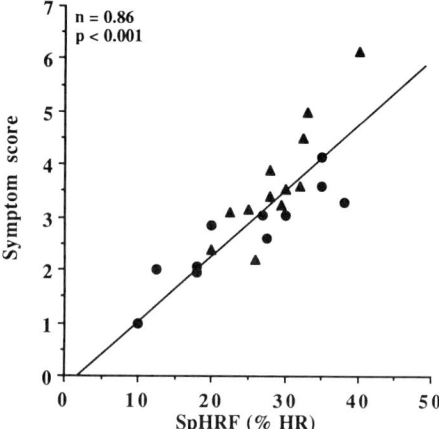

FIG. 7. The correlation between the change in PC20 for histamine and the change in spontaneous HRF production (prestudy versus poststudy) after 2 years of preseasonal immunotherapy in patients from the grass pollen-treated group. Prestudy measurements were performed before season and before the immunotherapy was initiated, whereas poststudy measurements were conducted in pollen season after 2 years of immunotherapy. (From Ref. 93.)

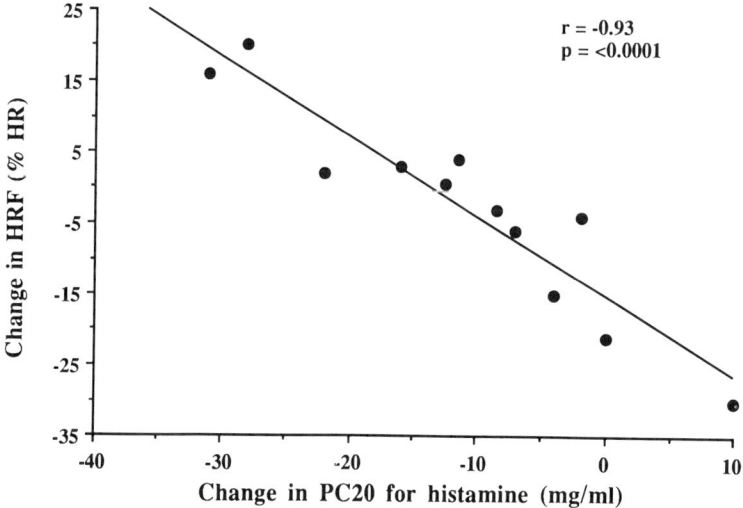

FIG. 8. Fitted curve illustrating the relationship between symptom scores and spontaneous HRF (SpHRF) production by MNCs from patients obtained from the placebo-treated group (circles) and grass pollen-treated group (triangles). Symptoms scores and SpHRF production were measured in the pollen season after preseasonal immunotherapy.

secretion of HRF-like activity into the supernatant (Fig. 7). A later study of immunotherapy (allegy injections) as a treatment for seasonal allergen rhinitis was performed to determine whether HRF production was affected. Using grass pollen-sensitive patients, specific immunotherapy was shown to diminish spontaneously produced HRF when the patients' mononuclear cells are cultured overnight, and the increase in HRF production seen upon incubation with grass pollen was blunted. The spontaneous HRF production correlated with those whose asthma improved, and all treated patients had a diminution in antigen-induced HRF production (93) (Fig. 8). An analogous study of subjects with food-dependent atopic dermatitis likewise revealed increased spontaneous HRF production associated with cutaneous symptoms that reversed when compulsive avoidance of skin test-positive foods (particularly egg allergen) was introduced. The diminution in HRF production correlated with improvement of skin rash (94).

References

1. Budzko, D. B., Bokisch, V. A., and Müller-Eberhard, H. J. (1971). A fragment of the third component of human complement with anaphylatoxin activity. *Biochemistry* **10**, 1166–1172.
2. Cochrane, C. G., and Müller-Eberhard, H. J. (1968). The deviation of two distinct anaphylatoxin activities from the third and fifth component of human complement. *J. Exp. Med.* **127**, 371–386.
3. Gorski, J. P., Hugli, T. E., and Müller-Eberhard, H. J. (1979). C4a: The third anaphylatoxin of the human complement system. *Proc. Natl. Acad. Sci. U.S.A.* **76**, 5299–5302.
4. Ross, G. D. (1986). Opsonization and membrane complement receptors. *In* "Immunobiology of the Complement System" (G. D. Ross, ed.), pp. 87–114. Academic Press, Orlando, Florida.
5. Thueson, D. O., Speck, L. S., Lett-Brown, M. A., and Grant, J. A. (1979). Histamine-releasing activity (HRA). II. Interaction with basophils and physicochemical characterization. *J. Immunol.* **123**, 633–639.
6. Thueson, D. O., Speck, L. S., Lett-Brown, M. A., and Grant, J. A. (1979). Histamine-releasing activity (HRA). I. Production by mitogen or antigen stimulated human mononuclear cells. *J.Immunol.* **122**, 623–632.
7. Sedgwick, B. D., Holt, P. G., and Turner, B. (1981). Production of a histamine releasing lymphokine by antigen or mitogen-stimulated human peripheral T cells. *Clin. Exp. Immunol.* **45**, 409–418.
8. Ezeumuzie, I. C., and Assem, E. S. K. (1985). Histamine releasing lymphokine characteristics of its production. *Agents Actions* **17**, 21–26.
9. Goetzl, E. J., Foster, D. W., and Payan, D. G. (1984). A basophil-activating factor from human T lymphocytes. *Immunology* **53**, 227–234.
10. Kaplan, A. P., Haak-Frendscho, M., Fauci, A., Dinarello, C., and Halbert, E. (1985).

A histamine-releasing factor from activated human mononuclear cells. *J. Immunol.* **135,** 2027–2032.
11. Schulman, E. S., Liu, P. C., Proud, D., MacGlashan, Jr., D. W., Lichtenstein, L. M., and Plaut, M. (1985). Human lung macrophages induce histamine release from basophils and mast cells. *Am. Rev. Respir. Dis.* **131,** 230–235.
12. Orchard, M. A., Kagey-Sobotka, A., Proud, D., and Lichtenstein, L. M. (1986). Basophil histamine release induced by a substance from stimulated platelets. *J. Immunol.* **136,** 2240–2244.
13. White, M. V., and Kaliner, M. A. (1987). Neutrophils and mast cells. I. Human neutrophil-derived histamine releasing activity (HRA-N). *J. Immunol.* **139,** 1624–1630.
14. White, M. V., Baer, H., Kubota, Y., and Kaliner, M. D. (1989). Neutrophils and mast cells. Characterization of cells responsive to neutrophil derived histamine releasing activity. *J. Allergy Clin. Immunol.* **84,** 773–780.
15. White, M. V., Kaplan, A. P., Haak-Frendscho, M., and Kaliner, M. (1989). Neutrophils and mast cells. Comparison of neutrophil-derived histamine-releasing activity with other histamine-releasing factors. *J. Immunol.* **141,** 3575–3583.
16. MacDonald, S. M., Lichtenstein, L. M., Proud, D., Plaut, M., Naclerio, R. M., MacGlashan, D. W., and Kagey-Sobotka, A. (1987). Studies of IgE-dependent histamine releasing factors: Heterogeneity of IgE. *J. Immunol.* **139,** 506–512.
17. Warner, J. A., Pienkowski, M. M., Plaut, M., Norman, P. S., and Lichtenstein, L. M. (1986). Identification of histamine releasing factor(s) in the late phase of cutaneous IgE-mediated reactions. *J. Immunol.* **136,** 2583–2587.
18. Lett-Brown, M. A., Thueson, D. O., Plank, D. E., Duffy, L., and Grant, J. A. (1984). Histamine-releasing activity. V. Characterization and purification using high-performance liquid chromatography. *Cell. Immunol.* **87,** 445–451.
19. Lett-Brown, M. A., Thueson, D. O., Plank, D. E., Langford, M. P., and Grant, J. A. (1984). Histamine-releasing activity. IV. Molecular heterogeneity of the activity from stimulated human thoracic duct lymphocytes. *Cell. Immunol.* **87,** 434–444.
20. Baeza, M. L., Reddigari, S., Haak-Frendscho, M., and Kaplan, A. P. (1989). Purification and further characterization of human mononuclear cell histamine-releasing factor. *J. Clin. Invest.* **83,** 1204–1210.
21. Baeza, M. L., Reddigari, S. R., Kornfeld, D., Ramani, N., Smith, E. M., Hossler, P. A., Fischer, T., Castor, C. W., Gorevic, P. G., and Kaplan, A. P. (1990). Relationship of one form of human histamine releasing factor to connective tissue activating peptide-III. *J. Clin. Invest.* **85,** 1516–1521.
22. Castor, C. W., Miller, J. W., and Walz, D. A. (1983). Structural and biological characteristics of connective tissue activating peptides (CTAP-III), a major human platelet derived growth factor. *Proc. Natl. Acad. Sci. U.S.A.* **80,** 765–769.
23. Walz, A., and Baggiolini, M. (1989). A novel cleavage product of β-thromboglobulin formed in cultures of stimulated mononuclear cells activates human neutrophils. *Biochem. Biophys. Res. Commun.* **159,** 969–975.
24. Walz, A., Dewald, B., Tscharner, V., and Baggiolini, M. (1989). Effects of the neutrophil activating peptide NAP-2, platelet basic protein, connective tissue activating peptide III, and platelet factor 4 on human neutrophils. *J. Exp. Med.* **170,** 1745–1750.
25. Walz, A., and Baggiolini, M. (1990). Generation of the neutrophil activating peptide NAP-2 from platelet basic protein or connective tissue-activating peptide III through monocyte protease. *J. Exp. Med.* **171,** 449–454.
26. Reddigari, S. R., Miragliotta, G. F., Kuna, P., Kornfeld, D., Baeza, M. L., Castor, C. W., and Kaplan, A. P. (1991). Connective tissue activating peptide III and its

derivative neutrophil activating peptide-2 release histamine from human basophils. Submitted for publication.
27. Castor, C. W., Walz, D. A., Johnson, P. H., Hossler, P. A., Smith, E. M., Bignall, M. C., Aaron, B. P., Underhill, P., Lazar, J. M., Hudson, D. H., Cole, L. A., Perini, F., and Mountjoy, K. (1990). Connective tissue activation 34. Effects of proteolytic processing on the biologic activities of CTAP III. *J. Lab. Clin. Med.* **116**, 516–526.
28. Levine, R. B., and Rabellino, E. M. (1986). Platelet glycoproteins IIb and IIIa associated with blood monocytes are derived from platelets. *Blood* **67**, 207–213.
29. Holt, J. C., Harris, M. E., Holt, A. M., Lange, E., Henschen, A., and Niewiarowski, S. (1986). Characterization of human platelet basic protein, a precursor form of low affinity platelet factor 4 and β thromboglobulin. *Biochemistry* **25**, 1988–1996.
30. Yoshimura, T., Matsushima, Y., Tanaka, S., Robinson, E. A., Appella, E., Oppenheim, J. J., and Leonard, E. J. (1987). Purification of a human monocyte-derived neutrophil chemotactic factor that has peptide sequence similarity to other host defense cytokines. *Proc. Natl. Acad. Sci. U.S.A.* **84**, 9233–9237.
31. Sugano, S., Stoeckle, M. Y., and Hanafusa, H. (1987). Transformation by Rous sarcoma virus induces a novel gene with homology to a mitogenic platelet protein. *Cell (Cambridge, Mass.)* **49**, 321–328.
32. Anisowicz, A., Bardwell, L., and Sager, R. (1987). Constitutive expression of a growth-regulated gene in transformed Chinese hamster and human cells. *Proc. Natl. Acad. Sci. U.S.A.* **84**, 7188–7192.
33. Richmond, A., and Thomas, H. G. (1988). Melanoma growth stimulating activity: Isolation from human melanoma tumors and characterization of tissue distribution. *J. Cell. Biochem.* **36**, 185–198.
34. Schmid, J., and Weissman, C. (1987). Induction of an mRNA for a serine protease and a β-thromboglobulin-like protein in mitogen stimulated human leukocytes. *J. Immunol.* **139**, 250–256.
35. Fisher, R. H., Kagey-Sobotka, A., Proud, D., Naclerio, R. M., and Lichtenstein, L. M. (1987). Histamine releasing factor: Release mechanisms and responding population. *J. Allergy Clin. Immunol.* **79**, 248 (abstr.).
36. Haak-Frendscho, M., Sarfati, M., Delespesse, G., and Kaplan, A. P. (1988). Comparison of mononuclear cell and B-lymphoblastoid histamine-releasing factor and their distinction from an IgE-binding factor. *Clin. Immunol. Immunopathol.* **49**, 72–82.
37. Alam, R., Forsythe, P. A., Lett-Brown, M. A., and Grant, J. A. (1989). Cellular origin of histamine-releasing factor produced by peripheral blood mononuclear cells. *J. Immunol.* **142**, 3951–3956.
38. Baeza, M. L., Haak-Frendscho, M., Satnick, S., and Kaplan, A. P. (1988). Responsiveness to human mononuclear cell derived histamine releasing factor: Studies of allergic status and the role of IgE. *J. Immunol.* **141**, 2688–2692.
39. Liu, M. C., Proud, D., Lichtenstein, L. M., MacGlashan, D. W. J., Schleimer, R. P., Adkinson, N. F. J., Kagey-Sobotka, A., Schulman, E. S., and Plaut, M. (1986). Human lung macrophage-derived histamine releasing activity is due to IgE-dependent factors. *J. Immunol.* **136**, 2588–2595.
40. MacDonald, S. M., White, J. M., Kagey-Sobotka, A., and Lichtenstein, L. M. (1988). Is glycosylation the basis of IgE heterogeneity. *Clin. Res.* **36**, 602 (abstr.).
41. McDonald, S. M. (1990). IgE-dependent histamine releasing factors and IgE heterogeneity purification and synthesis. *Proc. Coll. Int. Allergol.*, Funchal, Madeira, *1990*.
42. Alam, R., Forsythe, P. A., Rankin, J. A., Boyaro, M. C., Lett-Brown, M. A., and Grant, J. A. (1990). Sensitivity of basophils to histamine releasing factor(s) of various origin: Dependency on allergic phenotypes of the donor and surface-bound IgE. *J. Allergy Clin. Immunol.* **86**, 73–81.

43. Haak-Frendscho, M., Arai, N., Arai, K.-I., Baeza, M. L., Finn, A., and Kaplan, A. P. (1988). Human recombinant granulocyte-macrophage colony-stimulating factor and interleukin 3 cause basophil histamine release. *J. Clin. Invest.* **82**, 17–20.
44. MacDonald, S. M., Schleimer, R. P., Kagey-Sobotka, A., Gillis, S., and Lichtenstein, L. M. (1989). Recombinant IL-3 induces histamine release from human basophils. *J. Immunol.* **142**, 3527–3532.
45. Alam, R., Welter, J. B., Forsythe, P. A., Lett-Brown, M. A., and Grant, J. A. (1989). Comparative effect of recombinant IL-1, -2, -3, -4, and -6, IFN-gamma, granulocyte–macrophage-colony-stimulating factor, tumor necrosis factor-alfa, and histamine releasing factors on the secretion of histamine from basophils. *J. Immunol.* **142**, 3431–3435.
46. Subramanian, N., and Bray, M. A. (1987). Interleukin 1 releases histamine from human basophils and mast cells *in vitro*. *J. Immunol.* **138**, 271–275.
47. Haak-Frendscho, M., Dinarello, C., and Kaplan, A. P. (1988). Recombinant human interleukin-1β causes histamine release from human basophils. *J. Allergy Clin. Immunol.* **82**, 218–223.
48. White, M. V., Yoshimura, T., Hook, W., Kaliner, M. A., and Leonard, E. J. (1989). Neutrophil attractant/activation protein-1 (NAP-1) causes human basophil histamine release. *Immunol. Lett.* **22**, 151–154.
49. Lopez, A. M., Lyons, A. B., Eglinton, J. M., Park, L. S., To, L. B., Clark, S. C., and Vadas, M. A. (1990). Specific binding of human interleukin-3 and granulocyte–macrophage colony-stimulating factor to human basophils. *J. Allergy Clin. Immunol.* **85**, 99–102.
50. Robertson, M. W., Albrand, T. K., Keller, D., and Liu, F. T. (1990). Human IgE-binding protein: A soluble lectin exhibiting a highly conserved interspecies sequence and differential recognition of IgE glycoforms. *Biochemistry* **29**, 8093–8100.
51. Tung, R., and Lichtenstein, L. M. (1980). *In vitro* histamine release from basophils of asthmatics and atopic individuals in D_2O. *J. Immunol.* **128**, 2067–2072.
52. Kazimierczak, W., Plaut, M., Knauer, K. A., Meier, H. L., and Lichtenstein, L. M. (1984). Deuterium-oxide-induced histamine release from basophils of allergic subjects. (1) Responsiveness to deuterium oxide requires an activation step. *Am. Rev. Resp. Dis.* **129**, 592–596.
53. Ida, S., Hooks, J. J., Siraganian, R. P., and Notkins, A. L. (1977). Enhancement of IgE-mediated histamine release from human basophils by viruses: Role of interferon. *J. Exp. Med.* **145**, 892–906.
54. Busse, W. W., Swenson, C. A., Borden, E. C., Treuhaft, M. W., and Dick, E. C. (1983). Effect of influenza A virus on leucocyte histamine release. *J. Allergy Clin. Immunol.* **71**, 382–388.
55. Hernandez-Asensio, M., Hooks, J. J., Ida, S., Siraganian, R. P., and Notkins, A. L. (1979). Interferon-induced enhancement of IgE-mediated histamine release from human basophils requires RNA synthesis. *J. Immunol.* **122**, 1601–1603.
56. Ida, S., Hooks, J. J., Siraganian, R. P., and Notkins, A. L. (1980). Enhancement of IgE-mediated histamine release from basophils by immune-specific lymphokines. *Clin. Exp. Immunol.* **41**, 380–387.
57. Hirai, K., Morita, Y., Misaki, Y., Ohta, K., Takaishi, T., Suzuki, S., Motoyoshi, K., and Miyamoto, T. (1988). Modulation of human basophil histamine release by hemopoietic growth factors. *J. Immunol.* **141**, 3958–3964.
58. Kurimoto, Y., de Weck, A. L., and Dahandin, C. A. (1989). Interleukin 3-dependent mediator release in basophils triggered by C5a. *J. Exp. Med.* **170**, 467–479.
59. Massey, W. A., Randall, T. C., Kagey-Sobotka, A., Warner, J. A., MacDonald, S. M., Gillis, S., Allison, A. C., and Lichtenstein, L. M. (1989). Recombinant human IL-1α

and -1β potentiate IgE-mediated histamine release from human basophils. *J. Immunol.* **143**, 1875–1880.
60. Bischoff, S. C., Brunner, T., de Weck, A. L., and Dahinden, C. A. (1990). Interleukin 5 modifies histamine release and leukotriene generation by human basophils in response to diverse agonists. *J. Exp. Med.* **172**, 1577–1582.
61. Dahinden, C. A., Kurimoto, Y., de Weck, A. L., Lindley, I., Dewald, B., and Baggiolini, M. (1989). The neutrophil-activating peptide NAF/NAP-1 induces histamine release and leukotrine release by interleukin 3-primed basophils. *J. Exp. Med.* **170**, 1787–1792.
62. Alam, R., Grant, J. A., and Lett-Brown, M. A. (1988). Identification of a histamine releasing inhibitory factor produced by human mononuclear cells *in vitro*. *J. Clin. Invest.* **82**, 2056–2062.
63. Lichtenstein, L. M., and Margolis, S. (1968). Histamine release in vitro. Inhibition by catecholamines and methylxanthines. *Science* **161**, 902–903.
64. Kuna, P., Reddigari, S. R., Kornfeld, D., and Kaplan, A. P. (1991). Interleukin 8 inhibits histamine release from human basophils induced by histamine releasing factors, connective tissue activating peptide III, and interleukin 3. Submitted for publication.
65. Dvorak, A. M., Lett-Brown, M. A., Thueson, D. O., Pyne, K., Raghuprasad, P. K., Galli, S. J., and Grant, J. I. (1984). Histamine Releasing Activity (HRA) III. HRA induces human basophil histamine release by provoking non-cytotoxic granule exocytosis. *Clin. Immunol. Immunopathol.* **32**, 142–150.
66. Ezeumuzie, I. C., and Assem, E. S. K. (1982). Histamine and SRS release from leukocytes by lymphokines. 2nd International Conference on Immunopharmacology. *Int. J. Immunopharmacol.* **4**, 378 (abstr.).
67. Ezeumuzie, I. C., and Assem, E. S. K. (1983). A study of histamine release from human basophils and lung mast cells by products of lymphocyte stimulation. *Agents Actions* **13**, 222–230.
68. Gruber, B., Poznansky, M., Boss, E., Partin, J., Gorevic, P., and Kaplan, A. P. (1986). Characterization and functional studies of rheumatoid synovial mast cells: Activation by secretagogues, anti-IgE, and a histamine-releasing lymphokine. *Arthritis Rheum.* **29**, 944–955.
69. Dolovich, J., Hargreave, F. E., Chalmers, R., Shier, K. J., Gauldie, J., and Bienenstock, J. (1973). Late cutaneous allergic responses in isolated IgE-dependent reactions. *J. Allergy Clin. Immunol.* **52**, 38–46.
70. Solley, G. O., Gleich, G. J., Jordan, R. E., and Schroeter, A. L. (1976). The late phase of the immediate wheal and flare skin reaction. Its dependence upon IgE antibodies. *J. Clin. Invest.* **58**, 408–420.
71. Dvoracek, J. E., Yunginger, J. W., Kern, E. B., Hyatt, R. E., and Gleich, G. J. (1984). Induction of nasal late-phase reactions by insufflation of ragweed-pollen extract. *J. Allergy Clin. Immunol.* **73**, 363–368.
72. Nagy, L., Lee, T. H., and Kay, A. B. (1982). Neutrophil chemotactic activity in antigen-induced late asthmatic reactions. *N. Engl. J. Med.* **306**, 497–501.
73. Durham, S. R., Lee, T. H., Cromwell, O., Shaw, R. J., Merrett, T. G., Merrett, J., Cooper, P., and Kay, A. B. (1984). Immunologic studies in allergen-induced late phase asthmatic reactions. *J. Allergy Clin. Immunol.* **74**, 49–60.
74. Frew, A. J., and Kay, A. B. (1990). Eosinophils and T lymphocytes in late-phase allergic reactions. *J. Allergy Clin. Immunol.* **85**, 533–539.
75. Pienkowski, M. M., Adkinson, N. F., Jr., Plaut, M., Norman, P. S., and Lichtenstein, L. M. (1988). Prostaglandin D2 and histamine during the immediate and late phase components of allergic cutaneous responses. *J. Allergy Clin. Immunol.* **82**, 95–100.

76. Dorsch, W., Ring, J., Reimann, H. J., and Geiger, R. (1982). Mediator studies in skin blister fluid from patients with dual skin reactions after intradermal allergen injection. *J. Allergy Clin. Immunol.* **70**, 236–242.
77. Roberts, N. M., Page, C. P., Chung, K. F., and Barnes, P. J. (1988). Effect of a PAF antagonist, BN52063, on antigen-induced, acute, and late onset cutaneous responses in atopic subjects. *J. Allergy Clin. Immunol.* **82**, 236–241.
78. Reshef, A., Kagey-Sobotka, A., Adkinson, N. F., Jr., Lichtenstein, L. M., and Norman, P. S. (1989). The pattern and kinetics in human skin of erythema and mediators during the acute and late phase response (LPR). *J. Allergy Clin. Immunol.* **84**, 678–687.
79. Naclerio, R. M., Proud, D., Togias, A. G., Adkinson, N. F., Jr., Meyers, D. A., Kagey-Sobotka, A., Plaut, M., Norman, P. S., and Lichtenstein, L. M. (1985). Inflammatory mediators in late antigen-induced rhinitis. *N. Engl. J. Med.* **313**, 65–70.
80. Charlesworth, C. N., Hood, A. F., Soter, N. A., Kagey-Sobotka, A., Norman, P. S., and Lichtenstein, L. M. (1989). Cutaneous late-phase response to allergen. Mediator release and inflammatory cell infiltration. *J. Clin. Invest.* **83**, 1519–1526.
81. Bascom, R., Wachs, M., Naclerio, R. M., Pipkorn, U., Galli, S. J., and Lichtenstein, L. M. (1988). Basophil influx occurs after nasal antigen challenge: Effects of topical corticosteroid treatment. *J. Allergy Clin. Immunol.* **81**, 580–589.
82. Pipkorn, U., Karlsson, G., and Enerback, L. (1988). The cellular response of the human allergic mucosa to natural allergen exposure. *J. Allergy Clin. Immunol.* **82**, 1046–1054.
83. Shalit, M., Shcwartz, L. B., von Allmen, C., Atkins, P. C., Lavker, R. M., and Zweiman, B. (1990). Release of histamine and tryptase during cutaneous and interrupted cutaneous challenge with allergen in humans. *J. Allergy Clin. Immunol.* **86**, 117–125.
84. Castells, M. C., Irani, A. A., and Schwartz, L. B. (1987). Evaluation of human peripheral blood leukocytes for mast cell tryptase. *J. Immunol.* **138**, 2184–2189.
85. Broide, D. H., Smith, C. M., and Wasserman, S. I. (1990). Mast cells and pulmonary fibrosis. Identification of histamine releasing factor in bronchoalveolar lavage fluid. *J. Immunol.* **145**, 1838–1844.
86. Natbony, S. F., Phillips, M. E., Elias, J. M., Godfrey, H. P., and Kaplan, A. P. (1983). Histologic studies of chronic idiopathic urticaria. *J. Allergy Clin. Immunol.* **77**, 177–183.
87. Elias, J., Boss, E., and Kaplan, A. P. (1986). Studies of the cellular infiltrate of chronic idiopathic urticaria: Prominence of T-lymphocytes, monocytes, and mast cells. *J. Allergy Clin. Immunol.* **78**, 914–918.
88. Crisp, A. J., Chapman, C. M., Kirkham, S. E., Schiller, A. L., and Krane, S. M. (1984). Articular mastocytosis in rheumatoid arthritis. *Arthritis Rheum.* **27**, 845–851.
89. Godfry, H. P., Ilardi, C., Engbar, W., and Graziano, F. M. (1984). Quantitation of human synovial mast cells in rheumatoid arthritis and other rheumatic diseases. *Arthritis Rheum.* **27**, 852–856.
90. Hawkins, R. A., Claman, H. N., Clark, R. A. F., and Steigerwald, J. C. (1985). Increased dermal mast cell populations in progressive systemic sclerosis: A link in chronic fibrosis? *Ann. Intern. Med.* **102**, 182–186.
91. Seibold, J. R., Giorno, R. C., and Claman, H. N. (1990). Dermal mast cell degranulation in systemic sclerosis. *Arthritis Rheum.* **33**, 1702–1709.
92. Alam, R., and Rozniecki, J. (1985). A mononuclear cell-derived histamine releasing factor in asthmatic patients. *Allergy* **40**, 124–129.
93. Kuna, P., Alam, R., Kuzminska, B., and Rozniecki, J. (1989). The effect of preseasonal immunotherapy on the production of histamine releasing factor (HRF) by mononu-

clear cells from patients with seasonal asthma: Results of a double blind placebo-controlled, randomized study. *J. Allergy Clin. Immunol.* **83**, 816–824.
94. Sampson, H. A., Broadbent, K. R., and Bernhisel-Broadbent, J. (1989). Spontaneous release of histamine from basophils and histamine-releasing factor in patients with atopic dermatitis and food hypersensitivity. *N. Engl. J. Med.* **321**, 228–232.

This article was accepted for publication on 4 March 1991.

Immunologic Interactions of T Lymphocytes with Vascular Endothelium

JORDAN S. POBER AND RAMZI S. COTRAN

Department of Pathology, Brigham and Women's Hospital, Boston, Massachusetts 02115 and Department of Pathology, Harvard Medical School, Boston, Massachusetts 02115

I. Introduction

Inflammatory cell-mediated immune reactions, such as delayed hypersensitivity, may be locally elicited by introduction of specific foreign antigen into a peripheral tissue site of a previously sensitized host. In the events that follow, antigen-specific memory T cells first recognize the foreign antigen, become activated, and then release cytokines that recruit circulating leukocytes, including many T cells not specific for the antigen, into the local site. This inflammatory reaction normally functions to protect the host by eliminating the foreign antigen. Consideration of tissue anatomy leads to the conclusion that vascular endothelium, which forms a barrier between the blood and the tissue, must be traversed early by blood-borne T cells that are specific for the eliciting antigen and at later times by circulating T cells independent of their antigen specificity. In this article, we will review the growing body of data that suggests that the vascular endothelium is an active participant in both of these steps. Specifically, we will discuss the potential roles of endothelium in antigen presentation to and activation of specific memory T cells, as well as in antigen-independent recruitment of T cells into inflammatory sites. We will compare this latter process with T cell homing to lymphoid organs; this has been recently reviewed in this series (1). Other leukocyte–endothelial interactions, which are important in acute and nonimmunologic inflammatory responses, have been subjects of other recent reviews (2,3) and will not be covered here. Finally, although the proposed roles of endothelium in immune reactions are most relevant *in vivo*, we emphasize at the outset that experimental data have been largely acquired from *in vitro* experiments. *In vivo* confirmation of these proposed events is by no means conclusive, and we will refer to relevant *in vivo* observations when available.

II. Specific Antigen Presentation by Vascular Endothelial Cells

T cells specifically recognize foreign protein antigens only when these antigens are associated with self MHC molecules on the surface of an antigen-presenting cell (4). In general, endogenously synthesized proteins (such as viral antigens) are bound to class I MHC molecules and are presented to $CD8^+$ T cells, whereas exogenously acquired (endocytosed) proteins are bound to class II MHC molecules and are presented to $CD4^+$ T cells. Association of protein with class I MHC molecules is proposed to occur soon after translation of the MHC polypeptides in the cisternae of the endoplasmic reticulum, whereas association with class II molecules is proposed to occur in other intracellular compartments. In both cases, globular proteins must be processed to produce unfolded regions or proteolytic fragments (peptides) that can fit into the MHC molecule's cleft. The processing of protein antigen to peptide and the association of peptide with an MHC molecule are functions that must be performed by the cell presenting the antigen.

Most cell types are able to present antigens associated with class I MHC molecules. This is thought to benefit the host because class I-associated viral antigens serve as the molecular target recognized by $CD8^+$ cytolytic T lymphocytes, a defense mechanism that can eradicate virally infected cells. In contrast, presentation of antigens in association with class II MHC molecules is largely limited to a special group of antigen-presenting cells, such as bone marrow-derived B lymphocytes, mononuclear phagocytes, and mononuclear cells with dendritic morphology (e.g., epidermal Langerhans cells). This limitation of antigen presentation to special cell types is thought to provide greater control over activation of $CD4^+$ helper T lymphocytes, the cells responsible for initiating many immune reactions. However, almost all cell types can be induced to express class II MHC molecules under certain circumstances (see below) and experiments with T cell lines and hybridomas have revealed that virtually all cells possess the "machinery" to process both endogenous and exogenous antigens appropriately (although efficiency may vary). As a consequence, the major determinant of whether a cell can display a signal capable of interacting with a T cell antigen receptor is the ability of that cell to synthesize and express class I or class II MHC molecules. Therefore, we will begin our discussion of antigen presentation by endothelium with a discussion of regulated MHC molecule expression.

A. REGULATION OF MHC MOLECULE EXPRESSION ON VASCULAR ENDOTHELIUM

The expression of class I and of class II MHC molecules is independently regulated and the details of regulation vary among different cell types. Much of our knowledge of MHC molecule expression by endothelium is derived from cell culture experiments, and the most extensively studied system is human endothelial cells (ECs). Human ECs, under standard culture conditions, constitutively synthesize and express class I molecules but not class II molecules (5). The failure of early studies to detect class II molecules was initially surprising because human ECs are able to stimulate proliferation of class II molecule-specific primed typing T cells (6). The apparent explanation of this paradox is that the T cells induce class II molecule expression by the ECs during coculture, possibly through secretion of interferon-γ (IFN-γ) (7). These data imply that autologous T cells, perhaps activated by antigen, are responsible for physiologic regulation of MHC molecule expression on ECs. However, it should be noted that induction of class II molecules *in vitro* has largely been studied by use of T cells allogeneic to the ECs.

Induction of ECs to express class II molecules by coculture with allogeneic T cells is inhibited by cyclosporine A, supporting the notion that induction is mediated by T cell-secreted cytokines (8–10). (Cyclosporin A, which inhibits cytokine synthesis, does not block the response of cultured ECs to IFN-γ.) It has been further observed that whereas class I MHC expression is increased by coculture with purified allogeneic $CD4^+$ or $CD8^+$ T cell subsets, class II MHC expression is increased only by coculture with $CD8^+$ T cells (9–11). This difference is attributable to the fact that $CD4^+$ T cells secrete an inhibitor of class II MHC molecule expression (10). The identity of this inhibitor is as yet unknown, but it appears to be a stable protein mediator (i.e., a cytokine). $CD4^+$ T cells can induce class II expression on ECs if monocytes are added to the culture system (9), suggesting that monocytes induce $CD4^+$ T cells to secrete different profiles of cytokines from those induced by ECs. These studies imply that class II molecule induction on ECs *in vivo* may be initiated by infiltrating $CD8^+$ T cells and subsequently limited in their expression by interactions with infiltrating $CD4^+$ T cells.

By means of antibody neutralization, the induction of class II molecules by $CD8^+$ T cells has been shown to be mediated largely by secretion of IFN-γ (9). [Natural killer (NK) cells also can induce EC expression of class II MHC molecules, but this process is thought to be cell contact dependent rather than cytokine mediated (9).] The actions of IFN-γ on cultured human ECs have been studied in considerable detail. Recombinant IFN-γ increases the rate of class I MHC gene transcription and mRNA accumulation (12,13). It also induces *de novo* appearance of class II MHC transcripts, including HLA-DR, DP, and DQ α and β chains (12), probably also through activation of transcription. No other characterized cytokine studied to date can induce class II MHC mRNA accumulation or protein expression in ECs (14, but see Ref. 15). Class I transcription and mRNA accumulation can also be increased by IFN-α or IFN-β and by tumor necrosis factor (TNF) or lymphotoxin (LT, also called TNF-β) (14,16). TNF (or LT) acts synergistically with IFN-α/β or IFN-γ to increase class I transcription, whereas IFN-α/β and IFN-γ act through similar mechanisms, showing interactions that are at most additive (13,16). Both IFN-α/β and TNF/LT inhibit IFN-γ-mediated induction of class II MHC molecule expression (16–19).

In general, vascular ECs cultured from species other than human show similar patterns of modulation of MHC molecule expression by cytokines (20–22). One difference has been reported in ECs cultured from rat heart, wherein TNF augments class II expression (23). Furthermore, in these same cells interleukin-1 (IL-1), which has little effect on human EC expressin of MHC molecules (16), has been reported to inhibit expression of both class I and class II MHC molecules (23). It is also worth noting that several other cytokines have no discernible effect on MHC molecule expression by ECs. These include IL-2, IL-3, IL-4, IL-5, IL-6, IL-8, transforming growth factor-β (TGF-β), granulocyte colony-stimulating factor (G-CSF), and granulocyte/macrophage colony-stimulating factor (GM-CSF). [Rat heart endothelial cells once again apparently differ from other ECs by displaying class I down-regulation in response to GM-CSF (24).] Several of these other cytokines (e.g., IL-4, TGF-β, and the CSFs) do have other well-described actions on ECs, for example, as growth regulators; they simply do not regulate MHC molecule synthesis. The two major conclusions of these cytokine studies are (1) that the predominant regulators of MHC molecule expression by endothelial cells seem to be the interferons, including IFN-α, IFN-β, and especially IFN-γ, and TNF as well as the related cytokine LT and (2) that all of these cytokines increase class I molecule expression whereas only IFN-γ increases class II molecule expression.

In cell culture, many of the effects of these cytokines appear neither specific nor selective for ECs. For example, the same concentrations of IFN-γ that induce class II MHC molecule expression by dermal ECs also act on other skin cell populations such as fibroblasts (25), vascular smooth muscle cells (26), or keratinocytes (27) to produce comparable induction of class II molecule expression. Nevertheless, ECs are more strikingly sensitive to IFN-γ *in situ*. In short-term (3-day) human skin organ culture, ECs increase class II MHC expression in response to IFN-γ within 24 hours so that class II expression on ECs can be observed throughout the microvasculature, involving arterioles, capillaries, and venules (28). Under these same conditions, keratinocytes, fibroblasts, and smooth muscle cells do not respond, although epidermal Langerhans cells do. Thus, the *in situ* milieu of the various responding cell types appears to influence cytokine regulation of class II MHC molecule expression.

There are limited data regarding the regulation of EC expression of MHC molecules *in vivo*, as detected by immunocytochemical staining. Endothelia basally express class I molecules in all species and tissues examined. Human (29–31), nonhuman primate (31), and dog (32) venular and venous ECs basally express class II molecules *in situ;* mouse, rabbit, and guinea pig ECs generally do not (31). Certain human capillary ECs (e.g., in kidney and lung) may also basally express class II molecules (29–31, 33,34). In those species in which it has been observed, basal expression of class II molecules by ECs may not mean constitutive expression. For example, in dog, basal expression of class II molecules is reduced by cyclosporin A, suggesting that basal expression is actually a response to low levels of constitutive cytokine production (32). In baboon skin, class II molecules are readily increased in expression on venular ECs and induced on arteriolar ECs by intradermal injection of IFN-γ (35). In contrast, IFN-γ does not seem to induce class II molecules on murine ECs (36,37).

Inflammatory immune reactions cause increased class II MHC molecule expression on the venular ECs in most species. This upregulation phenomenon was first observed in mouse skin allograft rejection (38), but the most clear-cut demonstration is in guinea pig brain, where ECs express class II molecules as the first detectable sign of the development of experimental allergic encephalitis (39). Arterial ECs of man, which usually do not bear class II molecules, will express class II MHC molecules in settings of immune-mediated arterial reactions, such as Kawasaki disease (40) or in allograft coronary arteries postcardiac transplantation (41). Rat ECs appear to resist induction *in vivo* and usually remain class II MHC negative until relatively late in the course of inflammatory reactions (42). Overall, these data suggest that, although there are species differences and differences among endothelial cell types, most ECs can express class I and class II MHC molecules *in vivo*, probably in response to IFN-γ.

B. Protein Antigen Presentation by Cultured Endothelial Cells

The expression of class I and class II MHC molecules by ECs imply that these cells should be able to express processed protein antigens in a form that is recognizable by T cells. At the present time, the only assay for T cell recognition of a specific antigen is activation of a T cell by that antigen. Studies with mouse brain-derived ECs (20) or human umbilical vein-derived ECs (43) have shown that IFN-γ-treated ECs can indeed present protein antigens to and thereby activate T cell clones, confirming an ability to process antigen. (In mouse experiments, such T cells are syngeneic to the ECs, whereas in human experiments, the clones are usually allogeneic to the ECs but share an HLA-DR allele that serves as a restricting element for antigen recognition.) This property is not unique to ECs. For example, even dermal fibroblasts, treated with IFN-γ to express HLA-DR, will present antigen so as to activate human T cell clones (44). However, human umbilical vein-derived ECs can also present protein antigens to HLA-DR-matched resting T cells freshly isolated from peripheral blood of previously sensitized hosts (45,46). This capacity to stimulate proliferation of resting T cells distinguishes ECs from many other cell types with inducible class II MHC molecule expression, including dermal fibroblasts, which cannot present soluble antigens in a way that can activate resting T cells (47). It is not known if ECs can present protein antigens to T cells *in vivo*, but early experiments involving rat bone marrow chimeras were consistent with there being a functional antigen-presenting cell population not of bone marrow origin, possibly endothelium (48). This conclusion has since been disputed by others (49).

C. ALLOANTIGEN PRESENTATION BY CULTURED ENDOTHELIAL CELLS

Antigen-specific T cell responses are usually weak and require prior *in vivo* immunizations before they can be measured *in vitro*. The ability of ECs to present a signal recognized by T cells has also been studied by allogeneic reactions (50). Allogeneic reactions are fundamentally similar to specific antigen reactions in that the same populations of antigen-specific T cells are involved. It is now understood that the recognition of allogeneic MHC molecules, often with bound peptide, is actually a cross-reaction: the allogeneic MHC molecules mimic the three-dimensional structure of self MHC molecules plus bound peptides. A key difference between alloantigen proliferative responses and proliferative responses to protein antigens is that all individuals possess a remarkably high frequency of alloreactive T cells compared to T cells specific for any one foreign protein (i.e., many different antigen-specific T cells cross-react with each foreign MHC molecule). As a result, proliferative alloreactions may be studied without prior immunization and expansion of T cell populations.

EC class II MHC molecules can be recognized by allospecific T cell clones (25) or by alloreactive T cell populations (51), measured by cytolysis or adhesion, respectively. Just as in the case of presentation of foreign protein antigens, ECs are able to stimulate a resting T cell proliferative allogeneic response, whereas fibroblasts cannot (25,52), despite the fact that the same fibroblasts can stimulate proliferation of alloreactive helper T cell clones (44). Mouse brain ECs also strongly stimulate primary allogeneic responses, but, surprisingly, may be less potent than brain smooth muscle at presenting protein antigens (53).

The proliferative response to allogeneic ECs, like that to allogeneic leukocytes, is largely mediated by $CD4^+$ T cells and is largely directed against class II MHC molecules, especially HLA-DR (41). Removal of HLA-DR-expressing cells from untreated human umbilical vein EC cultures will prevent such cultures from stimulating a T cell response (54). The HLA-DR-bearing cells in the primary EC cultures are thought to initiate the response, and this observation raised the question of whether a contaminating, non-EC class II-bearing cell, presumably a leukocyte, was actually responsible for stimulating all of the allogeneic response of resting T cells. However, recent experiments, with better culture systems, support the idea that it is the endothelial cells that initiate and sustain the response. Serially passaged EC cultures, apparently devoid of leukocytes, are unable to activate resting $CD4^+$ T cells in their resting state. However, this capacity is generated if the ECs are treated with IFN-γ to induce HLA-DR expression (41,54a). This observation is not consistent with the proposal that constitutive HLA-DR-bearing cells, like dendritic cells, are responsible for stimulating the allogeneic response to ECs. Moreover, addition of class II-negative fixed ECs to fibroblast cultures has been shown to provide supplementing signals so that class II-bearing fibroblasts may activate resting T cells (55). These experiments point to the conclusion that although human ECs and other cells, such as dermal fibroblasts, are roughly comparable in producing a signal recognizable by a T cell antigen receptor, human ECs also provide additional costimulator activities needed for activation of resting T cells not provided by most other cell types. Such costimulator activities appear less important for activation of T cell clones. Rat brain ECs do not seem to express such costimulators (56), and guinea pig aortic ECs appear to express costimulators but not class II molecules (57). Both are necessary to activate $CD4^+$ T cells. The nature of the costimulator activities provided by human ECs to T cells is the subject of the next section.

III. Costimulator Activities of Vascular Endothelial Cells

Interleukin-2 has been identified as the key autocrine and paracrine growth progression factor for antigen-activated T cells, and insufficient production of IL-2 by antigen-activated T cells may lead to lack of a proliferative response or even a state of antigen-specific clonal anergy (58). Thus, a major function of costimulator activities on antigen-presenting cells is to increase the level of IL-2 produced by T cells in response to the antigen-specific signal. The inability of dermal fibroblasts or other cell types to cause T cell proliferation has been attributed to inadequate stimulation of IL-2 synthesis (47). Furthermore, augmentation of IL-2 in response to antigen can be used as a measure of costimulator activities. Experimentally, it is difficult to measure an effect of costimulators on IL-2 production in antigen-specific responses of resting T cells, because such antigen responses are usually very small, involving only a small number of T cells, until significant cytokine-mediated amplification has occurred. In practice, therefore, costimulators are often measured by their effects on polyclonal T cell mitogens such as anti-CD3 antibodies or mitogenic lectins such as phytohemagglutinin (PHA). We will begin by reviewing the effects of ECs on such polyclonal T cell responses.

A. EC COSTIMULATORS IN POLYCLONAL $CD4^+$ T CELL ACTIVATION

Purified resting human $CD4^+$ T cells secrete IL-2 and proliferate in response to high concentrations of PHA, with optimal stimulation occurring at 5 μg/ml or above. (These high concentrations of PHA are necessary to activate T cells if accessory cells, such as monocytes or B cells, are not present.) At all concentrations of PHA tested (0.1–10 μg/ml), the presence of ECs leads to augmented IL-2 synthesis by $CD4^+$ T cells, typically three- to eightfold or greater (59,60). The effect of ECs cannot be replaced by dermal fibroblasts or vascular smooth muscle cells. Interestingly, ECs are also not simply replacing monocytes as sources of costimulator activities, because (1) monocytes cannot substitute for ECs as a costimulator at high PHA concentrations and (2) ECs provide comparable degrees of costimulation whether or not monocytes are present (59,60). Similar augmentation of IL-2 synthesis by ECs may be observed upon stimulation of unfractionated peripheral blood mononuclear cells with soluble anti-CD3 antibody. We interpret these actions of ECs as indicative of strong costimulator functions that endow cultured human ECs with potent antigen-presenting capacity.

In the costimulator assay with optimal mitogen concentrations (10 μg/ml PHA), augmented IL-2 synthesis by T cells has been shown to depend upon contact between ECs and the responding T cells (60). Moreover, nearly equivalent levels of costimulation can be mediated by lightly paraformaldehyde-fixed ECs (61), consistent with the reported ability of fixed ECs to permit antigen presentation by fibroblasts (55). These observations suggest that the predominant EC costimulator activities that act on resting T cells are provided by cell surface ligands rather than by EC-secreted cytokines. Moreover, exogenous IL-1 and IL-6, two cytokines that are secreted by ECs (see below), fail to provide costimulation to resting blood CD4$^+$ T cells activated by PHA (60).

What is the identity of the EC ligands necessary for costimulation? Between 50 and 70% of EC costimulation can be inhibited by antibodies reactive with LFA-3 (CD58), the ligand for T cell CD2. Antibodies reactive with CD2 are equally effective, and inhibition by anti-LFA-3 and anti-CD2 is not additive (61). Moreover, at least 50% of the signal provided by ECs can be replaced by micelles of purified phosphatidylinositol-linked LFA-3 (PI-LFA-3) (61). EC costimulation is also more effective for CD4$^+$ T cells that express higher levels of CD2 than those selected to express lower levels, although there are numerous other differences between these T cell populations. Taken together, these observations establish that LFA-3 on the EC surface serves to provide costimulation to peripheral blood CD4$^+$ T cells.

Although EC expression of LFA-3 is important, LFA-3 cannot account for all of the costimulator activities of ECs (61). For example, addition of EC can further increase IL-2 synthesis at all concentrations of PI-LFA-3. Furthermore, in contrast to partial inhibition when costimulation is provided by ECs, costimulation by PI-LFA-3 is completely inhibited by antibody to LFA-3 or CD2. An additional piece of evidence to support the hypothesis that ECs use at least one ligand other than LFA-3 to provide costimulator activities is that PI-LFA-3 costimulation is strongly inhibited by cyclosporin A, whereas EC-mediated costimulation is not. The ligand(s) that mediates the remainder of the EC costimulator activity has not been identified. In antibody-blocking experiments, this signal appears to be independent of T cell LFA-1 (CD11a/CD18):EC–ICAM-1 (CD54) interactions, T cell VLA-4 (CD29/CD49d):EC–VCAM-1 interactions, T cell CD28:EC–B7/BB1 interactions, or any interactions involving CD44 on either cell. It is possible that EC-derived matrix molecules, such as fibronectin, provide costimulatory signals through T cell matrix receptors, such as VLA-4 and VLA-5. However, inhibitory antibodies to the common β chain of VLA-4 and VLA-5 (CD29) also fail to inhibit costimulation in the presence of PHA.

It is initially puzzling that so much of the costimulator activity provided by ECs can be attributed to LFA-3, because expression of this molecule is not restricted to ECs. Indeed, fibroblasts, which do not provide costimulation, express comparable levels of LFA-3, although the glycosylation patterns of the core polypeptide may differ (J. S.

Pober, unpublished observations). A possible explanation of these observations is that the non-LFA-3 signal is EC-specific and that these two signals interact to provide a greater than additive (i.e., synergistic) effect. This hypothesis would predict that cells that lack the EC-specific signal would contribute only a very weak LFA-3-dependent signal. The efficacy of purified LFA-3 to act as a costimulator does not refute this argument, as cells may never reach the same level of LFA-3-dependent signaling that can be achieved by optimal quantities of PI-LFA-3 micelles. Confirmation of this hypothesis must await identification of the other signal molecule.

Although the ligand(s) for the LFA-independent costimulator activity of EC is still unknown, a possible signaling pathway utilized by ECs to transduce costimulation in T cells has been uncovered. ECs increase steady-state levels of IL-2 mRNA of PHA-activated T cells, probably by increasing rates of IL-2 transcription. Several transcription factors that influence IL-2 synthesis have been identified, including AP-1 (62). AP-1 is usually composed of two protein subunits, one each encoded by the protooncogenes c-*fos* and c-*jun* (63). We have found that resting $CD4^+$ T cells express c-*jun* transcripts constitutively, and the level of c-*jun* mRNA is not influenced by PHA, by ECs, by LFA-3, or, in contrast to results with some T cell lines (62), by IL-1 (C. C. W. Hughes and J. S. Pober, unpublished observations). In contrast, c-*fos* transcripts are not detected in resting T cells. PHA induces transient synthesis and accumulation of the c-*fos* transcript, which peaks at around 30 minutes. ECs alone do not induce c-*fos* transcripts in T cells, but markedly augment peak c-*fos* mRNA levels induced by PHA. LFA-3 cannot mimic this effect and the EC effect on c-*fos* mRNA is not inhibited by cyclosporin A. Taken together, these data suggest that ECs utilize an LFA-3-independent signal to augment synthesis of AP-1 in activated T cells, leading to increased IL-2 transcription. This model is currently under investigation.

B. EC Costimulators in Polyclonal $CD8^+$ T Cell Activation

The studies described so far focus upon the activation requirements of $CD4^+$ T cells. Resting $CD8^+$ T cells also appear to have activation requirements in addition to signals provided by antigen recognition. The activation processes involved may be the entry of $CD8^+$ T cells into the cell cycle or the maturation of precytolytic T lymphocytes (pre-CTLs) into functional CTLs (64). Thus, one measure of costimulation of resting $CD8^+$ T cells is acquisition of cytolytic function in response to specific antigen. (Mature CTLs have different activation requirements, usually measured by target cell lysis; these are potentially relevant to EC injury *in vivo*, but will not be further discussed here.) *In vivo*, costimulation is often measured as successful allograft rejection by a previously unsensitized host. In many rodent transplantation models, class II MHC-expressing leukocytes present in an allograft are required to stimulate an effective rejection response (65). One interpretation of this result is that such "passenger" leukocytes in the graft activate $CD4^+$ T cells to produce cytokines, thereby providing costimulators for $CD8^+$ T cell maturation into CTLs. An alternative view is that the passenger leukocytes concomitantly but independently express both class II molecules, used to stimulate $CD4^+$ T cells, and costimulators, required for full activation of $CD4^+$ *and* $CD8^+$ T cells (66). In either model, elimination of class II-bearing leukocytes consequently removes all of the cells bearing costimulator. In human transplantation, passenger leukocytes have not been found to play a major role in sensitization, perhaps because human ECs express class II molecules and can activate $CD4^+$ T cells to produce costimulatory cytokines for $CD8^+$ T cells. Interestingly, rat ECs, induced to express class II MHC molecules by *in vitro* treatment with IFN-γ, can initiate a rejection response (67). Again, an alternative interpretation is that resting human ECs, in contrast to resting rodent ECs, could express costimulators required by $CD8^+$ T cells. Indeed, we have found that resting human ECs also increase IL-2 synthesis from polyclonally activated $CD8^+$ T cells (C. O. S. Savage and J. S. Pober, unpublished observations). The costimulators provided by ECs to $CD8^+$ T cells are as yet undefined.

C. EC COSTIMULATORS IN ALLOANTIGEN RESPONSES

The PHA-dependent costimulation assay, using a strong primary signal, was designed to minimize a need for additional accessory interactions between the T cell and an antigen-presenting cell (APC). In other culture systems (and presumably *in vivo*), T cell recognition of antigen on the surface of an APC critically depends upon accessory molecules that strengthen adhesion between the two cells. (PHA may bypass this requirement by directly cross-linking the two cells.) These additional accessory interactions can also be studied *in vitro* in models that utilize weaker, more "physiological" signals for T cell activation. As noted earlier, most responses to foreign protein antigens are very weak and not readily detected without prior *in vivo* immunization and often *in vitro* boosting (i.e., development of T cell lines). Again, a simple approach has been to examine the response of resting T cells to alloantigens, because the allogeneic response is sufficiently large to be measured without prior antigen exposure. We reiterate that cultured human ECs, unlike fibroblasts, are able to stimulate proliferation of allogeneic peripheral blood mononuclear cells or purified T cells (25,50,52) and this difference may be attributed to EC provision of costimulators to the alloreactive T cells. When $CD4^+$ T cells are used as responders, EC class II MHC molecules are major stimuli for proliferation (41; C. O. S. Savage, C. C. W. Hughes, and J. S. Pober, unpublished observations). This interpretation is based upon two key observations: (1) as noted earlier, activation of purified $CD4^+$ T cell subsets to proliferate depends upon pretreating the ECs with IFN-γ to induce class II expression; (2) the proliferative response of $CD4^+$ T cells to allogeneic IFN-γ-treated ECs can be blocked by antibody to class II MHC molecules or to T cell CD4, a molecule used to stabilize recognition of class I molecules, but not by control antibodies or antibodies to class I MHC molecules. In this system, we have found that $CD4^+$ T cell stimulation can also be inhibited by a variety of monoclonal antibodies to EC surface ligands, including anti-ICAM-1 (partial inhibition), anti-CD44, and anti-LFA-3. Antibodies to VCAM-1 do not inhibit. We presume LFA-3 is serving to provide costimulation, and although we have interpreted the role of ICAM-1 and CD44 as strengthening adhesion, it is possible that these molecules also provide additional costimulator signals. Our model of the molecules involved in antigen presentation, costimulation, and strengthening of adhesion is depicted in Fig. 1.

D. EC Production of Soluble Costimulators

Although our experimental systems appeared to emphasize contact-dependent signaling, ECs have been shown to produce a number of soluble costimulator activities for T cells (68–70), including IL-1 (71,72) and IL-6 (73,74). Resting ECs synthesize little IL-1 in the absence of cytokine treatment (71,72), although senescent ECs have been shown to spontaneously produce IL-1 (75). Analysis of mRNA species suggests that the predominant transcript in senescent ECs is IL-1α, and antisense oligonucleotides to IL-1α can reverse cell senescence (75)! [Surprisingly (and inexplicably), antibody to IL-1β can also reverse autocrine growth inhibition (76).] When early-passage ECs are treated with cytokines IL-1 (α or β), TNF, or LT, they accumulate IL-1 transcripts, again predominantly IL-1α (72). Lipopolysaccharide (LPS) induces similar accumulation of IL-1 transcrips (71,72). When assayed with specific IL-1-dependent cell lines (e.g., D10), relatively little IL-1 is released by ECs. However, the intact cells or their isolated membranes exhibit much more IL-1 activity, which is almost entirely IL-1α by serological neutralization (72). Although IL-1 is without effect in

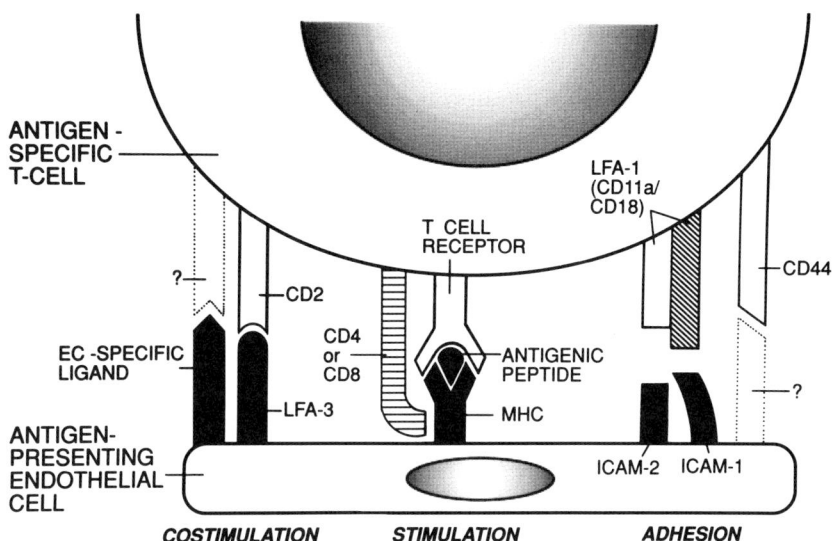

FIG. 1. Molecules involved in antigen presentation to T cells by vascular endothelial cells may serve to deliver a primary activation signal (stimulation), amplifying costimulatory signals (costimulation), or serve to stabilize cell contacts (adhesion).

our assays of IL-2 synthesis by peripheral blood of $CD4^+$ T cells activated by PHA, EC-expressed IL-1 could provide costimulation to bound T cells during antigen activation. It should also be noted that although ECs probably secrete less than 1% as much IL-1 as an activated macrophage, the surface levels of IL-1α may be more comparable.

ECs produce much more mouse thymocyte costimulator activator, a standard IL-1 assay, than can be accounted for by serologically neutralized IL-1 (69). The apparent explanation is that mouse thymocytes are also responsive to human IL-6 and that cytokine-treated ECs secrete copious quantities of IL-6 (73,74). The stimuli that induce IL-6 secretion appear similar to those that induce IL-1, namely IL-1 (α and β), TNF, LT, and LPS. IFN-γ may enhance IL-6 production (77). IL-1 appears far more potent than TNF for inducing IL-6 synthesis, but these mediators cooperate to induce greater than additive quantities of IL-6 (C. C. W. Hughes and J. S. Pober, unpublished observations). ECs are probably comparable to macrophages as sources of IL-6 and may account for much of the plasma IL-6 found in sepsis. Like IL-1, IL-6 cannot replace ECs as a source of costimulator in our PHA $CD4^+$ T cell assay, but IL-6 may also act as a costimulator of blood-borne T cells in other settings. In addition, IL-6 production may contribute to the ability of ECs to enhance B cell responses such as production of immunoglobulin (78) or proliferation to pokeweed mitogen (79,80).

In addition to these well-described T cell activators, ECs secrete several other cytokines, including IL-8 (81–83), monocyte chemotactic protein-1 (MCP-1) (84–86), GM-CSF (87–91), G-CSF (90,91), and TGF-β (92). ECs have not been found to secrete TNF. Human ECs can also synthesize various lipid mediators of inflammation, such as PGI_2 (93), PGE_2 (94), and platelet-activating factor (PAF) (95). The roles of these other molecules in human T cell activation are not fully defined.

In summary, the results described in the first two sections indicate that cultured human ECs can present antigen in an immunogenic form in association with MHC molecules and can provide costimulator activities that are needed to activate $CD4^+$ and (perhaps) $CD8^+$ T cells. Some of these costimulator activities depend on cell contact whereas others may involve secreted protein (cytokine) or lipid mediators. *In vivo*, these properties could allow ECs to present antigens to circulating memory T cells, providing both initial activation and recruitment of antigen-specific memory T cells into a site of antigenic challenge. Presentation of antigen by ECs would both initiate and localize delayed-type hypersensitivity responses. This view is teleologically appealing, but there are, to date, no convincing *in vivo* tests of this hypothesis. As noted in the introduction, endothelial cells may also recruit lymphocytes into inflammatory sites independent of specific antigens. The regulation of lymphocyte trafficking by endothelial cells is the subject of the next section.

IV. Antigen-Independent Recruitment of T Lymphocytes into Tissues by Endothelial Cells

T lymphocytes have a complex pattern of development that involves several migrations between blood and tissues. T cells must cross through EC barriers during three different phases of their life cycle: as pre-T cells, homing to the thymus; as circulating naive T cells, homing to lymph node; and as memory T cells, homing to sites of inflammation. These processes appear to be regulated by different lymphocyte surface structures interacting with different sets of EC surface molecules. Although our focus in this review will be on the homing of T cells to inflammatory sites, our understanding of this process is enhanced by comparisons with T cell migration in other circumstances, and we shall briefly review these here.

A. ENDOTHELIUM AND THE MIGRATION OF PRE-T CELLS TO THYMUS

The earliest cells committed to the T cell lineage, pre-T cells, arise in bone marrow, enter the circulation, and home to the thymus. It is not yet known what molecules are involved in the process of thymic homing. CD44, which can act as an adhesion molecule (see below), has been identified as a marker of early hematopoietic and lymphopoietic elements (96). CD44 is also expressed on a cell population in the medullary–cortical junction of the thymus, thought to represent newly arrived pre-T cells (97,98). The ECs in this region of the thymus are flat, and, in contrast to ECs in medullary vessels, do not express known markers of endothelial activation (98). To date, there are also no reports of expression of specific adhesion molecules by the endothelium of these thymic vessels.

Once in the thymus, pre-T cells undergo a series of ordered gene rearrangements that lead to expression of functional antigen receptors (99). The antigen receptors allow maturing T cells to be selected, both positively and negatively, so that the mature T cell antigen repertoire as expressed in the periphery is specific for various peptides bound to self alleles of MHC molecules. Other surface molecules concomitantly change in their level of expression during these selection processes. At the stage in which differentiated T cells are ready to leave the thymus, they express functional antigen receptors and either CD4 or CD8 corecognition molecules, which have been selected by interaction with class II or class I self MHC molecules, respectively. (In this review, we restrict our discussion to those T cells bearing α/β antigen receptors; the life cycle of γ/δ receptor-bearing T cells is not well worked out and their interactions with endothelial cells have not been described.) Such thymocytes then reenter the circulation as mature T cells.

B. ENDOTHELIUM AND THE MIGRATION OF NAIVE T CELLS TO LYMPH NODES

When T cells leave the thymus, they have functional antigen receptors but have not yet necessarily encountered their specific antigen. Such T cells are called "naive." In addition to antigen receptors and CD4 or CD8, naive T cells express the high-molecular-weight isoform of CD45 (called CD45RA); low levels of adhesion/signaling molecules such as CD2, LFA-1 (CD11a/CD18), VLA-4, VLA-5, VLA-6 (CD29, CD49d, e, and f), and CD44; and relatively high levels of a "peripheral lymph node homing receptor" molecule. This homing receptor molecule was originally identified in mouse by monoclonal antibody MEL-14 (100) and in man by antibodies TQ1 or Leu8 (101); it has also been called LAM-1 (leukocyte adhesion molecule-1) (102) and more recently, LECAM-1. This last name is an acronym for the structural motifs (103,104) that make up the extracellular portions of the molecule: an amino-terminal lectin domain (L), an adjacent epidermal growth factor-related domain (E), and two short consensus repeats characteristic of complement regulatory proteins (C). These extracellular regions function as a cell–cell adhesion molecule (AM). The extracellular domains are anchored to the lymphocyte surface by a short transmembrane region and an intracellular cytoplasmic region. LECAM-1 shares this overall structure with two other molecules: endothelial leukocyte adhesion molecule-1 (ELAM-1) (105), a cytokine-induced protein of vascular endothelial cells, and CD62, also called granule membrane protein-140 (GMP-140) (106) or platelet activation-dependent granule–external membrane protein (PADGEM), found in platelet α granules and endothelial cell Weibel–Palade bodies. The major structural differences among these molecules are that ELAM-1 and CD62 have six and nine short consensus repeats of complement regulatory proteins, respectively, instead of the two found in LECAM-1. These three molecules form a family that has been called LEC–CAMs (lectin–EGF–complement–cell adhesion molecules), or selectins. All three molecules are thought to mediate attachment to other cells by recognizing, in part, specific glycan structures borne on surface macromolecules of the other cells. Each selectin has its own carbohydrate specificity. The carbohydrate moiety recognized by LECAM-1 is not yet defined, but appears to require sialic acid (107). In addition, the interaction of LECAM-1-bearing cells with target cells (see below) can be efficiently inhibited by mannose phosphate polymers (108). More recent studies, using purified LECAM-1, have further supported the role of carbohydrate in the LECAM-1 ligand (109).

Naive T cells in the bloodstream preferentially migrate to secondary lymphoid organs, such as lymph nodes (1). It is thought that this emigration is a form of immune surveillance, checking to see if the antigen for which the particular T cell is specific has been carried to the node by afferent lymphatics draining epidermal or mucosal surfaces. If the specific antigen is not present, the naive T cells are thought to migrate through the nodal stroma to the efferent lymphatic and return to the bloodstream through the thoracic duct. Indeed, thoracic duct lymphocytes consist predominantly of naive T cells (110). Naive T cells are said to "recirculate" in search of specific antigen.

The key step in this recirculation is preferential transmigration through the postcapillary venules of the secondary lymphoid organs (111,112). In normal lymph nodes, the postcapillary venules that serve as the site of transmigration assume a characteristic appearance. Specifically, the cells become "tall" and protrude into the lumen. The altered morphology, which has given rise to the name "high endothelial venule" (HEV), is not an intrinsic property of these ECs. HEVs appear developmentally only after the lymph node is stimulated (113). Several markers associated with HEVs, including synthesis of a sulfate-rich proteoglycan (114) and HEV-associated cytoplasmic antigens (115), can be induced in cultured endothelial cells by treatment with cytokines (116,117). Tallness itself can be induced *in vitro* by treatment of EC cultures with certain combinations of cytokines (118). The morphological changes of the HEVs are thought to reflect changes that facilitate naive T cell trafficking across the venule. However, HEV morphology is not essential for extravasation; increased trafficking may be induced in germ-free mice by intraperitoneal injection of IFN-γ without inducing HEV morphology (119).

The most critical factor for T cell extravasation across an HEV appears to be increased adhesion to the vascular endothelium, allowing the T cell to resist the shear force of flowing blood long enough to find a site of exit (120). The adhesive interaction of naive T cells with lymph node venular endothelium can be studied *in vitro* by incubating lymphocytes with frozen sections of lymph node (121). At 37°C or room temperature, lymphocytes adhere to many structures in the node, more or less indiscriminately. This broad pattern of adhesion is largely mediated by T cell LFA-1 (CD11a/CD18) binding to its ligands, ICAM-1 and ICAM-2, and by various T cell very late activation (VLA) antigens, especially VLA-4, VLA-5, and VLA-6 (CD29/CD49d, e, and f) binding to extracellular matrix proteins and perhaps also to cellular receptors (see below). However, at 8°C, LFA-1- and VLA-dependent adhesion is minimal, and those lymphocytes that bear high levels of LECAM-1 will preferentially bind to the endothelium of the HEV. Furthermore, antibody to LECAM-1 blocks this adhesion (100). A soluble form of LECAM-1 has been used to show that ligands for LECAM-1 are expressed in high density by postcapillary venular endothelial cells of lymph nodes (122). It is likely that several different macromolecules carry the relevant carbohydrate epitope, but the endothelial core protein(s) may also participate in the interaction. In mouse, a monoclonal antibody, MECA-79, has been raised that recognizes the relevant epitope expressed by the endothelium and blocks naive T cell homing to lymph node *in vivo* (123). Studies with this antibody have shown that the LECAM-1-binding epitope is expressed in lymph node as a consequence of tissue-specific differentiation. MECA-79 staining can also be seen in some peripheral tissues with chronic inflammation (124), but these sites take on the appearance of ectopic lymph nodes. Molecules that bear the MECA-79 epitope, presumably conferred by expression of a glycosyl transferase in lymph node endothelium, have been called lymph node addressins; such molecules are not yet fully characterized.

C. ENDOTHELIUM AND MIGRATION OF MEMORY T CELLS TO INFLAMMATORY SITES

When a naive T cell finally encounters its specific antigen on the surface of an antigen-presenting cell, it undergoes a number of short-lived and long-lived changes in phenotype (125). In a matter of hours, antigen-activated T cells enter the cell cycle and express acute markers of activation, such as transferrin receptor, the 55-kDa subunit of the IL-2 receptor [called IL-2Rα or T activation (Tac) antigen], and HLA-DR. Activated T cells also increase the avidity of surface LFA-1, VLA-4, VLA-5, and VLA-6 for their ligands (126). The expression of acute activation markers and the increased avidity of adhesion molecules persist only as long as the T cells continue to be activated by specific antigen, and when antigen is cleared, these acute changes subside. However, concomitant with these markers of acute activation, antigen-activated T cells also undergo a series of more long-lived alterations. These latter changes include a profound reduction in the level of LECAM-1 expression with parallel increased expression of CD2, LFA-1, VLA-4,5,6, CD44 and LFA-3, among other surface structures (125,127,128). In addition, these cells express predominantly the lower molecular weight isoforms of CD45 (CD45RO) instead of CD45RA. Since these phenotypic changes are induced by encounter with antigen, T cells that show this phenotype are said to have changed from naive to "memory" cells. It is not clear whether the memory phenotype is stable or whether T cells may eventually revert to the so-called "naive" state (128). Memory T cells in the lymph node also eventually return to the circulation through the efferent lymphatics. However, since memory T cells express little if any LECAM-1, they do not preferentially return to lymph node. Instead, these cells, which may be recovered from peripheral sites of inflammation (129) or from afferent lymphatics draining sites of inflammation (110), remain in the circulation or preferentially home to sites of inflammation. Thus memory T cells become adhesive for endothelium at sites of inflammation.

1. Endothelial Leukocyte Adhesion Molecules

The preferential recognition of inflamed endothelium by memory T cells is thought to involve recognition of specific adhesion molecules on venular endothelial cells that are induced at sites of inflammation (2,3). Based primarily on studies with cultured human ECs, three cytokine-induced endothelial leukocyte adhesion molecules have been molecularly characterized: ELAM-1, VCAM-1, and ICAM-1. These molecules will each be described in more detail.

a. ELAM-1. ELAM-1 is a 110-kDa glycoprotein expressed by IL-1- or TNF-treated ECs, but not by resting ECs (130,131). These cytokines induce ELAM-1 expression by causing *de novo* appearance of the ELAM-1 mRNA, presumably by activating transcription (100). ELAM-1 transcripts reach maximal levels within 2 hours of cytokine treatment and then spontaneously decline, even in the presence of persistent cytokine. ELAM-1 protein surface expression peaks at about 4 hours. At later times, ELAM-1 is internalized and ultimately degraded. IL-1 and TNF in combination can increase the mean level of ELAM-1 expressed per cell, but the maximal level of expression per cell is unchanged (132). IFN-γ, which by itself does not induce ELAM-1, can accelerate the response to TNF and prolong ELAM-1 protein, but not mRNA, expression (132). IL-4 is reported to suppress IL-1- and TNF-induced ELAM-1 expression (133), but this effect is small.

Structurally, ELAM-1 is related to LECAM-1 (see above). It contains an amino-terminal lectin domain, an EGF-related domain, and six short consensus repeats of complement regulatory proteins. Recently, several laboratories have identified the glycan sialylated Lewis X antigen as a ligand for ELAM-1 (134–136). This moiety is expressed on several cell surface macromolecules of neutrophils and monocytes. Although T lymphocytes do not express sialylated Lewis X antigen, some T cell subpopulations do bind to ELAM-1 (137a,b,c), presumably through other as yet unknown ligands. It is currently in dispute whether all memory T cells can bind to ELAM-1 (137b) or whether only a small subset of memory T cells can recognize this endothelial cell ligand (137c). In general, however, ECs that express ELAM-1, but not other adhesion molecules, are most adhesive for resting neutrophils.

b. VCAM-1. A second inducible leukocyte adhesive glycoprotein of ECs is VCAM-1. VCAM-1 was originally detected as an IL-1- or TNF-inducible 110-kDa structure that mediated adhesion of certain tumor cells to the endothelium (and was originally called inducible cell adhesion molecule-110, or INCAM-110) (138). VCAM-1 was subsequently and independently identified by expression cloning of a lymphocyte and monocyte adhesion molecule on cytokine-activated endothelium (139). VCAM-1 is present on resting ECs at low levels and is inducible by IL-1, TNF, and IL-4 (138–140). VCAM-1 is also expressed on certain leukocyte populations, especially follicular dendritic cells (141). Inducible expression on ECs is slower than that of ELAM-1, with onset at 4–6 hours and peak expression at 24 hours. IFN-γ has no direct effect on VCAM-1, although it may enhance TNF-mediated induction.

Structural studies have shown that VCAM-1 is a member of the immunoglobulin (Ig) gene superfamily (139). Its extracellular amino terminus consists of a tandem array of six or seven Ig domains, depending on alternative splicing of the gene (141a). A leukocyte receptor for VCAM-1 has been identified as the β1 integrin VLA-4 (CD29/CD49d) (142,143). VLA-4 is present on monocytes and lymphocytes, but not neutrophils, and, as noted above, VLA-4 is expressed at higher levels on recently activated or previously activated (memory) T cells than on naive T cells, which have not been previously stimulated by antigen (126,127). T cell activation causes increased affinity of VLA-4 for fibronectin (126); it is not known if it also causes increased affinity for VCAM-1. In general, VCAM-1-expressing ECs or transfected COS cells preferentially bind mononuclear leukocytes (T cells and monocytes).

c. ICAM-1. A third inducible leukocyte adhesive glycoprotein of ECs is ICAM-1. ICAM-1 was identified as the ligand recognized by LFA-1 (CD11a/CD18) involved in phorbol ester-induced homotypic adhesion of lymphocytes (144). Resting ECs express low levels of ICAM-1, which can be increased by treatment with IL-1 or TNF, with onset at 4–6 hours and reaching plateau levels by 24 hours (145). ICAM-1 regulation differs from that of VCAM-1 in that it is directly inducible by IFN-γ (145) and is not inducible by IL-4 (133); in fact, IL-1- and TNF-mediated induction may be slightly inhibited by IL-4. ICAM-1 expression is widely distributed and may be present or induced on almost all cell types (146).

Structurally, ICAM-1, like VCAM-1, is a member of the Ig gene superfamily (147,148). It contains five extracellular Ig domains. The two amino-terminal domains are highly homologous to another molecule constitutively expressed on ECs, ICAM-2; ICAM-2 has only these two extracellular Ig domains (149). Also, in contrast to ICAM-1, ICAM-2 expression on ECs is apparently not subject to cytokine regulation. Both ICAM-1 and ICAM-2 can serve as ligands for recognition by the β2 integrin on lymphocytes, LFA-1 (CD11a/CD18) (149,150). ICAM-1 may also be recognized by Mac-1 (CD11b/CD18), a β2 integrin expressed on monocytes and neutrophils but not on lymphocytes (151).

Leukocyte recognition of ICAM-1 is regulated. Activated neutrophils adhere to surfaces more efficiently than resting neutrophils by up-regulating CD18-dependent mechanisms, which partly involve ICAM-1 (or ICAM-2) recognition (150). Moreover, as noted earlier, memory T cells express more LFA-1 than do naive T cells, and T cell activation increases the affinity of LFA-1 for ICAM-1 (152,153). ICAM-1 has been directly implicated in transmigration of neutrophils across cultured EC monolayers (154,155). T lymphocyte transmigration can be blocked by antibodies to LFA-1 (156,157), similarly suggesting a special role of ICAM-1 in EC interactions with activated, motile leukocytes. It should be pointed out, however, that congenital deficiency in CD18 molecules (LFA-1, Mac-1, and p150,95) produces more severe defects in neutrophil-mediated inflammation than in lymphocyte-mediated inflammation (158), suggesting that lymphocytes can avail themselves of alternative receptor–ligand interactions. In general, ICAM-1-expressing ECs or transfected COS cells preferentially bind activated leukocytes, including granulocytes and lymphocytes.

d. Other Endothelial Leukocyte Adhesion Molecules. There may be yet other endothelial leukocyte adhesion molecules with specificity for T lymphocytes. IFN-γ, which does not induce ELAM-1 or VCAM-1 and only weakly increases expression of ICAM-1, can make ECs more adhesive for T cells (159–163). One early report identified this as an interaction between T cell CD4 and class II MHC molecules on ECs (159). One other study has reported preferential adhesion of CD4$^+$ T cells (162), but another claimed preferential adhesion of CD8$^+$ T cells to IFN-γ-treated monolayers (163)! The basis for these differences is not clear and the molecules involved are poorly defined.

2. Involvement of Endothelial Leukocyte Adhesion Molecules in Lymphocyte Binding in Vitro

The presumptive function of increased endothelial leukocyte adhesion molecule expression is increased binding of lymphocytes at sites of inflammation. To date, the evidence linking EC expression of adhesion molecules, defined *in vitro*, with *in vivo* adhesion and subsequent extravasation of lymphocytes is largely indirect. Much of this evidence has come from *in vitro* studies in which antibodies to adhesion molecules have been shown to inhibit lymphocyte binding to resting or cytokine-activated human ECs (137,143,150,156,157,164–166). The use of inhibitory antibodies to endothelial leukocyte adhesion molecules *in vivo* to modify lymphocyte homing has not yet been reported, except for anti-ICAM-1 inhibition of transplant rejection (167). However, in these experiments, anti-ICAM-1 may be blocking antigen recognition by T cells rather than homing of memory T cells.

Adhesion assays with lymphocytes and cultured human endothelial cells pose certain difficulties in interpretation. The avidity of the interaction is usually inferred from the percentage of lymphocytes added that resist washing off under controlled but undefined conditions. In the earliest such experiments, performed by DeBono, it was noted that although freshly opened pig or human blood vessels were nonadhesive for lymphocytes, lymphocytes stuck avidly to ECs cultured from these same vessels (168). This result differs from adhesion of other leukocytes, which is relatively low in the absence of *in vitro* activation of either the endothelial cell or the leukocyte. Thus, culture itself appears to cause EC changes that selectively induce lymphocyte adhesion. Cytokine treatment of the ECs with IL-1, TNF, IFN-γ, or IL-4 does increase the level of T cell adhesion even further, but these increments in adhesion are generally smaller than the basal level of adhesion to cultured cells (137,143,150,156,157,159–163,169–171). Moreover, increased adhesion caused by IL-1 or TNF treatment of EC cultures appears maximal by 4 hours of cytokine activation (168,169), yet these same cytokines do not produce lymphocyte adhesion and extravasation *in vivo* until much later times (9–12 hours) (35).

Despite these technical issues, adhesion assays with cultured ECs have provided some important information about the molecules involved in adhesion. Resting ECs preferentially bind $CD8^+$ T cells and $CD16^+$ NK cells; little adhesion of $CD4^+$ T cells is noted (172). This adhesion to resting ECs can be blocked by monoclonal antibody to LFA-1 on the T cell and to ICAM-1 (and perhaps to ICAM-2) on the EC (150,154,155,165). Importantly, TNF- or IL-1-activated ECs appear to preferentially bind memory T cells of both $CD4^+$ and $CD8^+$ subsets (173). Indeed, cytokine-activated EC cultures can selectively adsorb those rat T cells that preferentially home to inflammatory sites (171). The adhesion of human T cells to TNF- or IL-1-activated ECs can be blocked by antibodies to VLA-4 on the T cell and to VCAM-1 on the EC (142,143,165). Antibodies to ELAM-1 and to LFA-1 or ICAM-1 have less individual inhibitory action (137,150,164,165), but can further block T cell adhesion to cytokine-treated ECs in the presence of anti-VCAM-1 antibody (137,165). These data suggest that there may be a primary contribution of VCAM-1 and secondary contributions of ELAM-1 and ICAM-1 in T lymphocyte adhesion to TNF-treated ECs. IL-4-treated ECs also bind T lymphocytes through VCAM-1 (140). As noted earlier, the molecules that mediate attachment to IFN-γ-treated ECs have not been well defined.

An alternative model for study of lymphocyte adhesion to ECs is the frozen-section binding assay first used to study naive T cell binding to HEVs. Cytokine-activated ECs at sites of inflammation take on morphological characteristics of HEVs (174–176) and share some biochemical and antigenic features with lymph node HEVs (124,176,177). When inflamed tissues are used in this assay, it has been possible to demonstrate specific lymphocyte adhesion to these altered venules (176,178–180). However, the molecular basis of the adhesion and the phenotype of the T cells that adhere (e.g., memory versus naive) have not yet been reported.

3. Expression of Endothelial Leukocyte Adhesion Molecules in Vivo

A second kind of experiment that supports the role of endothelial leukocyte adhesion molecules in lymphocyte homing is the immunocytochemical analysis in man and primates of experimental or pathological tissues. For example, ELAM-1, VCAM-1, and ICAM-1 are all increased in EC expression in elicited delayed hypersensitivity reactions in the skin of humans and baboons (141,172,181–183). In baboon (172), the onset of ELAM-1 expression in response to TNF (2–4 hours) correlates with onset of neutrophil adhesion and accumulation, whereas the onset of lymphocyte accumulation is delayed until there is upregulation of ICAM-1 (9–12 hours). (ELAM-1, although diminished, is by no means absent at the time of lymphocyte infiltration.) The onset of ELAM-1 expression also corresponds to the onset of inflammation in elicited late-phase reactions of immediate hypersensitivity in human (184). In general, these experimental studies provide a strong correlation of increased ELAM-1 expression with neutrophil accumulation, and support the role of other molecules, such as ICAM-1 and VCAM-1, in lymphocyte homing.

The results from elicited reactions are also supported by examination of pathological tissues. For example, in a pilot study of cardiac allograft biopsies, EC expression of ICAM-1 and VCAM-1 but not of ELAM-1 was highly correlated with presence of $CD3^+$ T cell infiltrates (185). Moreover, in these biopsies VCAM-1 was largely confined to the ECs of the postcapillary venules surrounded by the infiltrating T cells. In general, ELAM-1 is most commonly noted in diseases associated with acute inflammation or in the vasculature of tumors associated with high levels of cytokine production (e.g., T cell lymphoma, Hodgkin's disease, or Kaposi's sarcoma) (184,186,187). VCAM-1 is more abundant in subacute inflammation (141). However, these correlations are not absolute, and a major lesson from the examination of pathological tissues is that expression of endothelial leukocyte adhesion molecules cannot be sufficient to explain the specificity of leukocyte infiltration (188). The contribution of other factors, particularly the state of activation of the leukocytes, is also of critical importance.

D. ENDOTHELIUM AND TISSUE-SPECIFIC HOMING OF LYMPHOCYTES

In addition to the homing of memory T cells to sites of inflammation, there is evidence that different populations of circulating T cells may preferentially migrate to specific tissues. The best demonstration of tissue-specific homing has been provided by frozen-section binding assays. Certain lymphocyte cell lines strongly adhere to HEVs of gut-associated lymphoid tissue (GALT), especially Peyer's patch HEVs of the ileum, whereas others do not (189,190). Monoclonal antibodies MECA-89 and MECA-376 have tentatively identified a murine gut-specific endothelial antigen as the "addressin" molecule that confers Peyer's patch-specific recognition (191). This molecule is a glycoprotein of 58 to 66 kDa and is preferentially expressed by venules in the gut.

The lymphocyte receptor for the gut addressin(s) is not known with certainty, and various candidate gut lymphocyte homing receptors have been proposed. The first such candidate, identified in the rat by an antibody, is as yet molecularly uncharacterized (192). The two best studied candidates are VLA-4 (or sometimes a molecule composed of the VLA-4α chain complexed to a unique β chain distinct from CD29) in mouse (193,194) and CD44 in man (195,196). However, both of these molecules are also involved in interactions with cytokine-activated endothelial cells (142,143,157), and their expression could be markers of memory T cells rather than of tissue specificity. It is possible that inflammation is a "normal" feature of gut mucosa and that apparent tissue-specific homing is an artifact of such inflammatory changes in gut endothelium. However, there is also evidence, by the same kinds of assays, for distinct tissue-specific homing of lymphocytes to bronchial-associated lymphoid tissue (BALT) (197), to skin-associated lymphoid tissue (SALT) (178), and to ectopic lymphoid tissue that forms in rheumatoid synovium (179). It has recently been reported that lymphocytes that accumulate in skin express a common surface marker (originally noted on lymph node HEV) called HECA-452 (198), and it has been proposed that this molecule may identify T cells that preferentially home to skin (although it is equally possible that the expression of this marker is related to prolonged activation in the skin rather than homing). A different antigen (HML-1) has been reported to be preferentially expressed by mucosal lymphocytes in the gut (199). It is not known if either of these molecules is a homing receptor, nor have their putative ligands been identified. In our judgment, tissue-specific endothelial differentiation and expression of "addressins" as a mechanism for the recruitment of specific subsets of memory T cells, each expressing unique "homing receptors," is not yet established as fact.

E. ENDOTHELIUM AND EXTRAVASATION OF T LYMPHOCYTES

As commented on above, EC expression of endothelial leukocyte adhesion molecules does not entirely correlate with leukocytic infiltration into tissues (188), and several other coconditions probably must exist (2). For example, there must be adequate delivery of leukocytes to the tissue. In addition, the hemodynamic forces that serve to prevent leukocyte–endothelial adhesion must be sufficiently reduced in the local microcirculation. And most critically, the state of the circulating leukocyte will influence the interaction with microvascular EC. We have emphasized that for T cells, homing to inflammatory sites is the property of the memory subset. However, memory cells, although adhesive for cytokine-activated ECs (173), circulate in an immotile form that probably cannot extravasate into the tissues without a change in phenotype. It is presumed that lymphocytes undergo a transition while adherent to the venular EC surface, changing from the immotile circulating phenotype to a motile one. The full extent of this change is not defined and the signals that cause it are largely unidentified.

T cell activation independent of antigen, may result from interactions with surface ligands or by binding of cytokines, analogous to T cell costimulation. For example, activation of T cells through CD2 leads to increases in LFA-1 avidity for ICAM-1 (152), comparable to those seen in T cells activated through the CD3–antigen receptor complex (153). Alternatively, signals may be transmitted by other membrane events. For example, we have shown that T cells can form channels of intracellular communication with endothelial cells, permitting exchange of cytoplasmic signaling molecules (200). Gap junction-like connections between infiltrating T cells and ECs have been observed *in vivo* in the brains of mice with early experimental allergic encephalitis (201). Finally, ECs may synthesize and secrete molecules that deliver chemotactic signals to bound T cells, including lipid mediators and cytokines.

Perhaps the most potent lipid mediator of inflammation produced by ECs is platelet-activating factor (PAF). This alkyl ether lipid is made by vascular endothelial cells isolated from all tissues examined (95), and PAF can cause T cell transmigration through endothelial monolayers (202). In general, PAF is made by ECs in response to rapidly acting agonists, such as histamine. It has been reported by one laboratory that cytokines can also induce PAF synthesis (203,204), although we have not been able to confirm this (205). Little PAF is secreted by ECs, but PAF can be found in EC plasma membranes, where it is capable of activating bound leukocytes (206). It should be noted that PAF synthesis and activity is usually measured under serum-free conditions, and the ability of ECs to synthesize PAF *in vivo* remains to be shown.

EC cytokines that activate leukocytes fall into two classes. The first such cytokines to be described were the colony-stimulating factors. ECs activated by IL-1 or TNF secrete GM-CSF and G-CSF (87–91). (There are no reports as to whether ECs synthesize IL-3, IL-7, or other newly described CSFs.) GM-CSF, in addition to its role as a regulator of bone marrow, is a local activator of mature monocytes and granulocytes, but has not been shown to act on lymphocytes. More recently, ECs have been found to synthesize and secrete several members of the newly described low-molecular-weight inflammatory cytokine family, including IL-8 (81–83), MCP-1 (84–86), γ-IP-10 (207), and GRO (α and perhaps β/γ as well) (208). IL-8 synthesis by ECs is induced by IL-1, TNF, or LPS, and is not affected by IFN-γ (86) or IL-4 (209). Some laboratories have reported that IL-8 is less completely processed by ECs than by monocytes (82), whereas others have reported it is further processed (83), but there are no known biological differences among the various forms. IL-8 is a potent chemoattractant and possible activator of neutrophils, and has also been found to act on T cells as a chemoattractant (210). However, the details of IL-8-induced phenotypic changes in T cells are not well known. In particular, it is not known if IL-8 increases T cell LFA-1 or VLA-4 affinity for their ligands. MCP-1 (encoded by the JE gene) is an activator of monocytes, but not T cells. It is synthesized in response to IL-1 and TNF and, at later times, in response to IFN-γ and IL-4 as well (84–86, 211). EC-derived MCP-1 appears larger than monocyte-derived MCP-1, but it is functional and may be the major monocyte chemotactic factor produced by cytokine-activated ECs. Thus MCP-1 secretion by ECs may contribute to monocyte accumulation in delayed hypersensitivity. However, as for platelet-activating factor, it has not yet been established that ECs make any of these inflammatory protein molecules *in vivo*.

Lymphocyte migration through the endothelium can be observed *in vitro*, using monolayers of cultured ECs as a barrier to migration. Transmigration can be stimulated either by T cell activation, e.g., with phorbol esters or PAF, or by cytokine treatment of the endothelial cell monolayer (157,202,212). In these systems, antibody-blocking studies have revealed that LFA-1 : ICAM-1 interactions participate in transmigration (156,157); there are as yet no data on the role of VLA-4 : VCAM-1 interactions. However, as we previously noted, lymphocytes that do not express LFA-1 (due to a genetic defect) can extravasate *in vivo* into sites of inflammation (158); thus LFA-1-independent interactions in transmigration must also exist.

In summary, the observations reviewed in this section indicate that ECs may play a central role in the recruitment of lymphocytes into tissues independent of antigen. Different EC surface molecules can regulate T cell adhesion at different stages of the T cell life cycle. A model depicting these events and the molecules involved is represented in Fig. 2. Memory T cells are specifically adherent to the cytokine-activated endothelium at various sites of inflammation. A contribution of endothelium to tissue-specific homing in inflammation has been proposed, but not yet established. Finally, ECs may actively contribute to the conversion of adherent lymphocytes into motile cells as a necessary prelude to extravasation.

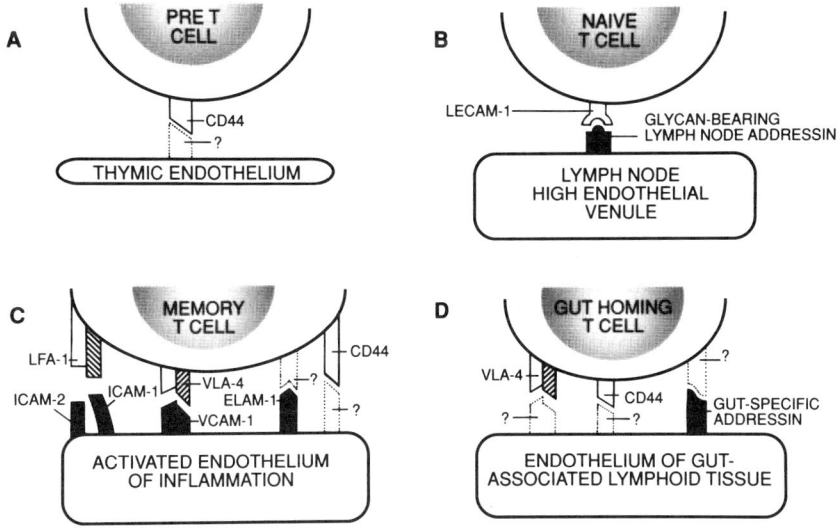

FIG. 2. Different combinations of molecules on lymphocytes and endothelial cells direct T cell migration at different stages of T cell maturation, such as in (A) the homing of pre T cells to thymus; (B) the homing of naive T cells to lymph nodes; and (C) the homing of memory T cells to peripheral sites of inflammation. Tissue specific homing, depicted for gut-associated lymphoid tissue in panel (D), may interact with the homing processes depicted in panels (B) and (C).

V. Summary

The data presented in this review establish that cultured human endothelial cells have the capacity to present antigens to T cells and to do so in the context of costimulators that lead to effective T cell activation. These activities raise the possibility that venular ECs, at sites of delayed hypersensitivity reactions, could be the primary antigen-presenting cell to circulating memory T cells. This putative role of ECs can explain the rapid rate of initiation of memory responses because ECs are uniquely positioned to have physical access to the pool of circulating memory T cells. Studies also suggest that ECs may present alloantigens to circulating T cells in the context of transplantation, thereby initiating rejection reactions. Nevertheless, we repeat our caveat that these proposed antigen-presenting functions of ECs have not been established *in vivo*.

Cytokine-mediated changes, particularly induction of adhesion molecules and synthesis of lymphocyte-activating cytokines, such as IL-8, provide ECs with the potential to recruit memory T cells to inflammatory sites independent of antigen specificity. Although these functions have also not been rigorously shown to occur *in vivo*, immunocytochemical studies of experimental and pathological tissues provide significant support for this proposal. Similar adhesive and activating functions of ECs may apply to preferential homing of pre-T cells to thymus and naive T cells to lymph node. We conclude by noting that the weight of evidence reviewed here supports the proposal that the vascular endothelium be considered an integral part of the *in vivo* immune system.

Acknowledgment

This work was supported by NIH Grant P01-HL36028. We thank Mr. David Lence for his expert assistance in preparation of the manuscript.

References

1. Yednock, T. A., and Rosen, S. D. (1989). Lymphocyte homing. *Adv. Immunol.* **44**, 313.
2. Pober, J. S., and Cotran, R. S. (1990). The role of endothelial cells in inflammation. *Transplantation* **50**, 537.
3. Osborn, L. (1990). Leukocyte adhesion to endothelium in inflammation. *Cell* **62**, (Cambridge, Mass.) 3.
4. Kourilsky, P., and Claverie, J.-M. (1989). MHC–antigen interaction: What does the T cell receptor see? *Adv. Immunol.* **45**, 107.
5. Pober, J. S., and Gimbrone, M. A., Jr. (1982). Expression of Ia-like antigens by human vascular endothelial cells is inducible *in vitro:* Demonstration by monoclonal antibody binding and immunoprecipitation. *Proc. Natl. Acad. Sci. U.S.A.* **79**, 6641.

6. Moen, T., Moen, M., and Thorsby, E. (1980). HLA-D region products are expressed in endothelial cells. *Tissue Antigens* **15**, 112.
7. Pober, J. S., Gimbrone, M. A., Jr.,Cotran, R. S., Reiss, C. S., Burakoff, S. J., Fiers, W., and Ault, K. A. (1983). Ia expression by vascular endothelium is inducible by activated T cells and by human γ-interferon. *J. Exp. Med.* **157**, 1339.
8. Groenewegen, G., Buurman, W. A., Jeunhomme,G. M. A. A., and van der Linden, C. J. (1985). Cyclosporin A affects MHC–class II mitogen expression by arterial and venous endothelium *in vitro*. *Transplantation* **40**, 21.
9. Pardi, R., Bender, J. R., and Engleman, E. G. (1987). Lymphocyte subsets differentially induce class II human leukocyte antigens on allogeneic microvascular endothelial cells. *J. Immunol.* **139**, 2585.
10. Doukas, J., and Pober, J. S. (1990). Lymphocyte-mediated activation of cultured endothelial cells (EC): $CD4^+$ T cells inhibit EC class II MHC expression despite secreting IFN-γ and increasing EC class I MHC and intercellular adhesion molecule-1 expression. *J. Immunol.* **145**, 1088.
11. Bender, J. R., Pardi, R., Kosek, J., and Engleman, E. G. (1989). Evidence that cytotoxic lymphocytes alter and traverse allogeneic endothelial cell monolayers. *Transplantation* **47**, 1047.
12. Collins, T., Korman, A. J., Wake, C. T., Boss, J. M., Kappes, D. J., Fiers, W., Ault, K. A., Gimbrone, M. A., Jr., Strominger, J. L., and Pober, J. S. (1984). Immune interferon activates multiple class II major histocompatibility complex genes and the associated invariant chain gene in human endothelial cells and dermal fibroblasts. *Proc. Natl. Acad. Sci. U.S.A.* **81**, 4917.
13. Johnson, D. R., and Pober, J. S. (1990). Tumor necrosis factor and immune interferon synergistically increase transcription of HLA class I heavy and light chain genes in vascular endothelium. *Proc. Natl. Acad. Sci. U.S.A.* **87**, 5183.
14. Collins, T., Lapierre, L. A., Fiers, W., Strominger, J. L., and Pober, J. S. (1986). Recombinant human tumor necrosis factor increases mRNA levels and surface expression of HLA-A,B antigens in vascular endothelial cells and dermal fibroblasts *in vitro*. *Proc. Natl. Acad. Sci. U.S.A.* **83**, 446.
15. Groenewegen, G., Buurman, W. A., de Ley, M., and Koostra, G. (1987). A new lymphokine different from interferon-gamma induces major histocompatibility complex class II antigen expression. *Transplant. Proc.* **19**, 190.
16. Lapierre, L. A., Fiers, W., and Pober, J. S. (1988). Three distinct classes of regulatory cytokines control endothelial cell MHC antigen expression: Interactions with immune (γ) interferon differentiate the effects of tumor necrosis factor and lymphotoxin from those of leukocyte (α) and fibroblast (β) interferons. *J. Exp. Med.* **167**, 794.
17. Wedgewood, J. F., Hakam, L., and Bonagura, V. R. (1988). Effect of interferon-gamma and tumor necrosis factor on the expression of class I and class II major histocompatibility molecules by cultured human umbilical vein endothelial cells. *Cell. Immunol.* **111**, 1.
18. Leeuwenberg, J. F., van Damme, J., Meager, T., Jeunhomme, T. M., and Buurman, W. A. (1988). Effects of tumor necrosis factor on the interferon-gamma-induced major histocompatibility complex class II antigen expression by human endothelial cells. *Eur. J. Immunol.* **18**, 1469.
19. Manyak, C. L., Tse, H., Fischer, P., Coker, L., Sigal, N. H., and Koo, G. C. (1988). Regulation of class II MHC molecules on human endothelial cells: Effects of IFN and dexamethasone. *J. Immunol.* **140**, 3817.
20. McCarron, R. M., Spatz, M., Kempski, O., Hogan, R. N., Muehl, L., and McFarlin,

D. E. (1986). Interaction between myelin basic protein-sensitized T lymphocytes and murine cerebral vascular endothelial cells. *J. Immunol.* **137**, 3428.
21. Male, D. K., Pryce, G., and Hughes, C. C. W. (1987). Antigen presentation in brain: MHC induction on brain endothelium and astrocytes compared. *Immunology* **60**, 453.
22. Launder, T. M., Gegen, N. W., Knedler, A., and Herbeck, R. J. (1987). The isolation and characterization of enriched microvascular endothelial cells from mouse adipose tissue. The induction of class II molecules of the major histocompatibility complex (MHC) by interferon-gamma (IFN-γ). *J. Immunol. Methods* **102**, 45.
23. Leszczynski, D. (1990). Interleukin-1 alpha inhibits the effects of gamma interferon and tumor necrosis factor alpha on the expression of the major histocompatibility antigens by the rat endothelium. *Am. J. Pathol.* **136**, 229.
24. Leszczynski, D., and Hayry, P. (1990). Granulocyte-macrophage colony-stimulating factor diminishes interferon-gamma-induced class I major histocompatibility complex antigen expression by endothelium with prostacyclin as intermediary. *Transplant. Proc.* **22**, 132.
25. Pober, J. S., Collins, T., Gimbrone, M. A., Jr., Cotran, R. S., Gitlin, J. D., Fiers, W., Clayberger, C., Krensky, A. M., Burakoff, S. J., and Reiss, C. S. (1983). Lymphocytes recognize human vascular endothelial and dermal fibroblast Ia antigens induced by recombinant immune interferon. *Nature (London)* **305**, 726.
26. Pober, J. S., Collins, T., Gimbrone, M. A., Jr., Libby, P., and Reiss, C. S. (1986). Overview: Inducible expression of class II major histocompatibility complex antigens and the immunogenicity of vascular endothelium. *Transplantation* **41**, 141.
27. Basham, T. Y., Nickoloff, B. J., Merigen, T. C., and Morhenn, V. B. (1984). Recombinant gamma interferon induces HLA-DR expression on cultured human keratinocytes. *J. Invest. Dermatol.* **83**, 88.
28. Messadi, D. V., Pober, J. S., and Murphy, G. F. (1988). Effects of recombinant gamma interferon on HLA-DR and DQ expression by skin cells in short term organ culture. *Lab. Invest.* **58**, 61.
29. Koyama, K., Fukunishi, T., Barcos, M., Tanigaki, N., andPressman, D. (1979). Human Ia-like antigens in non-lymphoid organs. *Immunology* **38**, 333.
30. Natali, P. G., de Martino, C., Quaranta, V., Nicotra, M. R., Frezza, F., Pellegrino, M., and Ferrone, S. (1981). Expression of Ia-like antigens in normal human non-lymphoid tissues. *Transplantation* **31**, 75.
31. Natali, P. G., Quaranta, V., Nicotra, M. R., Appolonj, C., Pellegrino, M. A., and Ferrone, S. (1981). Tissue distribution of Ia-like antigens in different species: Analysis with monoclonal antibodies. *Transplant. Proc.* **13**, 1026.
32. Groenewegen, G., Buurman, W. A., and von der Linden, C. J. (1985). Lymphokine dependence of *in vivo* expression of MHC class II antigens by endothelium. *Nature (London)* **316**, 361.
33. Baldwin, W. M., Claas, F. H. J., van Es, L. A., and van Rood, J. J. (1981). Distribution of endothelial-monocyte and HLA antigens on renal vascular endothelium. *Transplant. Proc.* **13**, 103.
34. Yamamoto, M., Shimokata, K., and Nagura, H. (1988). An immunohistochemical study on phenotypic heterogeneity of human pulmonary vascular endothelial cells. *Virchows Arch. A* **412**, 479.
35. Munro, J. M., Pober, J. S., and Cotran, R. S. (1989). Tumor necrosis factor and interferon-γ induce distinct patterns of endothelial activation and associated leukocyte accumulation in the skin of *Papio anubis*. *Am. J. Pathol.* **135**, 121.

36. Skoskiewicz, M. J., Colvin, R. B., Schneeberger, E. E., and Russell, P. S. (1985). Widespread and selective induction of major histocompatibility complex-determined antigens *in vivo* by γ interferon. *J. Exp. Med.* **162**, 1645.
37. Momburg, F., Koch, N., Moller, P., Moldenhauer, G., Butcher, G. W., and Hammerling, G. J. (1986). Differentiated expression of Ia and Ia-associatd invariant chain in mouse tissues after *in vivo* treatment with IFN-γ. *J. Immunol.* **136**, 940.
38. de Waal, R. M. W., Bogman, M. J. J., Maass, C. N., Cornelissen, L. M. H., Tax, W. J. M., and Koene, R. A. P. (1983). Variable expression of Ia antigens on the vascular endothelium of mouse skin allografts. *Nature (London)* **303**, 426.
39. Sobel, R. A., Blanchette, B. W., Bhan, A. K., and Colvin, R. B. (1984). The immunopathology of experimental allergic encephalomyelitis. II. Endothelial cell Ia increases prior to inflammatory cell infiltration. *J. Immunol.* **132**, 2402.
40. Terai, M., Kohno, Y., Namba, M., Unemiya, T., Niwa, K., Nakajima, H., and Mikata, A. (1990). Class II major histocompatibility antigen expression on coronary arterial endothelium in a patient with Kawasaki disease. *Hum. Pathol.* **21**, 231.
41. Salomon, R. N., Hughes, C. C. W., Schoen, F. J., Payne, D. D., Pober, J. S., and Libby, P. (1991). Human coronary transplantation-associated arteriosclerosis: Evidence for a chronic immune reaction to activated graft endothelial cells. *Am. J. Pathol.* **138**, 791.
42. Milton, A. D., and Fabre, J. W. (1985). Massive induction of donor type class I and class II major histocompatibility complex antigens in rejecting cardiac allografts in the rat. *J. Exp. Med.* **161**, 98.
43. Wagner, C. R., Vetto, R. M., and Burger, D. R. (1984). The mechanism of antigen presentation by endothelial cells. *Immunobiology* **168**, 453.
44. Umetsu, D. T., Pober, J. S., Jabara, H. J., Fiers, W., Yunis, E., Burakoff, S. J., Reiss, C. S., and Geha, R. S. (1985). Human dermal fibroblasts present tetanus toxoid antigen to antigen specific T cell clones. *J. Clin. Invest.* **76**, 254.
45. Hirschberg, H. (1981). Presentation of viral antigens by human vascular endothelial cells *in vitro*. *Hum. Immunol.* **2**, 235.
46. Wagner, C. R., Vetto, R. M., and Burger, D. R. (1985). Subcultured human endothelial cells can function independently as fully competent antigen-presenting cells. *Hum. Immunol.* **13**, 33.
47. Umetsu, D. T., Katzen, D., Jabara, H. J., and Geha, R. S. (1986). Antigen presentation by human dermal fibroblasts: Activation of resting T lymphocytes. *J. Immunol.* **136**, 440.
48. Standage, B. A., Vetto, R. M., Jones, R., and Burger, D. R. (1985). Vascular endothelial cells in cell-mediated immunity: Adoptive transfer with *in vitro* conditioned cells is genetically restricted at the endothelial cell barrier. *J. Cell. Biochem.* **29**, 45.
49. Hinrichs, D. J., Wegmann, K. W., and Dietsch, G. N. (1987). Transfer of experimental allergic encephalomyelitis to bone marrow chimeras. Endothelial cells are not a restricting element. *J. Exp. Med.* **166**, 1906.
50. Hirschberg, H., Evenson, S. A., Henricksen, T., and Thorsby, E. (1975). The human mixed lymphocyte–endothelium culture interaction. *Transplantation* **19**, 495.
51. Colson, Y. L., Markus, B. H., Zeevi, A., and Duquesnoy, R. J. (1990). Increased lymphocyte adherence to human arterial endothelial cell monolayers in the context of allorecognition. *J. Immunol.* **144**, 2975.
52. Geppert, T. D., and Lipsky, P. E. (1985). Antigen presentation by interferon-gamma-treated endothelial cells and fibroblasts: Differential ability to function as antigen-presenting cells despite comparable Ia expression. *J. Immunol.* **135**, 3750.

53. Fabry, Z., Waldschmidt, M. M., Moore, S. A., and Hart, M. N. (1990). Antigen presentation by brain microvessel smooth muscle and endothelium. *J. Neuroimmunol.* **28**, 63.
54. Nunez, G., Ball, E. J., and Stastry, P. (1983). Accessory cell function of human endothelial cells. I. A subpopulation of Ia positive cells is required for antigen presentation. *J. Immunol.* **131**, 666.
54a. Savage, C. O. S., Hughes, C. C. W., and Pober, J. S. (1991). In preparation.
55. Geppert, T. D., and Lipsky, P. E. (1987). Dissection of defective antigen presentation by interferon-gamma-treated fibroblasts. *J. Immunol.* **138**, 385.
56. Pryce, G., Male, D., and Sedgwick, J. (1989). Antigen presentation in brain: Brain endothelial cells are poor stimulators of T cell proliferation. *Immunology* **66**, 207.
57. Roska, A. K., Johnson, A. R., and Lipsky, P. E. (1984). Immunologic function of endothelial cells: Guinea pig aortic endothelial cells support mitogen-induced T lymphocyte activation, but do not function as antigen-presenting cells. *J. Immunol.* **132**, 136.
58. Schwartz, R. H. (1990). A cell culture model for T lymphocyte clonal anergy. *Science* **248**, 1349.
59. Guinan, E. C., Smith, B. R., Doukas, J. T., Miller, R. A., and Pober, J. S. (1989). Vascular endothelial cells enhance T cell responses by markedly augmenting IL-2 concentrations. *Cell. Immunol.* **118**, 166.
60. Hughes, C. C. W., Savage, C. O. S., and Pober, J. S. (1990). Endothelial cells augment T cell interleukin 2 production by a contact-dependent mechanism involving CD2/LFA-3 interaction. *J. Exp. Med.* **171**, 1453.
61. Savage, C. O. S., Hughes, C. C. W., Pepinsky, R. B., Wallner, B. P., Freedman, A. S., and Pober, J. S. (1991). Endothelial cell lymphocyte function-associated antigen-3 and an unidentified ligand act in concert to provide costimulation to human peripheral blood CD4$^+$ T cells. Submitted for publication.
62. Muegge, K., Williams, T. M., Kant, J., Karin, M., Chiu, R., Schmidt, A., Siebenlist, U., Young, H. A., and Durum, S. K. (1989). Interleukin-1 costimulatory activity on the interleukin-2 promoter via AP-1. *Science* **246**, 249.
63. Ransone, L. J., and Verma, I. M. (1990). Nuclear proto-oncogenes FOS and JUN. *Annu. Rev. Cell. Biol.* **6**, 539.
64. Nabholz, M., and MacDonald, H. R. (1983). Cytolytic T lymphocytes. *Annu. Rev. Immunol.* **1**, 273.
65. Lafferty, K. J., Prowse, S. J., Simeonovic, C. J., and Warren, H. S. (1983). Immunobiology of tissue transplantation: A return to the passenger leukocyte concept. *Annu. Rev. Immunol.* **1**, 143.
66. Weaver, C. T., and Unanue, E. R. (1990). The costimulatory function of antigen-presenting cells. *Immunol. Today* **11**, 49.
67. Ferry, B., Halttunen, J., Leszczynski, D., Schellekens, H., Meide, P. H. V. D., and Hayry, P. (1987). Impact of class II major histocompatibility complex antigen expression on the immunogenic potential of isolated rat vascular endothelial cells. *Transplantation* **44**, 499.
68. Wagner, C. R., Vetto, R. M., and Burger, D. R. (1985). Expression of I-region-associated antigen (Ia) and interleukin 1 by subcultured human endothelial cells. *Cell. Immunol.* **93**, 91.
69. Miossec, P., Cavender, D., and Ziff, M. (1986). Production of interleukin 1 by human endothelial cells. *J. Immunol.* **136**, 2486.
70. Nawroth, P. P., Bank, I., Handley, D., Cassimeris, J., Chees, L., and Stern, D.

(1986). Tumor necrosis factor/cachectin interacts with endothelial cell receptors to induce release of interleukin 1. *J. Exp. Med.* **163**, 1363.
71. Libby, P., Ordovas, J. M., Auger, K. R., Robbins, A. H., Birinyi, L. K., and Dinarello, C. A. (1986). Endotoxin and tumor necrosis factor induce interleukin 1 gene expression in adult human vascular endothelial cells. *Am. J. Pathol.* **124**, 179.
72. Kurt-Jones, E. A., Fiers, W., and Pober, J. S. (1987). Membrane interleukin 1 induction on human endothelial cells and dermal fibroblasts. *J. Immunol.* **139**, 2317.
73. Jirik, F. R., Podor, T. J., Hirano, T., Kishimoto, T., Loskutoff, D. J., Carson, D. A., and Lotz, M. (1989). Bacterial lipopolysaccharide and inflammatory mediators augment IL-6 secretion by human endothelial cells. *J. Immunol.* **142**, 144.
74. Sironi, M., Breviario, F., Proserpio, P., Biondi, A., Vecchi, A., Van Damme, J., Dejana, F., and Mantovani, A. (1989). IL-1 stimulates IL-6 production in endothelial cells. *J. Immunol.* **142**, 549.
75. Maier, J. A. M., Voulalas, P., Roeder, D., and Maciag, T. (1990). Extension of the life-span of human endothelial cells by an interleukin-1α antisense oligomer. *Science* **249**, 1870.
76. Cozzolino, F., Torcia, M., Aldinucci, D., Ziche, M., Almerigogna, F., Bani, D., and Stern, D. M. (1990). Interleukin 1 is an autocrine regulator of human endothelial cell growth. *Proc. Natl. Acad. Sci. U.S.A.* **87**, 6487.
77. Leeuwenberg, J. F. M., von Asmuth, E. J. V., Jeunhomme, T. M. A. A., and Buurman, W. A. (1990). IFN-γ regulates the expression of the adhesion molecule ELAM-1 and IL-6 production by human endothelial cells *in vitro. J. Immunol.* **145**, 2110.
78. Teitel, J. M., Shore, A., McBarron, J., Leary, P. L., and Schiavone, A. (1990). Endothelial cells modulate both T-cell-dependent and T-cell-independent plaque-forming cell generation *in vitro. Int. Arch. Allergy Appl. Immunol.* **91**, 66.
79. Shanahan, W. R., Jr., Weston, D. M., and Korn, J. H. (1985). Endothelial cells enhance the human pokeweed mitogen lymphocyte response. *J. Leukocyte Biol.* **37**, 305.
80. Hashimoto, Y., Nakano, K., Yoshinoya, S., Tanimoto, K., and Miyamoto, T. (1989). Accessory cell function of human vascular endothelial cells in pokeweed-mitogen-stimulated immunoglobulin production by peripheral blood lymphocytes. *Int. Arch. Allergy Appl. Immunol.* **89**, 11.
81. Strieter, R. M., Kunkel, S. L., Showell, H. J., Remick, D. G., Phan, S. H., Ward, P. A., and Marks, R. M. (1989). Endothelial cell gene expression of a neutrophil chemotactic factor by TNF-α, LPS and IL-1β. *Science* **243**, 1467.
82. Gimbrone, M. A., Jr., Obin, M. S., Brock, A. F., Luis, E. A., Hass, P. E., Hébert, C. A., Yip, Y. K., Leung, D. W., Lowe, D. G., Kohr, W. J., Darbonne, W. C., Bechtol, K. B., and Baker, J. B. (1989). Endothelial interleukin-8: A novel inhibitor of leukocyte–endothelial interactions. *Science* **246**, 1601.
83. Schroeder, J. M., and Christophers, E. (1989). Secretion of novel and homologous neutrophil-activating peptides by LPS-stimulated human endothelial cells. *J. Immunol.* **142**, 144.
84. Strieter, R. M., Wiggins, R., Phan, S. H., Wharram, B. L., Showell, H. J., Remick, D. G., Chensue, S. W., and Kunkel, S. L. (1989). Monocyte chemotactic protein gene expression by cytokine-treated human fibroblasts and endothelial cells. *Biochem. Biophys. Res. Commun.* **162**, 694.
85. Sica, A., Wang, J. M., Colotta, F., Dejana, E., Mantovani, A., Oppenheim, J. J.,

Larsen, C. G., Zachariae, C. O. C., and Matsushima, K. (1989). Monocyte chemotactic and activating factor gene expression induced in endothelial cells by IL-1 and tumor necrosis factor. *J. Immunol.* **144**, 3034.
86. Rollins, B. J., Yoshimura, T., Leonard, E. J., and Pober, J. S. (1990). Cytokine-activated human endothelial cells synthesize and secrete a monocyte chemoattractant, MCP-1/JE. *Am. J. Pathol.* **136**, 1229.
87. Broudy, V. C., Kaushansky, K., Segal, G. M., Harlan, J. M., and Adamson, J. W. (1986). Tumor necrosis factor type a stimulates human endothelial cells to produce granulocyte/macrophage colony-stimulating factor. *Proc. Natl. Acad. Sci. U.S.A.* **83**, 7467.
88. Munker, R., Gasson, J., Ogawa, M., and Koeffler, H. P. (1986). Recombinant human TNF induces production of granulocyte–monocyte colony-stimulating factor. *Nature (London)* **323**, 79.
89. Sieff, C. A., Tsai, S., and Faller, D. V. (1987). Interleukin 1 induces cultured human endothelial cell production of granulocyte–macrophage colony-stimulating factor. *J. Clin. Invest.* **79**, 48.
90. Broudy, V. C., Kaushansky, K., Harlan, J. M., and Adamson, J. W. (1988). Interleukin 1 stimulates human endothelial cells to produce granulocyte–macrophage colony-stimulating factor and granulocyte colony-stimulating factor. *J. Immunol.* **139**, 464.
91. Zsebo, K. M., Yuschenkoff, V. N., Schiffer, S., Chang, D., McCall, E., Dinarello, C. A., Brown, M. A., Altrock, B.,and Bagby, G. R., Jr. (1988). Vascular endothelial cells and granulopoiesis: Interleukin-1 stimulates release of G-CSF and GM-CSF. *Blood* **71**, 99.
92. Antonelli-Orlidge, A., Saunders, K., Smith, S., and D'Amore, P. A. (1989). An activated form of TGF-β is produced by co-cultures of endothelial cells and pericytes. *Proc. Natl. Acad. Sci. U.S.A.* **86**, 4544.
93. Weksler, B. B., Marcus, A. J., and Jaffe, E. A. (1977). Synthesis of prostaglandin I_2 (prostacyclin) by cultured human and bovine endothelial cells. *Proc. Natl. Acad. Sci. U.S.A.* **74**, 3922.
94. Charo, L. F., Shak, S., Karasek, M. A., Davison, P. M., and Goldstein, I. M. (1984). Prostaglandin I_2 is not a major metabolite of arachidonic acid in cultured endothelial cells from human foreskin microvessels. *J. Clin. Invest.* **74**, 914.
95. Whatley, R. E., Zimmerman, G. A., McIntyre, T. M., and Prescott, S. M. (1988). Endothelium from diverse vascular sources synthesizes platelet-activating factor. *Arteriosclerosis* **8**,321.
96. Lesley, J., Hyman, R., and Schulte, R. (1985). Evidence that the Pgp-1 glycoprotein is expressed on thymus-homing progenitor cells of the thymus. *Cell. Immunol.* **91**, 397.
97. Lesley, J., Trotter, J., and Hyman, R. (1985). The Pgp-1 antigen is expressed on early fetal thymocytes. *Immunogenetics* **22**, 149.
98. Horst, E., Meijer, C. J. L. M., Duijvestijn, A. M., Hartwig, N., van der Harten, K. J., and Pals, S. T. (1990). The ontogeny of human lymphocyte recirculation: High endothelial cell antigen (HECA-452) and CD44 homing receptor expression in the development of the immune system. *Eur. J. Immunol.* **20**, 1483.
99. Von Boehmer, H. (1988). The developmental biology of T lymphocytes. *Annu. Rev. Immunol.* **6**, 309.
100. Gallatin, W. M., Weissman, I. L., and Butcher, E. C. (1983). A lymphoid cell surface molecule involved in organ specific homing of lymphocytes. *Nature (London)* **304**, 30.
101. Camerini, D., James, S. P., Stamenkovic, I., and Seed, B. (1989). Leu8/TQ1 is the

human equivalent of the MEL-14 lymph node homing receptor. *Nature (London)* **342**, 78.
102. Tedder, T. F., Penta, A. C., Levine, H. B., and Freedman, A. S. (1990). Expression of the human leukocyte adhesion molecule, LAM-1. Identity with the TQ1 and Leu8 differentiation antigens. *J. Immunol.* **144**, 532.
103. Laskey, L. A., Singer, M. S., Yednock, A., Dowbenko, D., Fennie, C., Rodriguez, H., Nguyen, T., Stachel, S., and Rosen, S. D. (1989). Cloning of a lymphocyte homing receptor reveals a lectin domain. *Cell* **56**, (Cambridge, Mass.) 1045.
104. Siegelman, M. H., van de Rijn, M., and Weissman, I. L. (1989). Mouse lymph node homing receptor cDNA clone encodes a glycoprotein revealing tandem interaction domains. *Science* **243**, 1165.
105. Bevilacqua, M. P., Stengelin, S., Gimbrone, M. A., Jr., and Seed, B. (1989). Endothelial leukocyte adhesion molecule 1: An inducible receptor for neutrophils related to complement regulatory proteins and lectins. *Science* **243**, 1160.
106. Johnston, G. I., Cook, R. G., and McEver, R. P. (1989). Cloning of GMP-140, a granule membrane protein of platelets and endothelium: Sequence similarity to proteins involved in cell adhesion and inflammation. *Cell* **56**, (Cambridge, Mass.) 1033.
107. Rosen, S. D., Singer, M. S., Yednock, T. A., and Stoolman, L. M. (1985). Involvement of sialic acid on endothelial cells in organ-specific lymphocyte recirculation. *Science* **228**, 1005.
108. Stoolman, L. M., Tenforde, T. S., and Rosen, S. D. (1984). Phosphomannosyl receptors may participate in the adhesive interaction between lymphocytes and high endothelial venules. *J. Cell Biol.* **99**, 1535.
109. Imai, Y., True, D. D., Singer, M. S., and Rosen, S. D. (1990). Direct demonstration of the lectin activity of gp90 MEL, a lymphocyte homing receptor. *J. Cell Biol.* **111**, 1225.
110. Mackay, C. R., Marston, W. L., and Dudler, L. (1990). Naive and memory T cells show distinct pathways of lymphocyte recirculation. *J. Exp. Med.* **171**, 801.
111. Gowans, J. L., and Knight, E. J. (1964). The route of recirculation of lymphocytes in the rat. *Proc. R. Soc. London, Ser. B* **159**, 257.
112. Marchesi, V. T., and Gowans, J. L. (1964). The migration of lymphocytes through the endothelium of venules in lymph nodes: An electron microscopic study. *Proc. R. Soc. London, Ser. B* **159**, 283.
113. Miller, J. J., III (1969). Studies of the phylogeny and ontogeny of the specialized lymphatic venules. *Lab. Invest.* **21**, 284.
114. Andrews, P., Milsom, D. W., and Ford, W. L. (1982). Migration of lymphocytes across specialized vascular endothelium. V. Production of a sulphated macromolecule by high endothelial cells in lymph nodes. *J. Cell Sci.* **57**, 277.
115. Duijvestijn, A. M., Horst, E., Pals, S. T., Rouse, B. N., Steare, A. C., Picker, L. J., Meijer, C. J. L. M., and Butcher, E. C. (1988). High endothelial differentiation in human lymphoid and inflammatory tissues defined by monoclonal antibody HECA-452. *Am. J. Pathol.* **130**, 147.
116. Montesano, R., Mossaz, A., Ryser, J. E., Orci, L., and Vassalli, P. (1984). Leukocyte interleukins induce cultured endothelial cells to produce a highly organized, glycosaminoglycan-rich pericellular matrix. *J. Cell Biol.* **99**, 1706.
117. Horst, E., Pals, S. T., Duijvestijn, A. M., van Mourik, J., Kraal, G., Butcher, E. C., and Meijer, C. L. J. M. (1988). Expression and regulation of an antigen specific for endothelium involved in human lymphocyte homing. *Adv. Exp. Med. Biol.* **237**, 477.
118. Stolpen, A. H., Guinan, E. C., Fiers, W., and Pober, J. S. (1986). Recombinant tumor

necrosis factor and immune interferon act singly and in combination to reorganize human vascular endothelial cell monolayers. *Am. J. Pathol.* **123**, 16.
119. Manolios, N., Geczy, C. L., and Schreiber, L. (1988). High endothelial venule morphology and function are inducible in germ free mice: A possible role for interferon-γ. *Cell. Immunol.* **117**, 136.
120. Bjerknes, M., Cheng, H., and Ottaway, C. A. (1986). Dynamics of lymphocyte-endothelial interactions *in vivo*. *Science* **231**, 402.
121. Stamper, H. B., Jr., and Woodruff, J. J. (1976). Lymphocyte homing into lymph nodes: *In vitro* demonstration of selective affinity of recirculating lymphocytes for high-endothelial venules. *J. Exp. Med.* **144**, 828.
122. Watson, S. R., Imai, Y., Fennie, C., Geoffrey, J. S., Rosen, S. D., and Lasky, L. A. (1990). A homing receptor-IgG chimera as a probe for adhesive ligands of lymph node high endothelial venules. *J. Cell Biol.* **110**, 2221.
123. Streeter, P. R., Rouse, B. T. N., and Butcher, E. C. (1988). Immunohistologic and functional characterization of a vascular addressin involved in lymphocyte homing into peripheral lymph nodes. *J. Cell Biol.* **107**, 1853.
124. Berg, E. L., Goldstein, L. A., Jutila, M. A., Nakache, M., Picker, L. J., Streeter, P. R., Wu, N. W., Zhou, D., and Butcher, E. C. (1989). Homing receptors and vascular addressins: Cell adhesion molecules that direct lymphocyte traffic. *Immunol. Rev.* **108**, 5.
125. Sanders, M. E., Makgoba, M. W., and Shaw, S. (1988). Human naive and memory T cells: Reinterpretation of helper-inducer and suppressor-inducer subsets. *Immunol. Today* **9**, 195.
126. Shimizu, Y., van Seventer, G. A., Horgen, K. J., and Shaw, S. (1990). Regulated expression and binding of three VLA (β1) integrin receptors on T cells. *Nature (London)* **345**, 250.
127. Cerottini, J.-C., and MacDonald, H. R. (1989). The cellular basis of T cell memory. *Annu. Rev. Immunol.* **7**, 77.
128. Picker, L. J., Terstappen, L. W. M. M., Rott, L. S., Streeter, P. R., Stein, H., and Butcher, E. C. (1990). Differential expression of homing-associated adhesion molecules by T cell subsets in man. *J. Immunol.* **145**, 3247.
129. Issekutz, T. B., Stoltz, J. M., and Meide, P. (1988). Lymphocyte recruitment in delayed-type hypersensitivity. The role of gamma interferon. *J. Immunol.* **140**, 2989.
130. Pober, J. S., Bevilacqua, M. P., Mendrick, D. L., Lapierre, L. A., Fiers, W., and Gimbrone, M. A., Jr. (1986). Two distinct monokines, interleukin-1 and tumor necrosis factor, each independently induce biosynthesis and transient expression of the same antigen on the surface of cultured human vascular endothelial cells. *J. Immunol.* **136**, 1680.
131. Bevilacqua, M. P., Pober, J. S., Mendrick, D. L., Cotran, R. S., and Gimbrone, M. A., Jr. (1987). Identification of an inducible endothelial-leukocyte adhesion molecule. *Proc. Natl. Acad. Sci. U.S.A.* **84**, 9238.
132. Doukas, J., and Pober, J. S. (1990). IFN-γ enhances endothelial activation induced by TNF but not IL-1. *J. Immunol.* **145**, 1727.
133. Thornhill, M. H., and Haskard, D. O. (1990). IL-4 regulates endothelial cell activation by IL-1, tumor necrosis factor, or IFN-γ. *J. Immunol.* **145**, 865.
134. Lowe, J. B., Stoolman, L. M., Nair, R. P., Larsen, R. D., Berhend, T. L., and Marks, R. M. (1990). ELAM-1-dependent cell adhesion to vascular endothelium determined by a transfected human fucosyl transferase cDNA *Cell (Cambridge, Mass.)* **63**, 475.

135. Walz, G., Anuffo, A., Kolanus, W., Bevilacqua, M., and Seed, B. (1990). Recognition by ELAM-1 of the sialyl-Lex determinant on myeloid and tumor cells. *Science* **250**, 1132.
136. Goetz, S. E., Hession, C., Goff, D., Griffiths, B., Tizard, R., Newman, B., Chi-Rosso, G., and Lobb, R. (1990) ELFT: A gene that directs the expression of an ELAM-1 ligand. *Cell (Cambridge, Mass.)* **63**, 1349.
137a. Graber, N., Gopal, T. V., Wilson, D., Beall, L. D., Polte, T., and Newman, W. (1990). T cells bind to cytokine-activated endothelial cells via a novel, inducible sialoglycoprotein and endothelial leukocyte adhesion molecule-1. *J. Immunol.* **145**, 819.
137b. Shimizu, Y., Shaw, S., Graber, N., Gopan, T. V., Horgan, K. J., Van Seventer, G. A., and Newman, W. (1991). Activation-independent binding of human memory T-cells to adhesion molecule ELAM-1. *Nature (London)* **349**, 796.
137c. Picker, L. J., Kishimoto, T. K., Smith, C. W., Warnock, R. A., and Butcher, E. C. (1991). ELAM-1 is an adhesion molecule for skin-homing T-cells. *Nature (London)* **349**, 793.
138. Rice, G. E., and Bevilacqua, M. P. (1989). An inducible endothelial cell surface glycoprotein mediates melanoma adhesion. *Science* **246**, 1303.
139. Osborn, L., Hession, C., Tizard, R., Vassallo, C., Luhowskyj, S., Chi-Rosso, G., and Lobb, R. (1989). Direct cloning of vascular cell adhesion molecule 1, a cytokine-induced endothelial protein that binds to lymphocytes. *Cell (Cambridge, Mass.)* **59**, 1203.
140. Masinovsky, B., Urdal, D., and Gallatin, M. (1990). IL-4 acts synergistically with IL-1β to promote lymphocyte adhesion to microvascular endothelium by induction of vascular cell adhesion molecule-1. *J. Immunol.* **145**, 2886.
141. Rice, G. E., Munro, J. M., Corless, C., and Bevilacqua, M. P. (1991). Vascular and nonvascular expression of INCAM-110: A target for mononuclear leukocyte adhesion in normal and inflamed human tissue. *Am. J. Pathol.* **138**, 385.
141a. Cybulsky, M. I., Fries, J. W. V., Williams, A. J., Sultan, P., Davis, V. M., Gimbrone, M. A., Jr., and Collins, T. (1991). Alternative splicing of human VCAM-1 in activated vascular endothelium. *Am. J. Pathol.* **138**, 815.
142. Elices, M. J., Osborn, L., Takada, Y., Crouse, C., Luhowskyj, S., Hemler, M. E., and Lobb, R. R. (1990). VCAM-1, an activated endothelium, interacts with the leukocyte integrin VLA-4 at a site distinct from the VLA-4/fibronectin binding site. *Cell (Cambridge, Mass.)* **60**, 577.
143. Schwartz, B. R., Wagner, E. A., Carlos, T. M., Ochs, H. D., and Harlan, J. M. (1990). Identification of surface proteins mediating adherence of CD11/CD18-deficient lymphoblastoid cells to cultured human endothelium. *J. Clin. Invest.* **85**, 2019.
144. Rothlein, R., Dustin, M. L., Martin, S. D., and Springer, T. A. (1986). A human intercellular adhesion molecule (ICAM-1) distinct from LFA-1. *J. Immunol.* **137**, 1270.
145. Pober, J. S., Gimbrone, M. A., Jr., Lapierre, L. A., Mendrick, D. L., Fiers, W., Rothlein, R., and Springer, T. A. (1986). Overlapping patterns of activation by human endothelial cells by interleukin 1, tumor necrosis factor and immune interferon. *J. Immunol.* **137**, 1893.
146. Dustin, M. L., Rothlein, R., Bhan, A. K., Dinarello, C. A., and Springer, T. A. (1986). Induction by IL-1 and interferon, tissue distribution, biochemistry and function of a natural adherence molecule (ICAM-1). *J. Immunol.* **137**, 245.
147. Staunton, D. E., Marlin, S. D., Stratowa, C., Dustin, M. L., and Springer, T. A. (1988). Primary structure of intercellular adhesion molecule 1 (ICAM-1) demon-

strates interaction between members of the immunoglobulin and integrin supergene families. *Cell (Cambridge, Mass.)* **52**, 925.
148. Simmons, D., Makgoba, M. W., and Seed, B. (1988). ICAM, an adhesion ligand of LFA-1, is homologous to the neural cell adhesion molecule NCAM. *Nature (London)* **331**, 624.
149. Staunton, D. E., Dustin, M. L., and Springer, T. A. (1989). Functional cloning of ICAM-2, a cell adhesion ligand to LFA-1 homologous to ICAM-1. *Nature (London)* **339**, 61.
150. Dustin, M. L., and Springer, T. A. (1988). Lymphocyte function-associated antigen-1 (LFA-1) interaction with intercellular adhesion molecule-1 (ICAM-1) is one of at least three mechanisms for lymphocyte adhesion to cultured endothelial cells. *J. Cell Biol.* **107**, 321.
151. Diamond, M. S., Staunton, D. E., de Fougerolles, A. R., Stacker, S. A., Garcia-Aguilar, J., Hibbs, M. L., and Springer, T. A. (1990). ICAM-1 (CD54): A counter-receptor for Mac-1 (CD11b/CD18). *J. Cell Biol.* **111**, 3129.
152. Dustin, M. L., and Springer, T. A. (1989). T cell receptor cross-linking transiently stimulates adhesiveness through LFA-1. *Nature (London)* **341**, 619.
153. Van Kooyk, Y., Van de Wiel-van Kemenada, P., Weder, P., Kuijpers, T. W., and Figdor, C. G. (1989). Enhancement of LFA-1 mediated cell adhesion by triggering through CD2 or CD3 on T lymphocytes. *Nature (London)* **342**, 811.
154. Smith, C. W., Rothlein, R., Hughes, B. J., Marisclaco, M. M., Rudloff, H. E. Schmalstieg, F. C., and Anderson, D. C. (1988). Recognition of an endothelial determinant for CD18-dependent human neutrophil adherence and transendothelial migration. *J. Clin. Invest.* **82**, 1746.
155. Smith, C. W., Marlin, S. D., Rothlein, R., Toman, C., and Anderson, D. C. (1989). Cooperative interactions of LFA-1 and Mac-1 with intercellular adhesion molecule-1 in facilitating adherence and transendothelial migration of human neutrophils in vitro. *J. Clin. Invest.* **83**, 2008.
156. Van Epps, D. E., Potter, J., Vachula, M., Smith, C. W., and Anderson, D. C. (1989). Suppression of human lymphocyte chemotaxis and transendothelial migration by anti-LFA-1 antibody. *J. Immunol.* **143**, 3207.
157. Oppenheimer-Marks, N., Davis, L. S., and Lipsky, P. E. (1990). Human T lymphocyte adhesion to endothelial cells and transendothelial cell migration. Alteration of receptor use relates to the activation studies of both the T cell and the endothelial cell. *J. Immunol.* **145**, 140.
158. Anderson, D. C., Schmalstieg, E. C., Finegold, M. J., Hughes, B. J., Rothlein, R., Miller, L. J., Kohl, S., Tosi, M. F., Jacobs, R. L., Waldrop, T. C., Goldman, A. S., Shearer, W. T., and Springer, T. A. (1985). The severe and moderate phenotypes of heritable Mac-1, LFA-1 deficiency: Their quantitative definition and relation to leukocyte dysfunction and clinical features. *J. Infect. Dis.* **152**, 668.
159. Masuyama, J., Minato, N., and Kano, S. (1986). Mechanisms of lymphocyte adhesion to human vascular endothelial cells in culture. T lymphocyte adhesion to endothelial cells through endothelial HLA-DR antigens induced by γ interferon. *J. Clin. Invest.* **77**, 1596.
160. Yu, C.-L., Haskard, D., Cavender, D., and Ziff, M. (1986). Effects of bacterial lipopolysaccharide on the binding of lymphocytes to endothelial cell monolayers. *J. Immunol.* **136**, 569.
161. Hughes, C. C. W., Male, D. K., and Lantos, P. L. (1988). Adhesion of lymphocytes to cerebral microvascular cells: Effects of interferon-γ, tumour necrosis factor and interleukin-1. *Immunology* **64**, 677.

162. Thornhill, M. H., Williams, D. H., and Speight, P. M. (1989). Enhanced adhesion of autologous lymphocytes to γ-interferon-treated human endothelial cells *in vitro*. *Br. J. Exp. Pathol.* **70**, 59.
163. Nickoloff, B. J., Reusch, M. K., Bensch, K., and Karasek, M. A. (1988). Preferential binding of monocytes and Leu2$^+$ T lymphocytes to interferon-gamma treated cultured skin endothelial cells and keratinocytes. *Arch. Dermatol. Res.* **280**, 235.
164. Haskard, D. O., Cavender, D. E., Beatty, P., Springer, T. A., and Ziff, M. (1986). T lymphocyte adhesion to endothelial cells: Mechanisms demonstrated by anti-LFA-1 monoclonal antibodies. *J. Immunol.* **137**, 2901.
165. Rice, G. E., Munro, J. M., and Bevilacqua, L. P. (1990). Inducible cell adhesion molecule 110 (INCAM-110) is an endothelial receptor for lymphocytes. A CD11/CD18-independent adhesion mechanism. *J. Exp. Med.* **171**, 1369.
166. Thornhill, M. H., Kyan-Aurg, U., and Haskard, D. O. (1990). IL-4 increases human endothelial cell adhesiveness for T cells but not for neutrophils. *J. Immunol.* **144**, 3060.
167. Cosimi, A. B., Conti, D., Delmonico, F. L., Preffer, F. I., Wee, S.-L., Rothlein, R., Faanes, R., and Colvin, R. B. (1990). *In vivo* effects of monoclonal antibody to ICAM-1 (CD54) in nonhuman primates with renal allografts. *J. Immunol.* **144**, 4604.
168. DeBono, D. (1981). Lymphocyte interactions with vascular endothelium. In "Cellular Interactions" J. T. Dingle, and J. L. Gordon, eds.), pp. 97–105. Elsevier, New York.
169. Cavender, D. E., Haskard, D. O., Joseph, B., and Ziff, M. (1986). Interleukin 1 increases the binding of B and T lymphocytes to endothelial cell monolayers. *J. Immunol.* **136**, 203.
170. Cavender, D., Saegusa, Y., and Ziff, M. (1987). Stimulation of endothelial cell binding of lymphocytes by tumor necrosis factor. *J. Immunol.* **139**, 1855.
171. Issekutz, T. B. (1990). Effects of six different cytokines on lymphocyte adherence to microvascular endothelium and on *in vivo* lymphocyte migration in the rat. *J. Immunol.* **144**, 2140.
172. Bender, J., Pardi, R., Karasek, M., and Engleman, E. (1987). Phenotypic and functional characterization of lymphocytes that bind microvascular endothelial cells *in vitro*: Evidence for preferential binding of natural killer cells. *J. Clin. Invest.* **79**, 1679.
173. Damle, N. K., and Doyle, L. V. (1990). Ability of human T lymphocytes to adhere to vascular endothelial cells and to augment endothelial permeability to macromolecules is linked to their state of post-thymic maturation. *J. Immunol.* **144**, 1233.
174. Willms-Kretschmer, K., Flax, M. H., and Cotran, R. S. (1967). The fine structure of the vascular response in hapten-specific delayed hypersensitivity and contact dermatitis. *Lab. Invest.* **17**, 334.
175. Freemont, A. J., and Ford, W. L. (1985). Functional and morphological changes in post-capillary venules in relation to lymphocytic inflammation into BCG-induced granulomata in rat skin. *J. Pathol.* **147**, 1.
176. Freemont, A. J. (1988). Functional and biosynthetic changes in endothelial cells of vessels in chronically inflamed tissues: evidence for endothelial control of lymphocytes' entry into diseased tissues. *J. Pathol.* **155**, 225.
177. Duijvestijn, A. M., Kerkhove, M., Bargatze, R. F., and Butcher, E. C. (1987). Lymphoid tissue- and inflammation-specific endothelial cell differentiation defined by monoclonal antibodies. *J. Immunol.* **138**, 713.

178. Jalkanen, S., Steere, A. C., Fox, R. I., and Butcher, E. C. (1986). A distinct endothelial cell recognition system that controls lymphocyte traffic into inflamed synovium. *Science* **233**, 556.
179. Sachstein, R., Falanga, V., Streilein, J. W., and Chin, Y. H. (1988). Lymphocyte adhesion to psoriatic dermal endothelium is mediated by a tissue-specific receptor/ligand interaction. *J. Invest. Dermatol.* **91**, 423.
180. Renkonen, R., Turunen, J. P., Rapola, J., and Hayry, P. (1990). Characterization of high endothelial-like properties of peritubular capillary endothelium during acute renal allograft rejection. *Am. J. Pathol.* **137**, 643.
181. Cotran, R. S., and Pober, J. S. (1988). Endothelial activation: Its role in inflammatory and immune reactions. *In* "Endothelial Cell Biology" (N. Simionescu and M. Simionescu, eds.), p. 335. Plenum, New York.
182. Goerdt, S., Zwaldo, G., Schlegel, R., Hagemeier, H.-H., and Sorg, C. (1987). Characterization and expression kinetics of an endothelial activation antigen present *in vivo* only in acute inflammatory tissues. *Exp. Cell Biol.* **55**, 117.
183. Lewis, R. E., Buchsbaum, M., Whitaker, D., and Murphy, G. F. (1989). Intracellular adhesion molecule expression in the evolving human cutaneous delayed hypersensitivity reaction. *J. Invest. Dermatol.* **93**, 672.
184. Leung, D. Y. M., Pober, J. S., and Cotran, R. S. (1991). Expression of endothelial leukocyte adhesion molecule-1 (ELAM-1) in elicited late phase allergic reactions. *J. Clin. Invest.* (in press).
185. Briscoe, D. M., Schoen, F. J., Rice, G. E., Bevilacqua, M. P., Ganz, P., and Pober, J. S. (1991). Induced expression of endothelial–leukocyte adhesion molecules in human cardiac allografts. *Transplantation* (in press).
186. Cotran, R. S., Gimbrone, M. A., Jr., Bevilacqua, M. P., Mendrick, D. L., and Pober, J. S. (1986). Induction and detection of a human endothelial activation antigen *in vivo*. *J. Exp. Med.* **164**, 661.
187. Ruco, L. P., Pomponi, D., Pigott, R., Stoppacciaro, A., Monardo, F., Uccini, S., Boraschi, D., Tagliabue, A., Santoni, A., Dejana, E., Mantovani, A., and Baroni, C. D. (1990). Cytokine production (IL-1α, IL-1β and TNF-α) and endothelial cell activation (ELAM-1 and HLA-DR) in reactive lymphodenitis, Hodgkin's disease and in non-Hodgkin's lymphomas: An immunocytochemical study. *Am. J. Pathol.* **137**, 1173.
188. Pober, J. S., and Cotran, R. S. (1991). What can be learned from the expression of endothelial adhesion molecules in tissues? *Lab. Invest.* (in press).
189. Butcher, E. C., Scollay, R. G., and Weissman, I. L. (1980). Organ specificity of lymphocyte migration: Mediation by highly selective lymphocyte interaction with organ-specific determinants on high endothelial venules. *Eur. J. Immunol.* **10**, 556.
190. Chin, Y. H., Rasmussen, R., Cakiroglu, A. K., and Woodruff, J. J. (1984). Lymphocyte recognition of lymph node high endothelium. VI. Evidence for distinct structures mediating binding to high endothelial cells of lymph nodes and Peyer's patches. *J. Immunol.* **133**, 2961.
191. Streeter, P. R., Berg, E. L., Rouse, B. N., Bargatze, R. F., and Butcher, E. C. (1988). A tissue-specific endothelial cell molecule involved in lymphocyte homing. *Nature (London)* **331**, 41.
192. Chin, Y. H., Rasmussen, R. A., Woodruff, J. J., and Easton, T. G. (1986). A monoclonal anti-HEBF-PP antibody with specificity for lymphocyte surface molecules mediating adhesion to Peyer's patch high endothelium of the rat. *J. Immunol.* **136**, 2556.

193. Holzmann, B., McIntyre, B. W., and Weissman, I. L. (1989). Identification of a murine Peyer's patch-specific lymphocyte homing receptor as an integrin molecule with an α chain homologous to human VLA-4 α. *Cell (Cambridge, Mass.)* **56**, 37.
194. Holzmann, B., and Weissman, I. L. (1989). Peyer's patch-specific lymphocyte homing receptors consist of a VLA-4 like α chain associated with either of two integrin β chains, one of which is novel. *EMBO J.* **8**, 1735.
195. Jalkanen, S., Bargatze, R. F., de los Toyos, J., and Butcher, E. C. (1987). Lymphocyte recognition of high endothelium: Antibodies to distinct epitopes of an 85–95 kD glycoprotein antigen differentially inhibit lymphocyte binding to lymph node, mucosal or synovial endothelial cells. *J. Cell Biol.* **105**, 983.
196. Picker, L. J., de los Toyos, J., Telen, M. J., Haynes, B. F., and Butcher, E. C. (1989). Monoclonal antibodies against the CD44 [In (Lu)-related p80], and Pgp-1 antigens in man recognize the Hermes class of lymphocyte homing receptors. *J. Immunol.* **142**, 2046.
197. Van der Brugge-Gamelkoorn, G. J., and Kraal, G. (1985). The specificity of the high endothelial venule in bronchus-associated lymphoid tissue (BALT). *J. Immunol.* **134**, 3746.
198. Picker, L. J., Michie, S. A., Rott, L. S., and Butcher, E. C. (1990). A unique phenotype of skin-associated lymphocytes in humans: Preferential expression of the HECA 452 epitope by benign and malignant T cells at cutaneous sites. *Am. J. Pathol.* **136**, 1053.
199. Cerf-Bensussan, N., Jarry, A., Brousse, N., Lisowska-Grospierre, B., Guy-Grand, D., and Griscelli, C. (1987). A monoclonal antibody (HML-1) defining a novel membrane molecule present on human intestinal lymphocytes. *Eur. J. Immunol.* **17**, 1279.
200. Guinan, E. C., Smith, B. R., Davies, P. F., and Pober, J. S. (1988). Cytoplasmic transfer between endothelium and lymphocytes: Quantitation by flow cytometry. *Am. J. Pathol.* **132**, 406.
201. Raine, C. S., Cannella, B., Duijvestijn, A. M., and Cross, A. H. (1990). Homing to central nervous system vasculature by antigen-specific lymphocytes. II. Lymphocyte/endothelial cell adhesion during the initial stage of autoimmune demyelination. *Lab. Invest.* **63**, 476.
202. Renkonen, R., Matilla, P., Turenen, J. P., and Hayry, P. (1989). Lymphocyte binding to and penetration through vascular endothelium is stimulated by platelet-activating factor. *Scand. J. Immunol.* **30**, 673.
203. Bussolino, F., Breviario, F., Telta, C., Aglietta, M., Mantovani, A., and Dejana, E. (1986). Interleukin 1 stimulates platelet-activating factor production in cultured human endothelial cells. *J. Clin. Invest.* **77**, 2027.
204. Bussolino, F., Camussi, G., and Baglioni, C. (1988). Synthesis and release of platelet-activating factor by human vascular endothelial cells treated with tumor necrosis factor or interleukin 1a. *J. Biol. Chem.* **263**, 11856.
205. Zavoico, G. B., Ewenstein, B. M., Schafer, A. I., and Pober, J. S. (1989). Interleukin-1 and related cytokines enhance thrombin-stimulated PGI2 production in cultured endothelial cells without affecting thrombin-stimulated von Willebrand factor secretion or platelet-activating factor biosynthesis. *J. Immunol.* **142**, 3993.
206. Pober, J. S., and Cotran, R. S. (1990). The role of endothelial cells in inflammation. *Transplantation* **50**, 537.
207. Luster, A. D., Unkeless, J. C., and Ravetch, J. V. (1985). γ-interferon transcriptionally regulates an early-response gene containing homology to platelet proteins. *Nature (London)* **315**, 672.

208. Haskill, S., Peace, A., Morris, J., Sporn, S. A., Anisowicz, A., Lee, S. W., Smith, T., Martin, G., Ralph, P., and Sager, R. (1990). Identification of three related human GRO genes encoding cytokine functions. *Proc. Natl. Acad. Sci. U.S.A.* **87**, 7732.
209. Standiford, T. J., Strieter, R. M., Kasahara, K., and Kunkel, S. L. (1990). Disparate regulation of interleukin 8 gene expression from blood monocytes, endothelial cells, and fibroblasts by interleukin 4. *Biochem. Biophys. Res. Commun.* **171**, 531.
210. Larsen, C. G., Anderson, A. O., Oppenheim, J. J., and Matsushima, K. (1989). The neutrophil-activating protein (NAP-1) is also chemotactic for T lymphocytes. *Science* **243**, 1464.
211. Rollins, B. J., and Pober, J. S. (1991). IL-4 induces the synthesis and secretion of MCP-1/JE by human endothelial cells. *Am. J. Pathol.* (in press).
212. Male, D., Pryce, G., Hughes, C. C. W., and Lantos, P. (1990). Lymphocyte migration into brain modelled *in vitro:* Control by lymphocyte activation, cytokines, and antigen. *J. Cell. Immunol.* **127**, 1.

This article was accepted for publication on 4 March 1991.

Adoptive Transfer of Human Lymphoid Cells to Severely Immunodeficient Mice: Models for Normal Human Immune Function, Autoimmunity, Lymphomagenesis, and AIDS

DONALD E. MOSIER

Division of Immunology, Medical Biology Institute, La Jolla, California 92037

I. Introduction

The search for better animal models for human disease has led to the development of several new experimental systems that employ the transfer of human lymphoid cells or hematopoietic tissue to mice with profound immunodeficiency (Mosier *et al.*, 1988; McCune *et al.*, 1988; Kamel-Reid and Dick, 1988). This review will focus on experiments involving transfer of human lymphoid cells from normal donors or those suffering from autoimmune disorders to mice with severe combined immune deficiency (SCID); recent reviews (McCune, 1991; Dick, 1991) have dealt extensively with results of grafting SCID mice with fragments of human fetal tissue and hematopoietic precursor development in immunodeficient mice.

The concept of establishing a human-to-mouse xenograft model is firmly rooted in transplantation biology; what is needed is to avoid a rejection of the xenograft by the recipient or of the recipient by the xenograft (xenogeneic graft-versus-host disease). It is important to note at the outset that barriers to xenotransplantation are different from the more widely studied barriers to allotransplantation. In particular, rejection of cellular xenografts may involve antibody and natural killer (NK) cells to a larger extent than T cell-mediated allograft rejection (e.g., Sharabi *et al.*, 1990). Attempts at human-to-mouse xenotransplantation are not new. The discovery of the *nude* mutation, which rendered mice T cell deficient, led to a burst of studies involving transplantation of human tissues to nude mice. The general outcome of those studies was that transformed human cells could grow in nude mice, but that normal human cells could not (Fogh *et al.*, 1977; Nilsson *et al.*, 1977; Phillips *et al.*, 1989). The discovery of the *scid* mutation (Bosma *et al.*, 1983, 1988; Custer *et al.*, 1985; Schuler *et al.*, 1986, 1990; Malynn *et al.*, 1988) and the development of mice strains with multiple mutations affecting immune function (Andriole *et al.*, 1985) were critical antecedents to the observations that these severely immunodeficient mouse strains were permissive for the growth of normal as well as transformed human cells (Ware et al., 1985; Kamel-Reid et al., 1989). The goal of this review is to summarize the current status of experiments in this rapidly burgeoning field, and to explore the several unanswered issues raised by the success of these experiments.

II. Transfer of Normal or Autoimmune Human Cells to Severely Immunodeficient Mice

The first human cell suspensions transferred to SCID mice were normal adult peripheral blood leukocytes (PBLs) separated from whole blood by density centrifugation on Ficoll–Hypaque gradients (Mosier et al., 1988). Establishment of human immunoglobulin (Ig) production in these hu-PBL-SCID mice[1] was observed with intraperitoneal but not intravenous injection. Transfer of 10 to 100 × 10^6 PBLs routinely results in human Ig production in SCID recipients (Mosier et al., 1988, 1989a, 1989b; Bankert et al., 1989; Krams et al., 1989; Tighe et al., 1990; Duchosal et al., 1990; Markham et al., 1990), but injection of 5 × 10^6 PBLs reconstitutes only a small fraction of animals. This observation may mean that only a fraction of injected human cells persist, that collaboration between relatively rare human cells contributes to successful reconstitution, or that large numbers of human cells are required to overcome host resistance. All of these factors could contribute to the interaction between the transferred human cells and the SCID mouse host. With the injection of 20 × 10^6 PBLs or more, the success rate of engraftment approaches 100%. In the original experiments, we also observed that hu-PBL-SCID mice generated by the injection of 50 × 10^6 PBLs from donors seropositive for antibodies to Epstein–Barr virus (EBV) developed B cell lymphomas (or lymphoproliferative disease; see later) of human origin that expressed EBV genomic sequences (Mosier et al., 1988, 1989c; Rowe et al., 1991).

McCune et al. (1988) grafted fetal thymus fragments under the kidney capsule of SCID mice, and subsequently injected suspensions of 10 × 10^6 autologous fetal liver cells intravenously into the mice. Production of human T cells was transient with this protocol (McCune et al., 1988; Namikawa et al., 1990), and a more sustained generation of T cells was seen following grafting of adjacent fragments of fetal thymus

---

[1] A brief word on proposed nomenclature: we have called SCID mice injected with human peripheral blood leukocytes hu-PBL-SCID, following the convention of species–tissue–recipient. By the same convention, *beige.nude.xid* mice injected with human bone marrow would be called hu-BM-BNX mice rather than HID (human immunodeficient) mice, and SCID mice grafted with human fetal liver and thymus grafts would be called hu-FLT-SCID rather than SCID-hu. Until there is a wider agreement on these conventions, however, the current usage will be continued.

and fetal liver under the renal capsule (Namikawa et al., 1990). The success rate of human engraftment with this protocol is about 50%. Fetal lymph nodes have also been grafted into SCID mice (McCune, 1990, 1991; McCune et al., 1990a) and have demonstrated persistence of human cells and lymphoid architecture. Vessels of mouse origin invade the tissue grafts, and dendritic cells of mouse origin appear to seed human thymus grafts (Barry et al., 1991), so the human tissue is itself xenochimeric. The simultaneous expression of human and mouse major histocompatibility (MHC) antigens in this and other SCID transfer models has important implications for T cell development and antigen presentation.

Kamel-Reid and Dick (1988) injected intravenously suspensions of 7 to 10×10^6 human bone marrow cells into severely immunodeficient *beige.nude.xid* (BNX) mice previously conditioned with 400 cGy whole-body irradiation. This protocol resulted in the persistence of human cells and the recovery of human macrophage colony-forming unit precursors (hu-CFU-M) for at least 5 weeks. These experiments compared irradiated SCID recipients to irradiated BNX recipients and found BNX recipients to be superior for survival of human marrow progenitors. The authors argued that differences in NK or lymphokine-activated killer (LAK) cell function might explain their results, because BNX mice are more deficient in NK and LAK activity than are SCID mice (Andriole et al., 1985; Dorshkind et al., 1984; Hackett et al., 1986; Habu et al., 1981). This point will be further addressed below.

Bankert et al. (1989) reported successful transfer to SCID mice of both normal PBLs and tumor-infiltrating lymphocytes (TILs) from patients with adenocarcinoma of the lung. Though the levels of human Ig were similar to those seen in the original report from this laboratory (Mosier et al., 1988), most hu-PBL-SCID mice were reported to die of "graft-versus-host disease" (GVHD) at 12–14 weeks postinjection. The described histopathology and the timing of death were also consistent with the development of EBV-associated B cell lymphoproliferative disease, so it is difficult to unambiguously assign GVHD as the cause of death. In our laboratory, over 800 hu-PBL-SCID mice have been constructed using 20 to 50×10^6 PBLs from EBV-seronegative donors. We have observed no human B cell lymphomas or deaths in these animals, and clinical symptoms consistent with transient GVHD were observed on only one occasion, involving eight animals reconstituted with the same donor. Over 2000 hu-PBL-SCID mice have been constructed using EBV-seropositive donors, and GVHD has been similarly rare. Our experience indicates that clinically significant GVHD is a rare outcome in the standard hu-PBL-SCID model.

Pfeffer *et al.* (1989) reported in the same meeting that hu-PBL-SCID mice contained small numbers of functional human T cells in the peritoneal cavity as judged by limiting dilution estimates of cells responsive to anti-CD3 antibodies or phytohemagglutinin (PHA). They also presented flow cytometric analyses showing small numbers of human cells bearing HLA-DR and CD5 antigens appearing in the SCID spleen. At the same time, further experiments in our laboratory involving both flow cytometry and *in situ* hybridization with human *Alu*-specific probes demonstrated that only small numbers of human cells were present in the peritoneal cavity and spleen of hu-PBL-SCID mice, and a letter was published (Mosier *et al.*, 1989a) correcting the initial overestimate of human cells (Mosier *et al.*, 1988).

Krams *et al.* (1989) were the first to report generation of hu-PBL-SCID mice from donors suffering with autoimmune disease, in this instance primary biliary cirrhosis. Mice were injected intraperitoneally with 10 to 45×10^6 patient PBLs or 11 to 36×10^6 normal PBLs. Human Ig was observed in all hu-PBL-SCID mice, and mice derived from the patient's PBLs contained antibodies to human mitochondrial antigens. Analysis of hu-PBL-SCID mice (either normal or patient derived) showed varying levels of $CD3^+$ human T cells in the SCID mouse spleen, ranging from <1 to 63% of recovered cells. These $CD3^+$ cells were also shown to express either CD4 or CD8. The most interesting aspect of this report was the demonstration that hu-PBL-SCID mice developed hepatic lesions similar to those seen in primary biliary cirrhosis. This aspect of the disease model was potentially complicated by the appearance of less severe cellular infiltrates in the periportal space of some hu-PBL-SCID mice derived from normal donors, an observation consistent with microscopic GVHD. The normal donor in this case was a laboratory worker with a long history of mouse exposure, so the authors raised the issue of prior sensitization as a relevant factor for the development of GVHD. Tissues from hu-PBL-SCID mice generated in our laboratory are routinely submitted for histopathology, and similar hepatic periportal infiltrates have not been observed except in rare mice with EBV-associated lymphomas, so we conclude that microscopic GVHD is not a routine consequence of injection of human PBLs.

Tighe *et al.* (1990) recently reported autoantibody production in hu-PBL-SCID mice or SCID mice constructed using synovial fluid cells or synovial-infiltrating cells (hu-SL-SCID mice) from patients with rheumatoid arthritis (RA). The hu-SL-SCID mice produced less human IgG and IgM than did parallel normal hu-PBL-SCID mice, but the fraction of IgM with rheumatoid factor activity was much higher. In addition, autoantibodies to human nuclear antigens, cytoskeletal elements, and antibodies to exogenous antigens [e.g., EBV nuclear antigen-1 (EBNA-1)] were detected in hu-SL-SCID mice derived from RA patients, but rarely in normal hu-PBL-SCID mice. Rheumatoid factor production was observed to continue for as long as 20 weeks in some mice. Though not reported in the paper, no obvious joint or renal pathology was observed in these experiments, and no GVHD was observed.

Transfer of another human autoimmune disease, systemic lupus erythematosus (SLE), to hu-PBL-SCID mice has been reported (Duchosal et al., 1990). PBLs (15×10^6) were from healthy donors or SLE patients. All 15 hu-PBL-SCID mice showed human IgG and IgM production, although one SLE-hu-PBL-SCID animal, which developed a B cell lymphoma, had low IgG production and very high IgM levels. Human T and B cells were demonstrated in the peripheral blood of the hu-PBL-SCID mice, where they represented up to 15% of the total mononuclear cells. Human IgG production was observed to continue for at least 270 days, although IgG levels decreased from peak values at 2–3 months postreconstitution. Two SLE-hu-PBL-SCID mice died of B cell lymphomas; a higher incidence of EBV-associated B cell abnormalities might have been expected because of the higher frequency of latently infected B cells in disorders such as SLE (Tosato et al., 1985). The hu-PBL-SCID mice derived from SLE patients produced antinuclear antibodies, and human Ig was deposited in the renal glomeruli in six of seven mice examined. This deposition was not associated with mesangial proliferation or other indications of renal pathology, however. No clinical or histological evidence of GVHD was noted. These observations are clearly supportive of the use of hu-PBL-SCID mice as an *in vivo* model of human SLE.

A similar study has been reported in preliminary form (Geppert and Jasin, 1990). SLE-hu-PBL-SCID mice were reported to produce less total human Ig and more anti-DNA antibodies than normal hu-PBL-SCID mice. Provocative preliminary results of the independent transfer of T and B cells from normal or SLE donors were presented. First, both T and B cells were required for human Ig production, suggesting that virtually all of the Ig secretion in hu-PBL-SCID mice is T dependent. Second, the adoptive transfer of normal T cells and SLE B cells led to a great increase in both autoantibodies and total Ig, indicating a helper T cell deficit in SLE.

Davies et al. (1990) have shown that production of human thyroid autoantibodies can be transferred to hu-PBL-SCID mice. Both PBL and intrathyroidal lymphocytes from patients with Graves disease were used for reconstitution. As in the study of Tighe *et al.* (1990), hu-PBL-SCID mice constructed from patient cells produced less human IgG and IgM than did parallel normal hu-PBL-SCID mice. Both human T and B cells were detected by flow cytometry in the peripheral circulation and spleens of hu-PBL-SCID mice, and they were higher in normal hu-PBL-SCID mice than in Graves hu-PBL-SCID animals. Human T cells accounted for up to 30% of peripheral blood and spleen cells, and human B cells were found at 10% incidence. Six of 19 Graves hu-PBL-SCID mice contained detectable human autoantibody to thyroid peroxidase. Abnormal thyroid hormone levels or evidence of thyroiditis were not seen, nor was GVHD observed.

Saxon *et al.* (1991) have compared hu-PBL-SCID mice generated from normal donors and patients with common variable immunodeficiency (CVI) or X-linked agammaglobulinemia (XLA). SCID mice were injected with 10 to 20×10^6 PBLs and followed for up to 155 days postreconstitution. No human tumors or GVHD were seen in these experiments. Normal hu-PBL-SCID mice had serum Ig composed of all four human IgG subclasses, IgM, IgA, and low amounts of IgD and IgE in some mice. Human cells were analyzed by two-color immunofluorescence; $CD45^+$ human cells were generally found in the hu-PBL-SCID spleen, liver, and lung, with the highest level of human cells achieved at 35–50 days posttransfer. Human B cells accounted for 1–10% of recovered human cells at these time points; $CD3^+$ human T cells were the majority human population. CVI-hu-PBL-SCID mice produced human immunoglobulin levels that were virtually identical to that seen in normal hu-PBL-SCID mice with three of four CVI donors, despite the difficulty of initiating human Ig synthesis *in vitro* with PBLs from these same donors. Analysis of Ig gene rearrangements by polymerase chain reaction (PCR) amplification in both normal and CVI-hu-PBL-SCID mice suggested oligoclonal survival of B cells in long-term reconstituted mice; i.e., overutilization of selected V_H gene families was detected rather than the expected random V_H utilization. These observations raise the possibility that only a subset of memory B cells survives long term in hu-PBL-SCID mice, and that CVI patients have such memory B cells but not other B cells. How these B cells develop in CVI patients is unexplained, and an alternative hypothesis (that SCID mice somehow rescue defective human B cells) was neither excluded nor investigated in this report. Whereas this report does raise important issues about the selective survival of human B (and perhaps T) cells in hu-PBL-SCID at long intervals after PBL transfer, it does not address the important issue of short-term survival of primary IgM-secreting cells.

A recent study by Barry *et al.* (1991) presents interesting information about the engraftment of human postnatal thymic fragments in SCID mice pretreated with different immunosuppressive regimens. Untreated SCID mice grafted with thymic fragments maintained human thymic stromal cells as well as human thymocytes in 37% of recipients. Pretreatment of SCID mice with 400 cGy irradiation led to a much higher frequency of engraftment (83%), but the graft consisted mainly of human thymic epithelial cells and human thymocytes were not evident. A radiation-sensitive SCID mouse contribution to human thymocyte survival appears to be important. It should be noted that SCID mice have increased sensitivity to irradiation (Fulop and Phillips, 1986, 1990). Pretreatment of SCID mice with antiasialo-GM_1 antibodies led to successful engraftment of human thymic epithelium, but 38% of these animals were induced to produce *murine* Ig and contained *murine* T cells in the human thymic graft. These experiments point out the complexity of the interactions between the human xenograft and the SCID recipient, and the necessity to characterize completely cell populations in these animals. Treatment with antiasialo-GM_1 antibodies reduces mouse NK activity (Habu *et al.*, 1981), which appears to have a role in regulating differentiation of rare T and B cell precursors in the SCID mouse (Bosma *et al.*, 1988; Carroll *et al.*, 1989). We have recently observed (Riggs *et al.*, 1991) that transfer of neonatal mouse thymocytes to SCID recipients leads to the induction of "breakthrough" SCID Ig production (bearing the C.B-17 *scid* allotype) in every animal, and that this effect is abrogated by prior irradiation of the SCID recipients. It is thus possible that the neonatal human thymus graft placed into NK-depleted SCID recipients has an effect similar to transfer of neonatal mouse thymocytes.

These studies, taken together, demonstrate that human antibodies can be made in hu-PBL-SCID mice, and that both T and B cells survive for many weeks following adoptive transfer to the xenogeneic recipient. It appears likely that only a fraction of recirculating lymphocytes survive following intraperitoneal transfer to SCID recipients, and that high levels of human immunoglobulin occur despite persistence of relatively small numbers of human B cells. The occurrence of obvious GVHD is noted by a minority of groups using hu-PBL-SCID mice, but its sporadic occurrence requires some explanation. Though our laboratory has not observed evidence of GVHD in adult SCID mice reconstituted with adult PBLs, we have seen an interesting form of GVHD when adult PBLs were transferred in large numbers (50 to 100×10^6) to neonatal SCID.*beige* double mutant (Roder and Duwe, 1979; Lauzon *et al.*, 1986; MacDougall *et al.*, 1990) recipients. In this setting, high levels of human IgM are associated with red blood cell agglutination, anemia, and microvascular occlusion. The impression gained from these preliminary experiments is that xenoreactive antibodies are more important in the pathogenesis of this process than are xenoreactive T cells. This observation also suggests that the failure of GVHD to occur in adult SCID recipients is due in part to an NK cell-mediated resistance, and that interlaboratory variation in results could be explained by differences in the level of NK cell activation.

III. EBV-Associated Lymphoproliferative Disease in hu-PBL-SCID Mice

A fraction of SCID mice injected with human PBLs were observed to develop B cell immunoblastic lymphomas of human origin in the first series of transfer experiments (Mosier *et al.*, 1988). It was predicted that these tumors might be related to Epstein–Barr virus transformation based on the known association of EBV with B cell lymphoproliferative disease in immunosuppressed patients (Honto *et al.*, 1983; Cleary *et al.*, 1984; Shapiro *et al.*, 1988; Klein, 1989). Analysis of tumor DNA with a probe specific for the *Bam*W region of the EBV genome confirmed the presence of EBV sequences in all hu-PBL-SCID tumors examined (Mosier *et al.*, 1988, 1990a). These data have been extended to greater than 200 tumors, and all are EBV positive (Mosier *et al.*, 1990a). It should be noted, however, that older SCID mice have a high incidence of murine thymomas (Custer *et al.*, 1985), so the human origin of any lymphoma in a hu-PBL-SCID mouse must be directly confirmed.

The causative role of EBV in the generation of B cell lymphoproliferative disease in hu-PBL-SCID mice was examined by reconstituting mice with PBLs from either EBV-seropositive or -seronegative donors. Most adults are chronically infected with EBV and maintain high titers of antibody to EBV viral capsid antigen. No SCID mouse reconstituted with PBLs from an EBV-seronegative donor has ever developed a B cell lymphoma. Most EBV-positive donors have produced tumors in some hu-PBL-SCID mice, although the frequency with which tumors appear is dependent upon both the donor and the number of PBLs injected. To date, 88% of EBV-seropositive donors have produced lymphomas in some animals. The latent period until tumor detection does not vary much from donor to donor, but is related to the number of PBLs injected. The greater the number of human PBLs injected, the shorter the latent period. The mean latent period following injection of 5×10^7 PBLs is 9.5 ± 1.2 weeks. The relationship between the number of PBLs injected and the frequency of hu-PBL-SCID mice developing lymphomas follows a Poissonian distribution, which allows the estimate that approximately one to four B cells per million can give rise to a detectable tumor. This estimate is of interest because it is similar to the frequency of peripheral blood B cells latently infected with EBV (Kirchner et al., 1979; Yao et al., 1985; Katz et al., 1989; Klein, 1989). These early observations were consistent with the hypothesis that the lymphomas in hu-PBL-SCID mice represented outgrowth of B lymphocytes spontaneously transformed *in vivo* by EBV.

This hypothesis was addressed by comparing the properties of EBV-transformed lymphoblastoid cell lines (LCLs) established by *in vitro* transformation of PBLs with B95-8 strain EBV to lymphomas arising in hu-PBL-SCID mice derived from the same donor. Both LCLs and tumor cells expressed a similar array of B cell differentiation antigens and EBV-associated antigens, with a few interesting exceptions. The antigens expressed in common included CD19, CD20, CD21, CD23, CD39, LFA-1, MHC class I and II, EBV nuclear antigen-1, and EBV latent membrane protein (LMP). The tumors differed from the LCL in much diminished expression of cell adhesion proteins (LFA-1, LFA-3, and ICAM-1) and virtual absence of EBNA-2 expression. These changes are likely to be due to adaptation of the transformed B cells to *in vivo* growth, as transfer of LCLs to SCID mice results in similar alterations of adhesion molecule and EBNA-2 expression.

B cell lymphomas arising in immunosuppressed transplant recipients (Honto et al., 1983) or AIDS patients (Birx et al., 1986; Groopman et al., 1986) may be polyclonal or monoclonal (Cleary et al., 1984). Examination of DNA from tumors arising in hu-PBL-SCID mice by Southern blot analysis (Southern, 1975; Arnold et al., 1983) with a human J_H probe showed that most tumors were oligoclonal (two to four unique rearranged J_H bands), with the occasional monoclonal tumor. Southern blot analyses of the c-*myc* locus and the *bcl-2* locus were also performed. No changes in the germ-line context of *bcl-2* were seen, and the rare restriction-length polymorphisms seen at the c-*myc* locus were not associated with rearrangement to Ig loci (Cory, 1986; Pelicci et al., 1986), so their importance to lymphomagenesis is difficult to assess. Earlier preliminary reports (Mosier et al., 1989b, 1989c) from our laboratory emphasized these c-*myc* polymorphisms (mutations?) as potentially important in the causation of lymphomagenesis, but they have been limited to tumors derived from a small number of donors and probably represent secondary events in tumor evolution. The overall phenotype and genotype of the spontaneous immunoblastic B cell lymphomas in hu-PBL-SCID mice suggest that they are analogous to EBV-associated lymphomas arising in immunosuppressed or immunodeficient patients.

Cannon *et al.* (1990) have shown that a similar B cell lymphoproliferative disease can be induced by the injection of EBV into hu-tonsil- or hu-PBL-SCID mice generated from EBV-seronegative donors. These lymphomas were virtually identical to those appearing spontaneously in hu-PBL-SCID mice derived from EBV-seropositive donors. They were typed as high-grade immunoblastic lymphomas, and they could be serially passaged to new SCID recipients. Both karyotypic and Southern blot analysis failed to reveal evidence of chromosomal abnormalities or rearrangements of the c-*myc* or *bcl-2* loci. The phenotype of the *in vivo* EBV-induced transformants was that of activated B cells, with expression of LFA-3 and ICAM-1 documented. No data on expression of EBV-associated antigens were presented. This report also noted no evidence of GVH disease in SCID recipients of 5×10^7 PBLs or tonsil cells. This study leads to the conclusion that there is little difference in B cell lymphomas derived from spontaneous or induced transformation of human B cells *in vivo*. One observation reported by Cannon *et al.* (1990) differs from experience in our own laboratory. The transfer of *in vitro*-transformed LCLs to SCID mice was said not to result in tumors; in our studies (Picchio *et al.*, manuscript in preparation), transfer of LCLs does result in solid tumor formation within 3–6 weeks of i.p. injection. This result is in striking contrast to the inability of LCLs to proliferate in nude mice unless they are inoculated intracerebrally (Giovanella *et al.*, 1979).

A subsequent report (Okano *et al.*, 1990) confirmed the appearance of B cell lymphoproliferative disorders both in EBV-positive hu-PBL-SCID mice and in EBV-negative hu-PBL-SCID mice subsequently injected with EBV. These tumors were oligoclonal by cytoplasmic Ig expression, and expressed LMP, CD23, and cell adhesion molecules. Chromosomal analysis revealed no consistent karyotypic abnormality, although trisomy 11 was noted in a small fraction of mitoses. The authors concluded that these tumors resemble EBV-transformed B cells and not Burkitt's lymphoma. One of the donors used in this study was a patient with X-linked lymphoproliferative (XLP) syndrome; it was clear from the limited numbers of animals analyzed whether the progression of lymphoproliferative disease was accelerated in mice derived from this donor. Subsequent work from the same group (Purtilo *et al.*, 1991) has further compared normal and XLP donors. Little difference was seen in the incidence or rate of progression of B cell lymphoproliferative disease. It was reported, however, that symptoms attributable to GVHD occurred in some hu-PBL-SCID mice derived from normal donors, but in none derived from XLP donors. The symptoms of GVHD reported were mouse erythrocyte agglutination, enlarged spleens with erythrophagocytosis, and the presence of human T cells in the skin and liver of some animals. Because symptoms of GVHD are not consistently observed in hu-PBL-SCID mice, it may be worth noting some details of the experimental protocol that distinguish these from other studies. SCID recipients were maintained on Bactrim (sulfamethoxazole and trimethoprim) prophylaxis for *Pneumocystis carinii* infection, and 5 to 7×10^7 human PBLs were injected in RPMI 1640 medium (rather than HBSS) (Mosier et al., 1988); very low levels (<10 μg/ml) of human IgM or IgG were detected. The latter result is remarkable, because we have seen signs of antierythrocyte antibody formation in fewer than 5% of hu-PBL-SCID mice, and this has always been associated with human IgM levels exceeding 1 mg/ml.

A detailed analysis of B cell lymphomas arising in hu-PBL-SCID mice has recently been published by Rowe et al. (1991). They conclude from their analysis that these tumors are distinct from Burkitt's lymphoma and instead resemble LCLs in terms of their surface phenotype, with expression of the B cell activation antigen CD39, and cell adhesion molecules, CD54 and CD58, their normal karyotype, and their expression of EBV antigens associated with latent infection. They did observe the growth of EBV-transformed LCLs in SCID mice [in contrast to Cannon et al. (1990)], and they also noted that the rate of LCL growth was related to whether type 1 (A) or type 2 (B) EBV was used for transformation (Sixbey et al., 1989; Rowe et al., 1989), with type 1-transformed cells showing faster growth. They suggest that EBV transformation *per se* in the absence of secondary genetic changes is sufficient to establish B cell lymphomas in both SCID mice and immunosuppressed humans. No evidence of GVHD was observed in these studies, and enlarged spleens were shown to be due to tumor infiltration.

These studies, taken together, clearly indicate that the hu-PBL-SCID model will be useful for understanding EBV-related lymphomagenesis, particularly with regard to evaluating the loss of cellular control of latently infected B lymphocytes and addressing therapy of EBV infection to prevent the development of lymphomas in immunosuppressed patients.

IV. Other Tumor Models in SCID Mice

A few reports have appeared regarding the study of other human tumors by adoptive transfer to SCID mice. Though these studies do not involve transfer of normal PBLs to SCID mice yet, it is obvious that such cotransfer studies will soon be accomplished, so it is appropriate to briefly present these model systems here.

Kamel-Reid et al. (1989) have studied an acute lymphoblastic leukemia (ALL) cell line as well as bone marrow from ALL patients following transfer to SCID mice. The A-1 pre-B cell ALL line was injected either intravenously or intraperitoneally into irradiated SCID mice. Widespread dissemination of tumor cells was observed 12 weeks following i.v. but not i.p. injection, in contrast to the survival of normal PBLs (Mosier et al., 1988). Bone marrow cells from three patients with pre-B cell ALL were transferred to irradiated SCID recipients, and ALL cells were detectable 7–10 weeks later in mice derived from two of the three patients. Given that these ALL cells did not grow in nude or BNX mice, it was concluded that SCID mice engrafted with human tumors might form a unique model for following tumor progression and evaluating treatment strategies.

The SCID mouse has also been used as a recipient of a human melanoma cell line in work recently reported (Mueller and Reisfeld, 1990; Mueller et al., 1991). The malignant melanoma line M24met grows and metastasizes more rapidly in SCID mice than in nude recipients (Tsuchida et al., 1987). The M24met line has been shown to be capable of growth in hu-PBL-SCID mice challenged with tumor cells 4 weeks after PBL transfer, so model conditions have been established that may allow study of immunomodulators that affect the interaction between the PBLs and the tumor cells. Other such studies are in progress, but none has yet been published.

V. HIV Infection of hu-PBL-SCID Mice

A major incentive for the development of both the hu-PBL-SCID model and the SCID-hu model was to provide a small animal model for AIDS research. As soon as suitable biocontainment facilities could be constructed, studies of HIV-1 infection were initiated by McCune and co-workers and by our laboratory. Namikawa et al. (1988) have published experiments showing that direct injection of HIV-1 into the thymus or lymph node of SCID-hu mice results in a time-dependent increase in cells expressing HIV-1-specific mRNA, and we have completed studies (Mosier, 1990; Mosier et al., 1991) showing that hu-PBL-SCID mice can be infected by intraperitoneal injection either with cell-free HIV-1 or with HIV-1-infected syngeneic T lymphoblasts. In our studies, virus can be cultured from the peritoneal cavity, spleen, peripheral blood, and lymph nodes of infected mice, and *in situ* hybridization confirms infection of human cells in the SCID spleen. In addition, HIV-specific sequences can be amplified from hu-PBL-SCID spleen by the polymerase chain reaction. Virus infection is accomplished by direct injection under anesthesia into the peritoneal cavity of hu-PBL-SCID mice. By 2 weeks postinfection, HIV replication is detected by increases in viral DNA detected by PCR, and by the increasing numbers of cells expressing viral transcripts by *in situ* hybridization. In the case of the hu-PBL-SCID mice, *in situ* positive HIV-infected cells are detected in the spleen, lymph node, and peripheral blood, although the site of both human PBLs and HIV injection is the peritoneal cavity. In the SCID-hu model, virus replication appears to be confined to the human thymus or lymph node graft (McCune, 1990, 1991), and does not spread to T cells in the periphery. HIV can be isolated from the peritoneal cavity and spleen of hu-PBL-SCID mice by coculture of cells from those sites with fresh PHA-stimulated human lymphoblasts, establishing that replication-competent virus is spreading throughout the human lymphocyte graft in the animal.

It is clear from these studies that both hu-PBL-SCID mice containing mature human lymphocytes and SCID-hu mice containing human fetal thymus or lymph node can be infected with HIV-1. There are important differences between the two model systems, however. Chief among these differences is the susceptibility to infection with different strains of HIV-1. We initially used tissue culture-passed isolates of LAV-1/Bru or HTLV-IIIB to infect hu-PBL-SCID mice, and we have now extended the virus stocks used for infection to include MN (Gallo *et al.*, 1984), SF2, SF13 (Cheng-Mayer and Levy, 1988), the macrophage-tropic isolate ADA 5.1 (Gendelman *et al.*, 1988), fresh patient isolates, and several molecular clones. All tested isolates of HIV-1 are infectious for hu-PBL-SCID mice. In contrast, Namikawa *et al.* (1988) report infection using the molecular clones JR-CSF and JR-FL isolated directly from a single patient (Koyanagi *et al.*, 1987) and fresh clinical isolates, but find that "laboratory isolates" such as IIIB, MN, RF, and HXB2 are *not* infectious for either hu-FT-SCID or hu-FLN-SCID mice. What might be the explanation for this major difference? It is known that tissue culture passage of HIV-1 or SIV_{mac} can lead to selection of variants with reduced ability to replicate *in vivo* (Meyerhans *et al.*, 1989; Hirsch *et al.*, 1989). However, HIV-1_{IIIB} grows well in chimpanzees, and the minimal infectious dose in chimps and hu-PBL-SCID mice is virtually identical (4–10 $TCID_{50}$) (L. Arthur, personal communication). In addition, laboratory workers accidentally exposed to HIV-1_{IIIB} have shown evidence of persistent infection (Levy, 1988). These results make it difficult to argue that HIV-

1_{IIIB} and related strains of virus fail to infect SCID-hu mice because they are limited to *in vitro* replication. It is much more likely that the use of fetal lymph nodes containing relatively immature T cells imposes a constraint upon HIV replication not seen with adult T cells. This difference might be exploited to study issues related to perinatal transmission of HIV-1 infection, but it also underscores potentially fundamental differences in the biology of hu-PBL-SCID mice and SCID mice bearing fetal tissue grafts.

HIV infection of hu-PBL-SCID mice leads to alteration of the transfered human immune system. As noted above, reconstitution of SCID mice leads to the rapid and reproducible onset of human immunoglobulin synthesis. hu-PBL-SCID mice were reconstituted with 2×10^7 PBLs and infected 8 weeks later with 10^5 $TCID_{50}$ of HIV-1_{IIIB}. At biweekly intervals after HIV-1 infection, total human serum immunoglobulin levels of four animals from each group were determined from a small sample of blood. HIV infection led to a 200–300% increase in human immunoglobulin levels within 2 weeks of infection, and then most infected mice showed a substantial decrease to <100 $\mu g/ml$ at 6 or more weeks after infection. These changes in human B cell function were reproducible after injection of high concentrations of virus, but infection of hu-PBL-SCID mice with 10^2–10^4 $TCID_{50}$ of HIV-1_{IIIB} led to smaller increases in immunoglobulin and only a fraction of HIV-infected mice showed the later decrease. The fate of the small number of $CD4^+$ human T lymphocytes in the spleens of hu-PBL-SCID mice has been examined 8 weeks after infection with 10^4 $TCID_{50}$ of HIV-1_{IIIB}. Changes in CD4 T cell numbers correlated with human Ig concentrations, and many mice had low Ig levels and no detectable $CD4^+$ T cells. These results demonstrated that HIV infection impairs the immunologic function of hu-PBL-SCID mice. HIV infection led to the depletion of human $CD4^+$ T cells that regulate immunoglobulin synthesis, and this depletion might directly explain the observed immunodeficiency. Alternatively, HIV infection could alter T and B cell function by more indirect means, such as by changing human cytokine levels, enhancing the frequency of opportunistic infection (although no evidence for this has been noted), or possibly stimulating mouse NK cells to reject the human PBLs. These immunosuppressive effects of HIV-1 infection in hu-PBL-SCID mice are more profound at higher initial infectious virus doses and with more rapidly growing strains of virus (such as HIV-1_{IIIB}). Thus, these immunologic consequences of HIV infection of hu-PBL-SCID mice mimic human disease (e.g., Martinez-Maza *et al.*, 1987), but the pace of these changes is accelerated in mice compared to humans.

The development of the two complementary models for HIV infection of human cells in the context of the SCID mouse has led to studies of antiviral compounds and vaccine approaches. McCune et al. (1991) have reported that azidothymidine (AZT) administered in the drinking water at 160–200 mg/kg/day, a dose much higher than that used in humans, reduced the PCR signal to background levels at 2 weeks postinfection, but did not eliminate *in situ* positive cells. Preliminary reports (McCune, 1990) of treatment of SCID-hu mice with dideoxyinosine have appeared. Our laboratory has focused on the development of novel agents for HIV-1 therapy; results of these studies will be presented elsewhere.

The hu-PBL-SCID model has been used extensively for analysis of candidate HIV-1 vaccines both in our laboratory and in that of R. Hesselton, R. A. Koup, and J. L. Sullivan (personal communication). SCID mice have been reconstituted with PBLs from normal seronegative volunteers who were immunized either with recombinant gp160 or with vaccinia vectors expressing gp160. Some of these gp160-immune hu-PBL-SCID mice have been boosted with recombinant gp160 (from VacSyn and MicroGenSys), and all have been challenged with homologous strain HIV-1$_{IIIB}$ to assess protective immunity. In this case, hu-PBL-SCID mice are serving as an invaluable human surrogate for protection studies against a potentially lethal HIV infection. The results of the initial studies have been reported in preliminary form (Mosier et al., 1990b), and at least one immunization protocol has generated HIV-immune hu-PBL-SCID mice that are resistant to HIV-1 challenge. These mice were generated from donors who first had received immunization with vaccinia–gp160 constructs and who were later boosted with recombinant gp160. Five of six animals derived from two such donors were resistant to infection ($0.05 > p < 0.10$ by χ^2), compared to five of five control animals becoming infected. These preliminary data indicate that the hu-PBL-SCID model is likely to be very useful in the evaluation of candidate vaccines, but there are insufficient data to indicate the efficacy of current vaccines at the moment.

VI. Biosafety Issues: Viral Pseudotypes in HIV-Infected hu-PBL-SCID Mice

The use of the HIV-infected hu-PBL-SCID or SCID-hu mice requires careful attention to biohazard control. Both laboratories involved in these studies work under Biosafety Level 3 containment conditions, with additional safeguards to control HIV-infected mice. This concern for novel risk factors inherent in infecting mice with HIV anticipated a recently raised issue about xenotropic pseudotypes in these models (Lusso et al., 1990). It was found that phenotypic mixing and/or pseudotype formation between xenotropic murine viruses and HIV occurred following HIV infection *in vitro* of a human leukemic T cell line previously passaged in immunosuppressed nude mice. Pseudotypes occur when the envelope of one virus packages the genetic core of another virus in a dually infected cell. Phenotypic mixing occurs when a dually infected cell produces virions containing a mixture of envelope components. It was known previously (Tralka et al., 1983) that passage of human tumors in nude mice could lead to activation of endogenous xenotropic murine leukemia viruses (MuLV) and infection of human cells (xenotropic virus will not grow in mouse cells by definition). Previous evidence for pseudotyping between amphotropic MuLV and HIV *in vitro* had been presented by Spector et al. (1990). Such viruses have an expanded human cell tropism that makes them potentially more hazardous than HIV, and their generation could complicate the interpretation of HIV infection in both hu-PBL-SCID and SCID-hu mice.

The risks and biological consequences of HIV-murine retrovirus pseudotypes has been overemphasized, however. Amphotropic murine retroviruses, those capable of growing both in mouse and human cells, are common to wild mice but not to inbred mouse strains. Some mouse strains generate dualtropic MuLV somatically by a process of recombination between endogenous ectopic and xenotropic proviruses; such viruses can also infect human and mouse cells. Activation of these viruses in the SCID models for HIV infection would lead to the spread of HIV to mouse cells; neither our laboratory nor that of McCune (1990) has observed any evidence for this process. Pseudotyping of HIV with xenotropic MuLV would lead to a wider cell tropism for human cells because CD4 would no longer be involved in infection. All HIV-1 recovered from hu-PBL-SCID mice has been shown to depend upon CD4 for infection, so there is no evidence for altered cell tropism in either our laboratory or that of McCune.

The possibilities of pseudotype formation between HIV-1 and other human viruses present in the transferred PBLs do exist, however. EBV and cytomegalovirus (CMV) are the two human viruses most likely to be activated in hu-PBL-SCID mice, and pseudotypes between CMV and HIV-1 have been reported (Spector et al., 1990). Such pseudotypes may have relevance for human pathogenesis; their potential existence also underscores the need for appropriate biohazard containment of HIV-1-infected mice. A recent publication (Milman and D'Souza, 1990) summarizes the recommendations growing out of a workshop on biosafety issues in SCID models. The report concludes that the hazards of working with HIV-1 in SCID mouse models do not exceed those of human or *in vitro* studies with HIV-1. Our experience over the past 2 years supports this conclusion; nonetheless, anyone contemplating setting up these SCID model systems for HIV infection should read the Milman and D'Souza article and implement Biosafety Level 3 practices in a Biosafety Level 2 containment facility.

VII. Summary

Though the development of human-to-mouse xenotransplant models is in its infancy, astonishing progress has been made in a short period of time. Two experimental applications have been developed: short-term transfer of human lymphocytes to generate models for autoimmunity and infectious diseases, and long-term engraftment of tissues with self-renewal potential. Human PBL-SCID mice have been used by multiple laboratories to study normal and autoimmune antibody responses, and have been shown to be readily infectable with HIV-1. SCID mice grafted with fetal tissue have been developed for studies of HIV-1 infection and its therapy as well as for studies of human hematopoietic cell differentiation. Human tumors appear to grow better in SCID mice than in nude mice, and hu-PBL-SCID mice can develop EBV-related B cell lymphoproliferative disease that resembles the immunoblastic lymphomas appearing in immunosuppressed transplant recipients. There is some evidence of mouse NK cells responding to the human xenograft, and of human T and B cells responding to mouse xenoantigens in these models, but these responses are not generally strong enough to have a major impact on human immune function. The use of these surrogate human models is expected to have a major impact on the understanding and treatment of human disease.

ACKNOWLEDGMENTS

This work has been supported by NIH Grants AI-27703, AI-29182, and AI-30238. The experiments with hu-PBL-SCID mice have involved many collaborators, whose contributions to these studies are gratefully acknowledged. Collaborators include Stephen Baird, Dennis Carson, Neil Cooper, Larry Corey, Phil Greenberg, Ruth Hesselton, Shiu-Lok Hu, Tom Kipps, Scott Koenig, Rich Koup, Ryo Kuboyashi, Bonnie Mathieson, Gaston Picchio, Ray Ranken, Doug Richman, Jim Riggs, Gene Shearer, Deborah Spector, Stephen A. Spector, John L. Sullivan, Carol Tacket, Helen Tighe, Bruce Torbett, Darcy Wilson, and Flossie Wong-Staal. Dedicated technicians who contributed to the work were Roberta Barstad, Jeremy Bergsman, Rick Gulizia, Patti Healy, Marybeth Kirven, Ken Kleinhenz, Paul MacIsaac, and Lynn Stell. A number of investigators generously shared information prior to publication. These include John Dick, Tom Geppert, Bart Haynes, Richard Markham, Mike McCune, Bob Phillips, David Purtilo, Ralph Reisfeld, Andrew Saxon, and Len Shultz.

REFERENCES

Andriole, G. L., Mule, J. J., Hansen, C. T., Linehan, W. M., and Rosenberg, S. A. (1985). Evidence that lymphokine-activated killer cells and natural killer cells are distinct based on an analysis of congenitally immunodeficient mice. *J. Immunol.* 135, 2911–2913.

Arnold, A., Crossman, J., Bakhshi, A., Jaffe, E. S., Waldmann, T. A., and Korsmeyer, S. J. (1983). Immunoglobulin gene rearrangements as unique clonal markers in human lymphoid neoplasms. *N. Engl. J. Med.* 309, 1593–1599.

Bankert, R. B., Umemoto, T., Sugiyama, Y., Chen, F. A., Repasky, E., and Yokota, S. (1989). Human lung tumors, patients' peripheral blood lymphocytes and tumor infiltrating lymphocytes propagated in SCID mice. *Curr. Top. Microbiol. Immunol.* 152, 201–210.

Barry, T. S., Jones, D. M., Richter, C. B., and Haynes, B. F. (1991). Successful engraftment of human postnatal thymus in severe combined immune deficient (SCID) mice: Differential engraftment of thymic components with irradiation versus antiasialo GM-1 immunosuppressive regimens. *J. Exp. Med.* 173, 167–180.

Birx, D. L., Redfield, R. R., and Tosata, G. (1986). Defective regulation of Epstein–Barr virus infection in patients with acquired immunodeficiency syndrome (AIDS) or AIDS-related disorders. *N. Engl. J. Med.* 14, 8711–8718.

Bosma, G. C., Custer, R. P., and Bosma, M. J. (1983). A severe combined immunodeficiency mutation in the mouse. *Nature* 301, 527–530.

Bosma, G. C., Fried, M., Custer, R. R., Carroll, A., Gibson, D. M., and Bosma, M. J. (1988). Evidence of functional lymphocytes in some (leaky) SCID mice. *J. Exp. Med.*, 167, 1016–1033.

Cannon, M. J., Pisa, P., Fox, R. I., and Cooper, N. R. (1990). Epstein–Barr virus induces aggressive lymphoproliferative disorders of human B cell origin in SCID/hu chimeric mice. *J. Clin. Invest.* 85, 1333–1337.

Carroll, A. M., Hardy, R. R., and Bosma, M. J. (1989). Occurrence of mature B (IgM$^+$, B220$^+$) and T (CD3$^+$) lymphocytes in SCID mice. *J. Immunol.* 143, 1087–1093.

Cheng-Mayer, C., and Levy, J. A. (1988). Distinct biological and serological properties of human immunodeficiency virus from the brain. *Ann. Neurol.* 23, S58–61.

Cleary, M. L., Warnke, R., and Sklar, J. (1984). Monoclonality of lymphoproliferative diseases occurring after renal transplantation. *N. Engl. J. Med.* 310, 477–482.

Cory, S. (1986). Activation of cellular oncogenes in hemopoietic cells by chromosome translocation. *Adv. Cancer Res.* 47, 189–234.

Custer, R. P., Bosma, G. C., and Bosma, M. J. (1985). Severe combined immunodeficiency (SCID) in the mouse: Pathology, reconstitution, neoplasms. *Am. J. Pathol.* 120, 464–477.

Davies, T. F., Kimura, H., Fong, P., Martin, A. and Shultz, L. D. (1990). Establishment of human thyroid autoantibody secretion in SCID/SCID mice. *Clin. Res.* 38, 264 (abstr.).

Dick, J. E. (1991). Immune-deficient mice as models for human hematopoietic disease. In "Molecular Genetic Medicine" (T. Friedman, ed.), Vol. 1. Academic Press, New York (in press).

Dorshkind, K., Pollack, S. B., Bosma, M. J., and Phillips, R. A. (1984). Natural killer cells are present in mice with severe combined immunodeficiency (SCID). *J. Immunol.* 134, 3798–3801.

Duchosal, M. A., McConahey, P. J., Robinson, C. A., and Dixon, F. J. (1990). Transfer of human systemic lupus erythematosus in severe combined immunodeficient (SCID) mice. *J. Exp. Med.* **172**, 985–988.

Fogh, J., Fogh, J. M., and Orfeo, T. (1977). One hundred and twenty seven cultured tumor cell lines producing tumors in nude mice. *J. Natl. Cancer Inst.* **59**, 2216–2221.

Fulop, G. M., and Phillips, R. A. (1986). Full reconstitution of immune deficiency in SCID mice with normal stem cells requires low-dose irradiation of the recipients. *J. Immunol.* **136**, 4438–4443.

Fulop, G. M., and Phillips, R. A. (1990). The *scid* mutation in mice causes a general defect in DNA repair. *Nature (London)* **347**, 479–482.

Gallo, R. C., Salahuddin, S. Z., Popovic, M., Shearer, G. M., Kaplan, M., Haynes, B. F., Palker, T. J., Redfield, R., Oleske, J., Safai, B., White, G., Foster, P., and Markham, P. D. (1984). Frequent detection and isolation of cytopathic retroviruses (HTLV-III) from patients with AIDS and at risk for AIDS. *Science* **224**, 500–503.

Gendelman, H. E., Orenstein, J. M., Martin, M. A., Ferrua, C., Mitra, R., Phipps, T., Wahl, L. A., Lane, H. C., Fauci, A. S., and Burke, D. S. (1988). Efficient isolation and propagation of human immunodeficiency virus on recombinant colony stimulating factor-1 treated monocytes. *J. Exp. Med.* **167**, 1428–1441.

Geppert, T. D., and Jasin, H. E. (1990). Immunoglobulin (Ig) and autoantibody production in mice with severe combined immunodeficiency (SCID) injected with lymphocytes from patients with systemic lupus erythematosus (SLE). *Arth. Rheum.* **33**, SS99 (abstr.).

Giovanella, B., Nilsson, K., and Jack, L. (1979). Growth of diploid, Epstein–Barr virus-carrying human lymphoblastoid cell lines heterotransplanted into nude mice under immunologically privileged conditions. *Int. J. Cancer* **24**, 103–113.

Groopman, J. E., Sullivan, J. L., Mulder, C., Ginsberg, D., Orkin, S., O'hara, C. J., Falchuk, K., Wong-Staal, F., and Gallo, R. C. (1986). Pathogenesis of B cell lymphomas in patients with AIDS. *Blood* **67**, 612–615.

Habu, S., Hiroyasu, F., Shimamura, K., Kasai, M., Yoshituka, N., Okumura, K., and Norikazu, T. (1981). *In vivo* effects of anti-asialo GM-1. I. Reduction of NK activity and enhancement of transplanted tumor growth in nude mice. *J. Immunol.* **127**, 34–38.

Hackett, J., Bosma, G. C., Bosma, M. J., Bennett, M., and Kumar, V. (1986). Transplantable progenitors of natural killer cells are distinct from those of T and B lymphocytes. *Proc. Natl. Acad. Sci. U.S.A.* **83**, 3427.

Hirsch, V. M., Edmondson, P., Murphey-Corb, M., Arbeille, B., Johnson, P. R., and Mullins, J. I. (1989). SIV adaption to human cells. *Nature* **341**, 574–575.

Honto, D. W., Gajl-Peczalska, K. J., Frizzera, G., Arthur, D. C., Balfour, H. H., Jr., McClain, K., Simons, R. L., and Najarian, J. J. (1983). Epstein–Barr virus (EBV) induced polyclonal and monoclonal B-cell lymphoproliferative diseases occurring after renal transplantation. *Ann. Surg.* **198**, 356–369.

Kamel-Reid, S., and Dick, J. E. (1988). Engraftment of immune-deficient mice with human hematopoietic stem cells. *Science* **242**, 1706–1709.

Kamel-Reid, S., Letarte, M., Sirard, C., Doedens, M., Grunberger, T., Fulop, G., Freedman, M. H., Phillips, R. A., and Dick, J. E. (1989). A model of human acute lymphoblastic leukemia in immune-deficient SCID mice. *Science* **246**, 1597–1600.

Katz, B. Z., Raab-Traub, N., and Miller, G. (1989). Latent and replicating forms of Epstein–Barr virus DNA in lymphomas and lymphoproliferative diseases. *J. Infect. Dis.* **160**, 589–598.

Kirchner, H., Tosato, G., Blaese, R. M., Broder, S., and MaGrath, I. T. (1979). Polyclonal immunoglobulin secretion by human B lymphocytes exposed to Epstein–Barr virus *in vitro*. *J. Immunol.* **122**, 1310–1313.

Klein, G. (1989). Viral latency and transformation: The strategy of Epstein–Barr virus. *Cell* **58**, 5–8.
Koyanagi, Y., Miles, S., Mitsuyasu, R. T., Merrill, J. E., Vinters, H. V., and Chen, I. S. Y. (1987). Dual infection of the central nervous system by AIDS viruses with distinct cellular tropisms. *Science* **236**, 819–822.
Krams, S. M., Dorshkind, K., and Gershwin, M. E. (1989). Generation of biliary lesions after transfer of human lymphocytes into severe combined immunodeficient (SCID) mice. *J. Exp. Med.* **170**, 1919–1930.
Lauzon, R. J., Siminovitch, K. A., Fulop, G. M., Phillips, R. A., and Roder, J. C. (1986). An expanded population of natural killer cells in mice with severe combined immunodeficiency (SCID) lack rearrangement and expression of T cell receptors genes. *J. Exp. Med.* **164**, 1797–1802.
Levy, J. A. (1988). The transmission of AIDS: The case of the infected cell. *JAMA* **259**, 3037–3038.
Lusso, P., Veronese, F. D-M., Ensoli, B., Franchini, G., Jemma, C., DeRocco, S. E., Kalyanaraman, V. S., and Gallo, R. C. (1990). Expanded HIV-1 cellular tropism by phenotypic mixing with murine endogenous retroviruses. *Science* **247**, 848–851.
MacDougall, J. R., Croy, B. A., Chapeau, C., and Clark, D. A. (1990). Demonstration of a splenic cytotoxic effector cell in mice of genotype *SCID/SCID.BG/BG*. *Cell. Immunol.* **130**, 106–117.
Malynn, B. A., Blackwell, T. K., Fulop, G. M., Rathbun, G. A., Furley, A. J. W., Ferrier, P., Heinke, L. B., Phillips, R. A., Yancopoulos, G. D., and Alt, F. W. (1988). The *scid* defect affects the final step of the immunoglobulin VDJ recombinase mechanism. *Cell* **54**, 453–460.
Markham, R. B., Barber, J. P., and Donnenberg, A. D. (1990). Adoptive transfer of human immune memory responses to hu-PBL/SCID mice; effect of donor and recipient immunization. *Int. Conf. AIDS, 6th* (abstr.), Th.A.359.
Martinez-Maza, O., Crabb, E., Mitsuyasu, R. T., Fahey, J. L., and Giogi, J. V. (1987). Infection with the human immunodeficiency virus (HIV) is associated with an *in vivo* increase in B lymphocyte activation and immaturity. *J. Immunol.* **138**, 3720–3724.
McCune, J. M. (1990). The rational design of animal models for HIV infection. *Sem. Virol.* **1**, 229–235.
McCune, J. M. (1991). The SCID-hu mouse: A small animal model for HIV infection and pathogenesis. *Ann. Rev. Immunol.* **9**, 399–429.
McCune, J. M., Namikawa, R., Kaneshima, H., Shultz, L. D., Lieberman, M., and Weissman, I. L. (1988). The SCID-hu mouse: Murine model for the analysis of human hematolymphoid differentiation and function. *Science* **241**, 1632–1639.
McCune, J. M., Namikawa, R., Shih, C.-C., Rabin, L., and Kaneshima, H. (1990a). Suppression of HIV infection in AZT-treated SCID-hu mice. *Science* **247**, 564–566.
McCune, J. M., Namikawa, R., Shih, C.-C., Rabin, L., and Kaneshima, H. (1990b). Pseudotypes in HIV-infected mice. *Science* **250**, 1152–1153 (technical comments).
Meyerhans, A., Cheynier, R., Albert, J., Seth, M., Kwok, S., Sninsky, J., Morfeldt-Manson, L., Asjo, B., and Wain-Hobson, S. (1989). Temporal fluctuations in HIV quasispecies *in vivo* are not reflected by sequential HIV isolations. *Cell* **58**, 901–910.
Milman, G., and D'Souza, P. (1990). HIV infections in SCID mice: Safety considerations. *ASM News* **56**, 639–643.
Mosier, D. E. (1990). Immunodeficient mice xenografted with human lymphoid cells: New models for *in vivo* studies of human immunobiology and infectious diseases. *J. Clin. Immunol.* **10**, 185–191.
Mosier, D. E., Gulizia, R. J., Baird, S. M., and Wilson, D. B. (1988). Transfer of a

functional human immune system to mice with severe combined immunodeficiency. *Nature (London)* **335**, 256–259.

Mosier, D. E., Gulizia, R. J., Baird, S., and Wilson, D. B. (1989a). On the SCIDs? *Nature (London)* **338**, 211.

Mosier, D. E., Gulizia, R. J., Wilson, D. B., Baird, S. M., Spector, S. A., Spector, D. H., Richman, D. D., Fox, R. I., and Kipps, T. J. (1989b). Elements of the human immune system: Studies of mature lymphoid cells following xenotransplantation to SCID mice. *In* "Progress in Immunology VII" (F. Melchers *et al.*, eds.), pp. 1264–1271. Springer-Verlag, Berlin and New York.

Mosier, D. E., Gulizia, R. J., Baird, S. M., Richman, D. D., Wilson, D. B., Fox, R. I., and Kipps, T. J. (1989c). B cell lymphomas in SCID mice engrafted with human peripheral blood leukocytes. *Blood* **74**, 52a.

Mosier, D. E., Baird, S. M., Kirven, M. B., Gulizia, R. J., Wilson, D. B., Kubayashi, R., Picchio, G., Garnier, J. L., Sullivan, J. L., and Kipps, T. J. (1990a). EBV-associated B cell lymphomas following transfer of human peripheral blood lymphocytes to mice with severe combined immune deficiency. *Curr. Top. Microbiol. Immunol.* **166**, 317–324.

Mosier, D. E., Gulizia, R. J., MacIsaac, P. D., Newell, A., Corey, L., Hu, S.-L., Greenberg, P., and Koenig, S. (1990b). Evaluation of prototype vaccines in hu-PBL-SCID mice; nature of the protective immune response. *Cinquiéme Colloque des "Cent Gardes,"* **72** (abstr.).

Mosier, D. E., Gulizia, R. J., Baird, S. M., Wilson, D. B., Spector, D. H., and Spector, S. A. (1991). Human immunodeficiency virus infection of human-PBL-SCID mice. *Science* **251**, 791–794.

Mueller, B. M., and Reisfeld, R. A. (1990). Establishment of spontaneous melanoma metastasis in SCID/hu mice as a model for immunotherapy. *Proc. Am. Assoc. Cancer Res.* **31**, 299 (abstr.).

Mueller, B. M., Romerdahl, C. A., Trent, J. M., and Reisfeld, R. A. (1991). Suppression of spontaneous melanoma metastasis in SCID mice with an antibody to the epidermal growth factor receptor. *Cancer Res.* **51**, 2193–2198.

Namikawa, R., Kaneshima, H., Lieberman, M., Weissman, I. L., and McCune, J. M. (1988). Infection of the SCID-hu mouse by HIV-1. *Science* **242**, 1684–1686.

Namikawa, R., Weilbaecher, K. N., Kaneshima, H., Yee, E. J., and McCune, J. M. (1990). Long-term hematopoiesis in the SCID hu mouse. *J. Exp. Med.* **172**, 1055–1064.

Nilsson, K., Giovanella, B. C., Stehlin, J. S., and Klein, G. (1977). Tumorigenicity of human hematopoietic cell lines in athymic nude mice. *Int. J. Cancer* **19**, 337–344.

Okano, M., Taguchi, Y., Nakamine, H., Pirruccello, S. J., David, J. R., Beisel, K. W., Kleveland, K. L., Sanger, W. G., Fordyce, R. R., and Purtilo, D. T. (1990). Characterization of Epstein–Barr virus-induced lymphoproliferation derived from human peripheral blood mononuclear cells transferred to severe combined immunodeficient mice. *Am. J. Pathol.* **137**, 517–522.

Pelicci, P. G., Knowles, D. M., Arlin, Z. A., Wieczorek, R., Luciw, P., Dina, D., Basilico, C., and Dalla-Favera, R. (1986). Multiple monoclonal B cell expansions and c-*myc* oncogene rearrangements in acquired immune deficiency syndrome-related lymphoproliferative disorders. *J. Exp. Med.* **164**, 2049–2060.

Pfeffer, K., Heeg, K., Bubeck, R., Condradt, P., and Wagner, H. (1989). Adoptive transfer of human peripheral blood lymphocytes (PBL) into scid mice. *Curr. Top. Microbiol. Immunol.* **152**, 211–218.

Phillips, R. A., Jewett, M. A. S., and Gallie, B. L. (1989). Growth of human tumors in immune-deficient *scid* mice and *nude* mice. *Curr. Top. Microbiol. Immunol.* **152**, 259–263.

Purtilo, D. T., Falk, K., Pirruccello, S. J., Nakamine, H., Kleveland, K., Davis, J. R., Okano, M., Taguchi, Y., Sanger, W. G., and Beisel, K. W. (1991). SCID mouse model of Epstein–Barr virus-induced lymphomagenesis of immunodeficient humans. *Int. J. Cancer* **47**, 510–517.

Riggs, J. E., Stowers, R. S., and Mosier, D. E. (1991). Adoptive transfer of neonatal T lymphocytes rescues immunoglobulin production in mice with severe combined immune deficiency. *J. Exp. Med.* **173**, 265–268.

Roder, J. C., and Duwe, A. (1979). The beige mutation in mouse selectively impairs natural killer cell function. *Nature (London)* **278**, 451–452.

Rowe, M., Young, L. S., Cadwallader, K., Petti, L., Kieff, E., and Rickinson, A. B. (1989). Distinction between Epstein–Barr virus type A (EBNA 2A) and type B (EBNA 2B) isolates extends to the EBNA 3 family of nuclear proteins. *J. Virol.* **63**, 1031–1039.

Rowe, M., Young, L. S., Crocker, H., Stokes, S., Henderson, S., and Rickinson, A. B. (1991). Epstein–Barr virus (EBV)-associated lymphoproliferative disease in the SCID mouse model: Implications for the pathogenesis of EBV-positive lymphomas in man. *J. Exp. Med.* **173**, 147–158.

Saxon, A., Macy, E., Denis, K., Tary-Lehmann, M., Witte, O., and Braun, J. (1991). Limited B cell repertoire in SCID mice engrafted with peripheral blood mononuclear cells derived from immunodeficient or normal humans. *J. Clin. Invest.* **87**, 658–665.

Schuler, W., Weiler, I. J., Schuler, A., Phillips, R. A., Rosenberg, N., Mak, T., Kearney, J. F., Perry, R. P., and Bosma, M. J. (1986). Rearrangements of antigen receptor genes is defective in mice with severe combined immunodeficiency. *Cell* **46**, 963–972.

Schuler, W., Schuler, A., and Bosma, M. J. (1990). Defective V-to-J recombination of T cell receptor gamma chain genes in *scid* mice. *Eur. J. Immunol.* **20**, 545–550.

Shapiro, R. S., McClain, K., Frizzera, G., Gajl-Peczalska, K. J., Kersey, J. H., Blazer, B. R., Arthur, D. C., Patton, D. F., Greenberg, J. S., Burke, B., Ramsay, N. K. C., McGlave, P., and Filipovich, A. H. (1988). Epstein–Barr virus associated B cell lymphoproliferative disorders following bone marrow transplantation. *Blood* **71**, 1234–1243.

Sharabi, Y., Aksentijevich, I., Sundt, T. M. III, Sachs, D. H., and Sykes, M. (1990). Specific tolerance induction across a xenogeneic barrier: Production of mixed rat/mouse lymphohematopoietic chimeras using a lethal preparative regimen. *J. Exp. Med.* **172**, 195–202.

Sixbey, J. W., Shirley, P., Chesney, P. J., Buntin, D. M., and Resnick, L. (1989). Detection of a second widespread strain of Epstein–Barr virus. *Lancet* **2**, 761–765.

Southern, E. M. (1975). Detection of specific consequences among DNA fragments separated by gel electrophoresis. *J. Mol. Biol.* **98**, 503–517.

Spector, D. H., Wade, E., Wright, D. A., Koval, V., Clark, C., Jaquish, D., and Spector, S. A. (1990). Human immunodeficiency virus pseudotypes with expanded cellular and species tropism. *J. Virol.* **64**, 2298–2308.

Tighe, H., Silverman, G. J., Kozin, F., Tucker, R., Gulizia, R., Peebles, C., Lotz, M., Rhodes, G., Machold, K., Mosier, D. E., and Carson, D. A. (1990). Autoantibody production by severe combined immunodeficient mice reconstituted with synovial cells from rheumatoid arthritis patients. *Eur. J. Immunol.* **20**, 1843–1848.

Tosato, G., Blaese, R. M., and Yarchoan, R. (1985). Relationship between immunoglobulin production and immortalization by Epstein–Barr virus. *J. Immunol.* **135**, 959–964.

Tralka, T. S., Yee, C. L., Rabson, A. B., Wivel, N. A., Stromberg, K. J., Rabson, A. S., and Costa, J. C. (1983). Murine type C retroviruses and intracisternal A particles in human tumors serially passaged in nude mice. *J. Natl. Cancer Inst.* **71**, 591–600.

Tsuchida, T., Saxton, R. E., and Irie, R. F. (1987). Gangliosides of human melanoma: GM2 and tumorigenicity. *J. Natl. Cancer Inst.* **78**, 55–59.

Ware, C. F., Donato, N. J., and Dorshkind, K. (1985). Human, rat, or mouse hybridomas secrete high levels of monoclonal antibodies following transplantation into mice with severe combined immunodeficiency (SCID). *J. Immunol. Methods* **85**, 353–361.

Yao, Q. Y., Rickinson, A. B., and Epstein, M. A. (1985). A re-examination of the Epstein–Barr virus carrier state in healthy seropositive individuals. *Int. J. Cancer* **35**, 35–42.

This article was accepted for publication on 4 March 1991.

Index

A

Abnormalities in endogenous production of IL-2, 148
Activation-induced death of hybridomas, 67
Alloantigen presentation by cultured endothelial cells, 266–267
Alloantigen responses, EC costimulator in, 271–272
Amino acid coding sequences of chicken, 110
Anti-IL-2Rα mAbs, treatment with, 188–189
Anti-TCR1, -TCR2, -TCR3 antibodies, embryonic treatment with, 103–105
Antigen, enhancement of responses to by Vβ selective elements, 36–37
Antigen-independent recruitment of T lymphocytes into tissues by endothelial cells, 274–287
  endothelium and extravasation of T lymphocytes, 285–287
  endothelium and migration of memory T cells to inflammatory sites, 278–284
    endothelial leukocyte adhesion molecules, 279–281
    expression of endothelial leukocyte adhesion molecules $in\ vivo$, 283–284
    involvement of endothelial leukocyte adhesion molecules in lymphocyte binding $in\ vitro$, 281–283
  endothelium and migration of naive T cells to lymph nodes, 275–278
  endothelium and migration of pre-T cells to thymus, 275
  endothelium and tissue-specific homing of lymphocytes, 284–285
Antigens, differentiation, in avian T cells, 88–89
Apoptosis, morphology, 56–57
Apopotic T cell death, and IL-2, 166–167
Autoimmune attack, IL-2 and IL-2R at site of, 171–172
Autoimmune disease, elevated serum levels of soluble IL-2Rα in, 183–186
Autoimmune manifestations
  after $in\ vivo$ applications of recombinant IL-2, 189–191
  in mice transgenic for human IL-2 or IL-2R components, lack of, 192–193
Autoimmune manifestations, lack of in mice transgenic for human IL-2 or IL-2R components, 192–193
Autoimmune or normal human cells, transfer of to SCID mice, 304–310
Autoimmunity, role of IL-2-induced cytokines in, 210–213
  interferon-γ and, 212–213
  interleukin-6 and, 210–211
  tumor necrosis factor-α and autoimmunity, 211–212
Autoimmunity and IL-2, 171–187
  abnormalities in IL-2 expression and responsiveness on circulating lymphocytes, 186–187
  at site of autoimmune attack, 171–172
  elevated serum levels of soluble IL-2Rα in autoimmune disease, 183–186
  $in\ vitro$ IL-2 production in autoimmune disease, 172–178
    autoantibodies, 176
    decrease of IL-2 producing cells, 176
    low IL-2, 175–176
    signal transduction, 177–178
    suppressor macrophages, 176
    suppressor T cells, 176–177
  $in\ vivo$ IL-2 production in autoimmune disease, 178–183
Autolerance and IL-2, 165–171
  effect $in\ vivo$, 165–166
  interference of with clonal deletion, 166–167

nonspecific killing induced by IL-2, 170–171
the system, 167–170
Avian T cell ontogeny
conclusion, 112–114
development, interest, 87
differentiation antigens, 88–89
thymic attraction and origin, 89–90
T cell development, experimental manipulation, 102–105
embryonic treatment with anti-TCR1, -TCR2, -TCR3 antibodies, 103–105
thymectomy, 102
T cell migration to periphery, 96–101
homing preferences, 96–97
interstitial T cell surface glycoprotein, 97–101
migration of TCR1, TCR2 and TCR3 sub-populations, 96
T cell tumors, 105–108
Marek's disease (MDV) virus-induced tumors, 105–107
reticuloendotheliosis virus-induced tumors, 107–108
TCR genes, 108–112
conserved structural features, 111–112
genomic organization, 108–109
TCRβ diversity, 109–111
TCR1, TCR2, TCR3 cells, functional capabilities, 101–102
thymus, diversification in, 90–96
embryonic waves of thymocyte development, 95–96
intrathymic clonal selection, 94
ontogeny of TCR1, TCR2, and TCR3, 91–94

B

Bacterial toxic mitogens and murine Vβ gene segments, 15
B cells, programmed cell death, 63–66
clonal abortion, 64
faulty recombination, 63–64
growth arrest of WE III-231 B cell line, 64–65
somatic mutation and terminal differentiation, 65–66

Bacterial toxic mitogen, retroviral sequences bearing no homology to, 41
Bacterial toxic mitogens, 15–16
B cells, role of in response, 31
Biochemical pathways of programmed cell death, 60–61
Burkitt's lymphoma translocations, C-myc in, 124–128

C

Caenorhabditis elegans, 72
Calcium, role of in apoptosis, 61
C-myc in Burkitt's lymphoma translocations, 124–128
CD4 T cells, response, 30
CD8 T cells, response, 30–31
Cell death in the immune system, *see* Programmed cell death in immune system
Cell source of HRF, 243–244
Cells producing or expressing IL-2, detection and distribution, 152–154
Chromosomal abnormalities on adjacent oncogenes, consequences of formation, 124–132
C-myc in Burkitt's lymphoma translocations, 124–128
gene fusion resulting from chromosome translocation, 128–129
helix-loop-helix oncogenes in chromosome translocations, 129–130
oncogene location after translocation, 131–132
transcriptional disruption of LIM domain oncogenes by translocation, 130–131
Chromosomal abnormalities on lymphoid cells, effect, 132–140
differentiation-related translocation oncogenes, 139
functional chimerism involving DNA binding and protein dimerization, 133–138
other unusual proteins affected by chromosome translocations, 139–140

Chromosome translocation
  gene fusion resulting from, 128–129
  helix-loop-helix oncogenes in, 129–130
  other unusual proteins affected by, 139–140
  and timing of inversion, 122–124
Clonal abortion, 64
Clonal deletion, putative interference of IL-2 with, 166–167
Clonal inactivation by Vβ selective elements, 35–36
Common variable immunodeficiency (CVI), and SCID mice, 308
Connective tissue-activating peptide (CTAP III), 240–243
  comparison of with HRF, 242
  conversion of to NAP-2, 242–243
Costimulator activities of vascular endothelial cells, 267–274
  EC costimulators in alloantigen responses, 271–272
  EC costimulator in polyclonal CD4+ T cell activation, 268–270
  EC costimulator in polyclonal CD8+ T cell activation, 270–271
  EC Production of soluble costimulators, 272–274
Csa, effects of on development of autoimmune phenomena, 209–210
CTL targets, death, 69
Cultured endothelial cells, alloantigen presentation by, 266–267
Cyclosporin A and autoimmunity, 208–210
Cyclosporin treatment, 104–105
Cytokine network, position of IL-2 in, 160
Cytokines
  and histamine release, 247–248
  modulation of histamine release by, 248–250

D

Death genes, induction activation, 71–73
Death program, release 75–76
Dendritic cells, role of in response, 31
Diabetic patients, IL-2 deficiency of, 173–174
Differentiation-related translocation oncogenes, 139
DNA damage in programmed cell death, 57–60
  fragmentation, 57–58
  methods, 59–60
  observations in rodent thymocyte, 58–59
  pattern of fragments, 58

E

EBV, see Epstein-Barr virus
EC, see also Human endothelial cells
EC costimulators in alloantigen responses, 271–272
EC costimulator in polyclonal CD4+ T cell activation, 268–270
EC costimulator in polyclonal CD8+ T cell activation, 270–271
EC production of soluble costimulators, 272–274
Embryonic treatment with anti-TCR1, -TCR2, -TCR3 antibodies, 103–105
  thymectomy, 102
Endogenous retroviral insertions inseparable from Vbse, 38–39
Endothelial cells, antigen-independent recruitment of T lymphocytes into tissues by, 274–287
  endothelium and extravasation of T lymphocytes, 285–287
  endothelium and migration of memory T cells to inflammatory sites, 278–288
    endothelial leukocyte adhesion molecules, 279–281
    expression of endothelial leukocyte adhesion molecules *in vivo*, 283–284
    involvement of endothelial leukocyte adhesion molecules in lymphocyte binding *in vitro*, 281–283
  endothelium and migration of naive T cells to lymph nodes, 275–278
  endothelium and migration of pre-T cells to thymus, 275
  endothelium and tissue-specific homing of lymphocytes, 284–285

Epstein-Barr virus (EBV), associated lymphoproliferative disease in hu-PBL-SCID mice, 310–313
Escherichia coli, 162
Extravasation of T lymphocytes, and endothelium, 285–287

F

Fetal thymus, and synchronized thymocyte ontogeny, 158–159
Functional chimerism involving DNA binding and protein dimerization, 133–138

G

Gene fusion resulting from chromosome translocation, 128–129
Gene regulation of IL-2 and IL-2R, 154–157

H

Helix-loop-helix oncogenes in chromosome translocations, 129–130
Histamine, release of from human basophil or mast cells
 cell source of HRF, 243–244
 clinical considerations, 250–254
 cytokines and histamine release, 247–248
 factors, 237
 mechanism of action of HRF, 244–247
 modulation of histamine release by cytokines, 248–250
 purification of HRF, 238–243
Histamine releasing factor, *see* HRF
HIV-infection of hu-PBL-SCID mice, 314–317
 other tumor models in SCID mice, 313–314
Homing preferences, avian, 96–97
HRF
 cell sources of, 243–244
 mechanism of action, 244–247
 purification of HRF, 238–243
Hu-PBL-SCID mice
 EBV-associated lymphoproliferative disease in, 310–313

 HIV infection of, 314–317
Human follicular B cell lymphoma, 73
Human immune function, animal models for
 EBV-associated lymphoproliferative disease in hu-PBL-SCID mice, 310–313
 HIV-infection of hu-PBL-SCID mice, 314–317
 other tumor models in SCID mice, 313–314
 search for better models, 303–304
 summary, 319
 transfer of normal or autoimmune human cells to SCID mice, 304–310
 viral pseudotypes in HIV-infected hu-PBL-SCID mice, biosafety issues and, 317–319
Human T cell lymphotropic virus type I (HTLV-I), 180, 182
Hybridomas, activation-induced death, 67

I

I-A molecules, involvement, 20–21
I-E molecules, involvement, 18–19
IFN-γ and TNF-α, in synthesis, 213
IL-2, *see also* Interleukin-2
 responsiveness, 164–165
 production, 163–164
 bioavailability, 164
IL-2/vaccina virus (IL-2.VV)
 induction of manifest autoimmunity in athymic mice by, 203–208
 effect of on neonatally thymectomized mice, 206–208
 effect of on *nu/nu* mice, 204–206
Immune intervention, IL-2R-targeted, 187–189
Immune system, programmed cell death in, *see* Programmed cell death in immune system
Immunological systems, programmed cell death in, 63–70
 in B cells, 63–66
 clonal abortion, 64
 faulty recombination, 63–64

growth arrest of WE III-231 B cell
line, 64–65
somatic mutation and terminal
differentiation, 65–66
in T cells, 66–69
CTL targets, death, 69
deprived of growth factors, 68–69
hybridomas, activation-induced
death, 67
non-selected thymocyte, death,
67–68
thymus, negative selection, 66
interphase death of T and B
lymphocytes and nonspecific
damage, 69–70
splenic B cells, 65
In vitro IL-2 production in response to T
cell mitogenes Con a or PHA, 173
In vitro IL-2 production in autoimmune
disease, 172–178
autoantibodies, 176
decrease of IL-2 producing cells, 176
low IL-2, 175–176
signal transduction, 177–178
suppressor macrophages, 176
suppressor T cells, 176–177
In vitro binding of endothelial leukocyte
adhesion molecules in lymphocyte,
281–283
In vitro suppression induced by Vβ
selective elements, 37
In vivo, antitolerance effect of IL-2, 165
In vivo expression of endothelial
leukocyte adhesion molecules,
283–284
In vivo interventions of Li-2 in
autoimmunity, 187–210
autoimmune manifestations after *in
vivo* applications of recombinant
IL-2, 189–191
cyclosporin A and autoimmunity,
208–210
IL-2R-targeted immune intervention,
187–189
induction of manifest autoimmunity in
athymic mice by IL-2/vaccina
virus, 203–206
effect of IL-2.VV on neonatally
thymectomized mice, 206–208

effect of IL-2.VV on nu/nu mice,
204–206
lack of autoimmune manifestations in
mice transgenic for human IL-2 or
IL-2R components, 192–193
recombinant IL-2/vaccina virus
construct on SLE or
MRL/Mp-lpr/lpr mice, 193–203
effect of IL-2.VV on life expectancy
and autoimmune manifestations,
195–196
effect of IL-2.VV on T cell
compartments, 196–200
mechanism of beneficial effect of
IL-2 on lymphoproliferation and
autoimmunity, 200–203
In vivo IL-2 production in autoimmune
disease, 178–183
Incompetent MHC class II
molecules, 22
Induction, mechanism of action of death
genes, 71–73
Interferon-γ and autoimmunity, 212–213
Interleukin-2 (IL-2), autotolerance and
autoimmunity
autoimmunity, 171–187
abnormalities in IL-2 expression and
responsiveness on circulating
lymphocytes, 186–187
at site of autoimmune attack,
171–172
elevated serum levels of soluble
IL-2Rα in autoimmune disease,
183–186
in vitro IL-2 production in
autoimmune disease, 172–178
autoantibodies, 176
decrease of IL-2 producing cells,
176
low IL-2, 175–176
signal transduction, 177–178
suppressor macrophages, 176
suppressor T cells, 176–177
in vivo IL-2 production in
autoimmune disease, 178–183
autolerance, 165–171
effect *in vivo*, 165–166
interference of with clonal deletion,
166–167

nonspecific killing induced by IL-2, 170–171
the system, 167–170
conclusions, 214–217
history of IL-2, 147–149
in vivo interventions of Li-2 in autoimmunity, 187–210
  autoimmune manifestations after in vivo applications of recombinant IL-2, 189–191
  cyclosporin A and autoimmunity, 208–210
  IL-2R-targeted immune intervention, 187–189
  induction of manifest autoimmunity in athymic miceby IL-2/vaccina virus, 203–206
    effect of IL-2.VV on neonatally thymectomized mice, 206–208
    effect of IL-2.VV on *nu/nu* mice, 204–206
  lack of autoimmune manifestations in mice transgenic for human IL-2 or IL-2R components, 192–193
  recombinant IL-2/vaccina virus construct on SLE or MRL/Mp-lpr/lpr mice, 193–203
    effect of IL-2.VV on life expectancy and autoimmune manifestations, 195–196
    effect of IL-2.VV on T cell compartments, 196–200
    mechanism of beneficial effect of IL-2 on lymphoproliferation and autoimmunity,200–203
physiology of and its receptor, 149–165
  compartmentalization reduces pleiotropy, 163–165
    IL-2 bioavailability, 164
    IL-2 production, 163–164
    IL-2 responsiveness, 164–165
  detection and distribution of cells producing or expressing, 152–154
  effects on peripheral immune system and toxicity, 160–163
  gene regulation, 154–157

mechanics of IL-2/IL-2 system, 149–152
and thymocyte differentiations, 158–160
role of IL-2-induced cytokines in autoimmunity, 210–213
  interferon-γ and autoimmunity, 212–213
  interleukin-6 and autoimmunity, 210–211
  tumor necrosis factor-α and autoimmunity, 211–212
Interleukin-6 and autoimmunity, 210–211
Interstitial T cell surface glycoprotein, avian, 97–101
Intrahepatic cholestatis, severe, 190–191
Intrarhythmic induction of clonal anergy, examples, 169
Intrarhythmic development, role of Vβ selective elements in, 33
  negative selection, 34–35
  positive selection, 34
Irradiated thymocyte, biochemical events in, 70

K

Klebsiella pneumoniae, 162

L

Leukemia, T cell, *see* T cell leukemia
Leukocyte adhesion molecules, endothelial, 279–281
  in vivo expression of, 283–284
Leukocyte adhesion molecules *in vivo*, endothelial, expression of, 283–284
Leukocyte adhesion molecules in lymphocyte binding *in vitro*, endothelial, involvement of, 281–283
Life expectancy and autoimmune manifestations, effect of IL-2.VV, 195–296
LIM domain oncogenes by translocation, transcriptional disruption of, 130–131

Lipopolysaccharide (LPS), treatment with, 65
lpr phenomenon, and IL-2, 202
Lupus-prone inbred strains of mouse, 193
Lymphadenopathy and autoimmune symptoms in MRL/Mp-*lpr/lpr* mice, amelioration, 199
Lymphocytes, circulating, abnormalities in IL-2R expression and responsiveness on, 186–187
Lymphocytes, circulating, abnormalities in IL-2 expression and responsiveness, 186–187
Lymphocytes, T and B lymphocytes, interphase death of and nonspecific damage, 69–70
Lymphoid cells, chromosomal abnormalities, effect, 132–140
  differentiation-related translocation oncogenes, 139
  functional chimerism involving DNA binding and protein dimerization, 133–138
  other unusual proteins affected by chromosome translocations, 139–140
Lymphoid tumors, chromosomal translocation in consequences of formation of chromosomal abnormalities on adjacent oncogenes, 124–132
  C-*myc* in Burkitt's lymphoma translocations, 124–128
  gene fusion resulting from chromosome translocation, 128–129
  helix-loop-helix oncogenes in chromosome translocations, 129–130
  oncogene location after translocation, 131–132
  transcriptional disruption of LIM domain oncogenes by translocation, 130–131
  development of T cell leukemia, 141–142
  effect of chromosomal abnormalities on lymphoid cells, 132–140
  differentiation-related translocation oncogenes, 139
  functional chimerism involving DNA binding and protein dimerization, 133–138
  other unusual proteins affected by chromosome translocations, 139–140
  mechanism of translocation and inversion, 121–122
  presence of abnormalities, 119–121
  timing of chromosome translocation and inversion, 122–124
Lymphoproliferation and autoimmunity, mechanism of beneficial effect of IL-2 on, 200–203
Lymphoproliferative disease in hu-PBL-SCID mice, EBV-associated, 310–313

M

Major histocompatibility complex, *see* MHC
Marek's disease (MDV) virus-induced tumors, 105–107
Memory T cells, endothelium and migration of to inflammatory sites, 278–284
MHC molecule expression on vascular endothelium, regulation, 262–265
MHC molecules, and immunological interactions of T lymphocytes with vascular endothelium, 261–266
MHC molecules, involvement, 17–22
  I-A molecules, involvement, 20–21
  I-E molecules, involvement, 18–19
  incompetent MHC class II molecules, 22
Mice bearing a male-specific transgenic T cell receptor, effect of IL-2 on, 170–171
Mice, neonatally thymectomized, effect of IL-2.VV on, 206–108
Mice transgenic for given Mtv, 40
Mice with severe combined immune deficiency (SCID), 303
  EBV-associated lymphoproliferative

disease in hu-PBL–SCID mice, 310–313
HIV-infection of hu-PBL-SCID mice, 314–317
other tumor models in SCID mice, 313–314
transfer of normal or autoimmune human cells to, 304–310
Mls, 7–10
  definition, 7–9
  multiple loci, 9–10
Mls and bacterial toxic mitogen action, working model, 4, 5–6
Molecular basis of element action, 16–22
  MHC molecules, involvement, 17–22
  I-A molecules, involvement, 20–21
  I-E molecules, involvement, 18–19
  incompetent MHC class II molecules, 22
Mouse retention of Vbse sequences, 41–42
Mouse mammary tumor virus (MMTV), 39
MRL/Mp-*lpr*/*lpr* mice, recombinant IL-2/vaccina virus construct on 193–203
Mtv transcripts expressed in activated B cells, 40
Mtv in B cells with Vbse expression, 39
Multiple Mls loci, existence, 9–10
  current nomenclature for, 11
Murine genome, non-Mls elements in, 11–14
Murine Vbse not stimulating primary T-cell response, 11–13
Mycobacterium avium, 162
Myelopathy, HTLV-I associated, 182

N

Naive T cells, endothelium and migration of to lymph nodes, 275–278
NAP-2, conversion of CTAP III to, 242–243
Neonatally thymectomized mice,effect of IL-2.VV on mice, 206–208
Nephritis, interstitial, induced by IL-2, 200–201
Normal or autoimmune cells, transfer of to SCID mice, 304–310
Nu/*nu* mice, effect of IL-2.VV on, 204–206

O

Oncogene location after translocation, 131–132
Oncogenes, adjacent, chromosomal abnormalities, consequences of formation, 124–132
  C-*myc* in Burkitt's lymphoma translocations, 124–128
  gene fusion resulting from chromosome translocation, 128–129
  helix-loop-helix oncogenes in chromosome translocations, 129–130
  oncogene location after translocation, 131–132
  transcriptional disruption of LIM domain oncogenes by translocation, 130–131

P

Peripheral immune system, effects and toxicity, 160–163
Phagocytosis, surface changes resulting in, 62–63
Pleiotropy, reduction of by compartmentalization, 163–165
  IL-2 bioavailability, 164
  IL-2 production, 163–164
  IL-2 responsiveness, 164–165
Pneumonia, inflammatory, induction of in patients treated with recombinant IL-2
Polyclonal CD4+ T cell activation, EC costimulator in, 268–270
Polyclonal CD8+ T cell activation, EC costimulator in, 270–271
Postnatal thymic fragments, human, and SCID mice, 309
Pre-T cells, endothelium and migration of to thymus, 275

Programmed cell death in immune system
  conclusions, 76
  definitions, 55–56
  in immunological systems, 63–70
    in B cells, 63–66
      clonal abortion, 64
      faulty recombination, 63–64
      growth arrest of WE III-231 B cell
        line, 64–65
      somatic mutation and terminal
        differentiation, 65–66
    in T cells, 66–69
      CTL targets, death, 69
      deprived of growth factors, 68–69
      hybridomas, activation-induced
        death, 67
      non-selected thymocyte, death,
        67–68
      thymus, negative selection, 66
      interphase death of T and B
        lymphocytes and nonspecific
        damage, 69–70
      splenic B cells, 65
  mechanisms, 56–63
    apoptosis, morphology, 56–57
    biochemical pathways, 60–61
    DNA damage, 57–60
    phagocytosis, surface changes
      resulting in, 62–63
  triggering and regulation, 71–73
    induction activation of death genes,
      71–73
    release of death program, 75–76
    transduction, 73–75
Protein antigen presentation by cultured
  endothelial cells, 265–266

R

Rat chloroleukemia cell line,
  observation, 73
Recombinant IL-2, autoimmune
  manifestations after *in vivo*
  application of, 189–191
Recombinant IL-2 vaccina virus
  infection, 205
Recombinant IL-2/vaccina virus
  construct on SLE or
  MRL/Mp-*lpr/lpr* mice, 193–203

effect of IL-2.VV on life expectancy
  and autoimmune manifestations,
  195–196
effect of IL-2.VV on T cell
  compartments, 196–200
mechanism of beneficial effect of IL-2
  on lymphoproliferation and
  autoimmunity, 200–203
Responding cells, 29–31
  CD4 T cells, response, 30
  CD8 T cells, response, 30–31
Reticuloendotheliosis virus-induced
  tumors, 107–108
Retroviruses, 40–41
Retroviruses, endogenous or extrinsic,
  37–42
  endogenous retroviral insertions
    inseparable from Vbse, 38–39
  expressions of Mtv in B cells with
    Vbse expression, 39
  mice transgenic for given Mtv, 40
  mouse mammary tumor virus
    (MMTV), 39
  mouse retention of Vbse sequences,
    41–42
  Mtv transcripts expressed in activated
    B cells, 40
  other retroviruses, 40–41
  retroviral sequences bearing no
    homology to bacterial toxic
    mitogen, 41
Rodent thymocyte, and DNA damage,
  58–59

S

Servomodulator regulating functional
  tolerance, 167–170
Signal transduction, mechanism of
  IL-2R-mediated, 152
SLE mice, recombinant IL-2/vaccina
  virus construct on, 193–203
SLE patients, low IL-1 production in,
  175–176, 186–187
Soluble costimulators, EC production of,
  272–274
Somatic mutation and terminal
  differentiation, 65–66
Splenic B cells, 65

Stimulating cells, 31–32
    B cells, role of in response, 31
    dendritic cells, role of in response, 31
    transfer of Vbse among cells, 32
Suppressor macrophages, 176
Suppressor T cells, 176–177
Systemic lupus erythematosus (SLE), transfer of to hu-PBL-SCID mice, 307

T

T cell development, avian, experimental manipulation, 102–105
    cyclosporin treatment, 104–105
    embryonic treatment with antibodies, 103
    thymectomy, 102
T cell leukemia, development, 141–142
T cell migration to periphery, avian, 96–101
    homing preferences, 96–97
    interstitial T cell surface glycoprotein, 97–101
    migration of TCR1, TCR2 and TCR3 sub-populations, 96
T cell ontogeny, avian, see Avian T cell ontogeny
T cell receptor, selective elements for Vbeta region
    cellular basis of Vβ selective element action, 29–32
        responding cells, 29–31
            CD4 T cells, response, 30
            CD8 T cells, response, 30–31
        stimulating cells, 31–32
            B cells, role of in response, 31
            dendritic cells, role of in response, 31
            transfer of Vbse among cells, 32
    conclusions, 47
    elements, 2–3
        questions about, 3–4
    impact of on T cell development and function, 32–37
        impact of Vβ selection elements in periphery, 35–37
            clonal inactivation by Vβ selective elements, 35–36
            enhancement of responses to antigen by Vβ selective elements, 36–37
            suppression induced by Vβ selective elements *in vitro*, 37
        role of Vβ selective elements in intrarhythmic development, 33
            negative selection, 34–35
            positive selection, 34
    molecular basis of element action, 16–22
        MHC molecules, involvement, 17–22
            I-A molecules, involvement, 20–21
            I-E molecules, involvement, 18–19
            incompetent MHC class II molecules, 22
        T cell receptors and coreceptors, involvement, 23–28
            involvement of coreceptors in response, 27–28
            involvement of T cell receptor, 23–24
            involvement of Vβ region of T cell receptor, 24–25
            mapping relevant sites on region, 25–27
        working model, evidence, 28–29
    questions, 42–47
        biological function, 45
        distribution, 44–45
        molecular nature, 44
        validity, 42–44
    retroviruses, endogenous or extrinsic, 37–42
        endogenous retroviral insertions inseparable from Vbse, 38–39
        expressions of Mtv in B cells with Vbse expression, 39
        mice transgenic for given Mtv, 40
        mouse mammary tumor virus (MMTV), 39
        mouse retention of Vbse sequences, 41–42
        Mtv transcripts expressed in activated B cells, 40
        other retroviruses, 40–41

retroviral sequences bearing no homology to bacterial toxic mitogen, 41
selective elements, 6–16
  bacterial toxic mitogens, 15–16
  Mls, 7–10
    definition, 7–9
    multiple loci, 9–10
    non-Mls elements in murine genome, 11–14
    murine Vbse not stimulating primary T-cell response, 11–13
    Vbse, general nomenclature, 13–14
    Vbse, null alleles, 14
  working model, 4–6
T cell tumors, avian, 105–108
  Marek's disease (MDV) virus-induced tumors, 105–107
  reticuloendotheliosis virus-induced tumors, 107–108
T cells, programmed cell death, 66–69
  CTL targets, death, 69
  deprived of growth factors, 68–69
  hybridomas, activation-induced death, 67
  non-selected thymocyte, death, 67–68
  thymus, negative selection, 66
T lymphocytes, immunological interactions of with vascular endothelium
  antigen-independent recruitment of T lymphocytes into tissues by endothelial cells, 274–287
    endothelium and extravasation of T lymphocytes, 285–287
    endothelium and migration of memory T cells to inflammatory sites, 278–284
      endothelial leukocyte adhesion molecules, 279–281
      expression of endothelial leukocyte adhesion molecules in vivo, 283–284
      involvement of endothelial leukocyte adhesion molecules in lymphocyte binding in vitro, 281–283
    endothelium and migration of naive T cells to lymph nodes, 275–278
    endothelium and migration of pre-T cells to thymus, 275
    endothelium and tissue-specific homing of lymphocytes, 284–285
  costimulator activities of vascular endothelial cells, 267–274
    EC costimulators in alloantigen responses, 271–272
    EC costimulator in polyclonal CD4+ T cell activation, 268–270
    EC costimulator in polyclonal CD8+ T cell activation, 270–271
    EC Production of soluble costimulators, 272–274
  specific antigen presentation by vascular endothelial cells, 261–267
    alloantigen presentation by cultured endothelial cells, 266–267
    protein antigen presentation by cultured endothelial cells, 265–266
    regulation of MHC molecule expression on vascular endothelium, 262–265
  summary, 288
TCR genes, avian, 108–112
  conserved structural features, 111–112
  genomic organization, 108–109
  TCRβ diversity, 109–111
TCR1 cells, functional capabilities, 101–102
  antibodies, embryonic treatment, 103–105
  migration of subpopulations, 96
TCR2 cells, functional capabilities, 101–102
  antibodies, embryonic treatment, 103–105
  migration of subpopulations, 96
TCR3 cells, functional capabilities, 101–102
  antibodies, embryonic treatment, 103–105
  migration of subpopulations, 96
Thymectomy, avian, 102

Thymic attraction and origin, 89–90
Thymus, diversification in, 90–96
  embryonic waves of thymocyte development, 95–96
  intrathymic clonal selection, 94
  ontogeny of TCR1, TCR2, and TCR3, 91–94
Thymus, negative selection, 66
Thymus, endothelium and migration of pre-T cells to, 275
Thymocyte differentiation and IL-2, 158–160
Thymocyte, non-selected, death, 67–68
Tissue-specific homing of lymphocytes, and endothelium, 284–285
Transduction of programmed cell death, 73–75
Translocation and inversion, mechanism, 121–122
Translocations, chromosome, see Chromosome translocation
Tumor necrosis factor-α and autoimmunity, 211–212

V

Vaccina virus, see IL-2/vaccina virus
Vascular endothelial cells, specific antigen presentation by, 261–267
  alloantigen presentation by cultured endothelial cells, 266–267
  protein antigen presentation by cultured endothelial cells, 265–266
  regulation of MHC molecule expression on vascular endothelium, 262–265
Vβ selective element action, cellular basis of 29–32, see also T cell receptor, selective elements for Vβ region
  responding cells, 29–31
    CD4 T cells, response, 30
    CD8 T cells, response, 30–31
  stimulating cells, 31–32
    B cells, role of in response, 31
    dendritic cells, role of in response, 31
    transfer of Vbse among cells, 32
Vβ selective elements on T cell development and function, 32–37, see also T cell receptor, selective elements for Vβ region
  impact of Vβ selection elements in periphery, 35–37
    clonal inactivation by Vβ selective elements, 35–36
    enhancement of responses to antigen by Vβ selective elements, 36–37
    suppression induced by Vβ selective elements in vitro, 37
  role of Vβ selective elements in intrarhythmic development, 33
    negative selection, 34–35
    positive selection, 34
Vbse, general nomenclature, 13–14
Vbse, null alleles, 14
VDJ joints isolated from chicken thymus, diversity, 111
Viral pseudotypes in HIV-infected hu-PBL-SCID mice, biosafety issues and, 317–319

W

WE III-231 B cell line, growth arrest, 64–65

X

Xenotransplant models, human-to-mouse, 319
X-linked agammaglobulinemia (XLA), and SCID mice, 308

CONTENTS OF RECENT VOLUMES

Volume 40

Regulation of Human B Lymphocyte Activation, Proliferation, and Differentiation
  Diane F. Jeliner and Peter E. Lipsky

Biological Activities Residing in the Fc Region of Immunoglobulin
  Edward L. Morgan and William O. Weigle

Immunoglobulin-Specific Suppressor T Cells
  Richard G. Lynch

Immunoglobulin A (IgA): Molecular and Cellular Interactions Involved in IgA Biosynthesis and Immune Response
  Jiri Mestecky and Jerry R. McGhee

The Arrangement of Immunoglobulin and T Cell Receptor Genes in Human Lymphoproliferative Disorders
  Thomas A. Waldmann

Human Tumors Antigens
  Ralph A. Reisfeld and David A. Cheresh

Human Marrow Transplantation: An Immunological Perspective
  Paul J. Martin, John A. Hansen, Rainer Storb, and E. Donnall Thomas

Index

Volume 41

Cell Surface Molecules and Early Events Involved in Human T Lymphocyte Activation
  Arthur Weiss and John B. Imboden

Function and Specificity of T Cell Subsets in the Mouse
  Jonathan Sprent and Susan R. Webb

Determinants on Major Histocompatibility Complex Class I Molecules Recognized by Cytotoxic T Lymphocytes
  James Forman

Experimental Models for Understanding B Lymphocyte Formation
  Paul W. Kincade

Cellular and Humoral Mechanisms of Cytotoxicity: Structural and Functional Analogies
  John Ding-E Young and Zanvil A. Cohn

Biology and Genetics of Hybrid Resistance
  Michael Bennett

Index

Volume 42

The Clonotype Repertoire of B Cell Subpopulations
  Norman R. Klinman and Phyllis-Jean Linton

The Molecular Genetics of the Arsonate Idiotypic System of A/J Mice
  GARY RATHBUN, INAKI SANZ,
  KATHERYN MEEK,
  PHILIP TUCKER, AND
  J. DONALD CAPRA

The Interleukin 2 Receptor
  KENDALL A. SMITH

Characterization of Functional Surface Structures on Human Natural Killer Cells
  JEROME RITZ,
  REINHOLD E. SCHMIDT,
  JEAN MICHON,
  THIERRY HERCEND, AND
  STUART F. SCHLOSSMAN

The Common Mediator of Shock, Cachexia, and Tumor Necrosis
  B. BEUTLER AND A. CERAMI

Myasthenia Gravis
  JON LINDSTROM,
  DIANE SHELTON, AND
  YOSHITAKA FUJII

Alterations of the Immune System in Ulcerative Colitis and Crohn's Disease
  RICHARD P. MACDERMOTT AND
  WILLIAM F. STENSON

INDEX

Volume 43

The Chemistry and Mechanism of Antibody Binding to Protein Antigens
  ELIZABETH D. GETZOFF,
  JOHN A. TAINER,
  RICHARD A. LERNER, AND
  H. MARIO GEYSEN

Structure of Antibody-Antigen Complexes: Implications for Immune Recognition
  P. M. COLEMAN

The $\gamma\delta$ T Cell Receptor
  MICHAEL B. BRENNER,
  JACK L. STROMINGER, AND
  MICHAEL S. KRANGEL

Specificity of the T Cell Receptor for Antigen
  STEPHEN M. HEDRICK

Transcriptional Controlling Elements in the Immunological and T Cell Receptor Loci
  KATHRYN CALAME AND
  SUZANNE EATON

Molecular Aspects of Receptors and Binding Factors for IgE
  HENRY METZGER

INDEX

Volume 44

Diversity of the Immunoglobulin Gene Superfamily
  TIM HUNKAPILLER AND
  LEROY HOOD

Genetically Engineered Antibody Molecules
  SHERIE L. MORRISON AND
  VERNON T. OI

Antinuclear Antibodies: Diagnostic Markers for Autoimmune Diseases and Probes for Cell Biology
  ENG M. TAN

Interleukin-1 and Its Biologically Related Cytokines
  CHARLES A. DINARELLO

Molecular and Cellular Events of T Cell Development
  B. J. FOWLKES AND
  DREW M. PARDOLL

Molecular Biology and Function of CD4 and CD8
  JANE R. PARNES

Lymphocyte Homing
  TED A. YEDNOCK AND
  STEVEN D. ROSEN

INDEX

Volume 45

Cellular Interactions in the Humoral Immune Response
  ELLEN S. VITETTA,
  RAPAEL FERNANDEZ-BOTRAN,
  CHRISTOPHER D. MYERS, AND
  VIRGINIA M. SANDERS

MHC-Antigen Interactions: What Does the T Cell Receptor See?
  PHILIPPE KOURILSKY AND
  JEAN-MICHEL CLAVERIE

Synthetic T and B Cell Recognition Sites: Implications for Vaccine Development
  DAVID R. MILICH

Rationale for the Development of an Engineered Sporozoite Malaria Vaccine
  VICTOR NUSSENZWEIG AND
  RUTH S. NUSSENZWEIG

Virus-Induced Immunosuppression: Infections with Measles Virus and Human Immunodeficiency Virus
  MICHAEL B. MCCHESNEY AND
  MICHAEL B. A. OLDSTONE

The Regulators of Complement Activation (RCA) Gene Cluster
  DENNIS HOURCADE,
  V. MICHAEL HOLERS, AND
  JOHN P. ATKINSON

Origin and Significance of Autoreactive T Cells
  MAURICE ZAUDERER

INDEX

Volume 46

Physical Maps of the Mouse and Human Immunoglobulin-like Loci
  ERIC LAI, RICHARD K. WILSON,
  AND LEROY E. HOOD

Molecular Genetics of Murine Lupus Models
  ARGYRIOS N. THEOFILOPOULOS,
  REINHOLD KOPLER,
  PAUL A. SINGER, AND
  FRANK J. DIXON

Heterogeneity of Cytokine Secretion Patterns and Functions of Helper T Cells
  TIM R. MOSMANN AND
  ROBERT L. COFFMAN

The Leukocyte Integrins
  TAKASHI K. KISHMOTO,
  RICHARD S. LARSON,
  ANGEL L. CORBI,
  MICHAEL L. DUSTIN,
  DONALD E. STAUNTON, AND
  TIMOTHY A. SPRINGER

Structure and Function of the Complement Receptors, CR1 (CD35) and CR2 (CD21)
  JOSEPH M. AHEARN AND
  DOUGLAS T. FEAROW

The Cellular and Subcellular Bases of Immunosenescence
MARILYN L. THOMAN AND
WILLIAM O. WEIGLE

Immune Mechanisms in Autoimmune Thyroiditis
JEANNINE CHARREIRE

INDEX

Volume 47

Regulation of Immunoglobin E Biosynthesis
KIMISHIGE ISHIZAKA

Control of the Immune Response at the Level of Antigen-Presenting Cells: A Comparison of the Function of Dendritic Cells and B Lymphocytes
JOSHUA P. METLAY,
ELLEN PURÉ, AND
RALPH M. STEINMAN

The CD5 B Cell
THOMAS J. KIPPS

Biology of Natural Killer Cells
GIORGIO TRINCHIERI

The Immunopathogenesis of HIV Infection
ZEDA F. ROSENBERG AND
ANTHONY S. FAUCI

The Obeses Strain of Chickens: An Animal Model with Spontaneous Autoimmune Thyroiditis
GEORGE WICK,
HANS PETER BREZINSCHEK,
KAREL HÁLA,
HERMANN DIETRICH,
HUGO WOLF AND
GUIDO KROEMER

INDEX

Volume 48

Internal Movements in Immunoglobulin Molecules
ROALD NEZLIN

Somatic Diversification of the Chicken Immunoglobulin Light-Chain Gene
WAYNE T. MCCORMACK AND
CRAIG B. THOMPSON

T Lymphocyte-Derived Colony-Stimulating Factors
ANNE KELSO AND
DONALD METCALF

The Molecular Basis of Human Leukocyte Antigen Class II Disease Associations
DOMINQUE CHARRON

Neuroimmunology
E. J. GOETZL, D. C. ADELMAN,
AND S. P. SREEDHARAN

Immune Privilege and Immune Regulation in the Eye
JERRY Y. NIEDERKORN

Molecular Events Mediating T Cell Activation
AMNON ALTMAN,
K. MARK COGGESHALL,
AND TOMAS MUSTELIN

INDEX

Volume 49

Human Immunoglobulin Heavy-Chain Variable Region Genes: Organization, Polymorphism, and Expression
VIRGINIA PASCUAL AND
J. DONALD CAPRA

Surface Antigens of Human
Leucocytes
V. Hořejší

Expression, Structure, and Function
of the CD23 Antigen
G. Delespesse, U. Suter,
D. Mossalayi, B. Bettler,
M. Sarfati, H. Hoffstetter,
E. Kilcherr, P. Debre, and
A. Dalloul

Immunology and Clinical Importance
of Antiphospholipid Antibodies
H. Patrick McNeil,
Colin N. Chesterman, and
Steven A. Krilis

Adoptive T Cell Therapy of Tumors:
Mechanisms Operative in the
Recognition and Elimination of
Tumor Cells
Philip D. Greenberg

The Development of Rational
Strategies for Selective
Immunotherapy against Autoimmune
Demyelinating Disease
Lawrence Steinman

The Biology of Bone Marrow
Transplantation for Severe
Combined Immune Deficiency
Robertson Parkman

Index